Lecture Notes in Mathematics

Edited by A. Dold, F. Takens and B. Teissier

Editorial Policy
for the publication of monographs

1. Lecture Notes aim to report new developments in all areas of mathematics – quickly, informally and at a high level. Monograph manuscripts should be reasonably self-contained and rounded off. Thus they may, and often will, present not only results of the author but also related work by other people. They may be based on specialized lecture courses. Furthermore, the manuscripts should provide sufficient motivation, examples and applications. This clearly distinguishes Lecture Notes from journal articles or technical reports which normally are very concise. Articles intended for a journal but too long to be accepted by most journals, usually do not have this "lecture notes" character. For similar reasons it is unusual for doctoral theses to be accepted for the Lecture Notes series.

2. Manuscripts should be submitted (preferably in duplicate) either to one of the series editors or to Springer-Verlag, Heidelberg. In general, manuscripts will be sent out to 2 external referees for evaluation. If a decision cannot yet be reached on the basis of the first 2 reports, further referees may be contacted: the author will be informed of this. A final decision to publish can be made only on the basis of the complete manuscript, however a refereeing process leading to a preliminary decision can be based on a pre-final or incomplete manuscript. The strict minimum amount of material that will be considered should include a detailed outline describing the planned contents of each chapter, a bibliography and several sample chapters.

Authors should be aware that incomplete or insufficiently close to final manuscripts almost always result in longer refereeing times and nevertheless unclear referees' recommendations, making further refereeing of a final draft necessary.

Authors should also be aware that parallel submission of their manuscript to another publisher while under consideration for LNM will in general lead to immediate rejection.

3. Manuscripts should in general be submitted in English.
Final manuscripts should contain at least 100 pages of mathematical text and should include
– a table of contents;
– an informative introduction, with adequate motivation and perhaps some historical remarks: it should be accessible to a reader not intimately familiar with the topic treated;
– a subject index: as a rule this is genuinely helpful for the reader.

Continued on back inside cover

Lecture Notes in Mathematics 1744

Editors:
A. Dold, Heidelberg
F. Takens, Groningen
B. Teissier, Paris

Springer
Berlin
Heidelberg
New York
Barcelona
Hong Kong
London
Milan
Paris
Singapore
Tokyo

Markus Kunze

Non-Smooth Dynamical Systems

Springer

Author

Markus Kunze
Mathematisches Institut
Universität Köln
Weyertal 86
50931 Köln, Germany

E-mail: mkunze@mi.uni-koeln.de

Cataloging-in-Publication Data applied for

Die Deutsche Bibliothek - CIP-Einheitsaufnahme

Kunze, Markus:
Non-smooth dynamical systems / Markus Kunze. - Berlin ; Heidelberg ;
New York ; Barcelona ; Hong Kong ; London ; Milan ; Paris ; Singapore
; Tokyo : Springer, 2000
(Lecture notes in mathematics ; 1744)
ISBN 3-540-67993-6

Mathematics Subject Classification (2000): 34Cxx, 34Dxx, 37-xx, 70Exx, 70Kxx

ISSN 0075-8434
ISBN 3-540-67993-6 Springer-Verlag Berlin Heidelberg New York

Springer-Verlag Berlin Heidelberg New York
a member of BertelsmannSpringer Science+Business Media GmbH

© Springer-Verlag Berlin Heidelberg 2000
Printed in Germany

Typesetting: Camera-ready TEX output by the author
SPIN: 10724224   41/3142-543210 - Printed on acid-free paper

The course of true love never did run smooth.

Shakespeare, A midsummer night's dream

Preface

There are many concrete problems in life, and particularly in mechanics and in the engineering sciences, where non-smooth phenomena play an important role: one might think of the noise of a squeaking chalk on a black-board or, sometimes more pleasantly, the sounds of stringed instruments like a violin. More relevant applications include noise generation in railway wheels, the chattering of machine tools, grating brakes, impact print hammers, percussion drilling machines, etc. Physically speaking, these effects often are due to the fact that there are rigid bodies which are in contact (they "stick"), whereas these contact phases are interrupted by "slip" phases during which one of the bodies moves relative to another. In addition to such behaviour mainly induced by friction, there may also be impacts between different parts of the system.

From a mathematical viewpoint, problems of this kind are not easy to handle, since the resulting models are dynamical systems whose right-hand sides are not continuous or not differentiable. In many cases the solutions have to observe additional restrictions that frequently appear in the form of inequality constraints. Since many concepts from classical dynamical systems theory do rely on the smoothness of the underlying system or (semi-) flow, it was necessary to generalize those concepts to cover non-smooth dynamical systems as well, and it turned out that almost always such generalization is a non-trivial issue.

This book is devoted to the analysis of mathematical aspects of non-smooth dynamical systems, and since we are aiming to develop a rigorous theory, we will often have to restrict ourselves to investigating simple model problems for which, however, we will then obtain quite satisfactory results. Besides mathematicians who are interested in dynamical systems or applications, the book also intends to address researchers e.g. from mechanics or the engineering sciences who want to read more on mathematical techniques that are useful for the analysis of non-smooth dynamical systems, and it would be much reward if those people could profit from these notes.

My Habilitation thesis KUNZE [114] may be considered a first draft of the book, and I'm grateful to T. Küpper for giving me the opportunity to work on this promising and interesting subject of non-smooth dynamical systems. Sincere thanks are due to the priority research program "Dynamik: Analysis, effiziente Simulation und Ergodentheorie" of Deutsche Forschungsgemeinschaft, and to its coordinator B. Fiedler for initiating valuable exchanges between different parts of dynamical systems theory, through many conferences, workshops, etc. For their support and interest in my work I further wish to thank L. Arnold, J. Batt, W.-J. Beyn, M. Brokate, F. Colonius, K. Deimling, J.-P. Eckmann, Ch. Jones, A. Komech, A. Mielke, M. Monteiro Marques, F. Pfeiffer, K. Popp, J.-F. Rodrigues, J. Scheurle, H. Spohn, and E. Zeidler.

June 2000   *Markus Kunze*

Contents

1. Introduction

In the preface we have mentioned that non-smooth phenomena often result from the fact that contact phases, during which rigid bodies stick to each other, are interrupted by slip phases, i.e., one of the bodies moves relative to another. One of the simplest examples exhibiting such "slip-stick motion" is provided by a pendulum with dry friction (also called Coulomb friction) as shown in Fig. 1.1; see DEIMLING [65, p. 193], or older books on mechanics like ANDRONOV/VITT/KHAIKIN [9] and KAUDERER [107]. The study of such systems started as early as in the 1930's with the work of DEN HARTOG [70], and then has been continued in the 1950's through a series of papers by REISSIG [188, 189, 190, 191, 192] and with SZABLEWSKI [212]; see also the review GURAN/FEENY/HINRICHS/POPP [96].

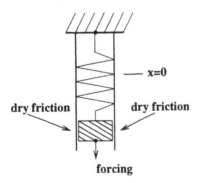

Fig. 1.1. A pendulum with dry friction

The pendulum consists of a mass being attached to a spring, and the mass, which is sinusoidally forced, moves in a straight tube and has contact to the wall of the tube. Depending on the size of the dry friction between mass and wall, and depending on the strength of the forcing, the mass moves up or down ("slip phase"), or it sticks to the wall.

To derive the governing equation for this system, we have to take into account three different kinds of forces:

1. the force through excitation, which is chosen to be $\gamma \sin(\eta\tau)$, with time τ and parameters $\gamma, \eta \geq 0$,
2. the restoring force of the spring, being (by Hooke's law) proportional to $-x(\tau)$, and
3. the friction force due to dry friction between the mass and the tube. By Coulomb's law, this friction force equals the negative of the normal force to the wall (which is constant), multiplied by the friction coefficient and by the signum of the velocity $\dot{x}$ of the mass.

Normalizing the mass of the pendulum and all other physical constants (besides γ and η) equal to one, Newton's force equation hence yields the discontinuous equation

$$\ddot{x}(\tau) + x(\tau) + \operatorname{sgn}\dot{x}(\tau) = \gamma \sin(\eta\tau). \tag{1.1}$$

Here we also assumed that on the left-hand side there is no additional viscous damping term $\delta\dot{x}(\tau)$ (e.g., through a fluid in the tube), but this is only for simplicity.

Note that (1.1) only makes sense if $\dot{x}(\tau) \neq 0$, and $\operatorname{sgn}(0) := 0$ is not the right choice. To see this, consider first the unforced equation with $\gamma = 0$. The corresponding phase portrait is given in Fig. 1.2.

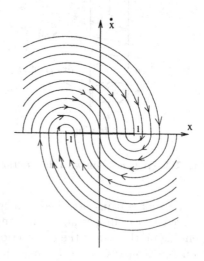

Fig. 1.2. Phase portrait of (1.1) with $\gamma = 0$

It can be seen that the system is "dissipative", and as $\tau \to \infty$, all trajectories approach the interval $[-1, 1] \times \{0\}$ in the $(x, \dot{x})$-phase plane. In particular, every $(x_0, 0)$ with $x_0 \in [-1, 1]$ is an equilibrium, but this effect is lost by defining $\text{sgn}(0) := 0$. This choice of $\text{sgn}(0)$ turns out to be too restrictive also in the presence of a forcing $(\gamma > 0)$, since a solution can have dead-zones, i.e., $\dot{x}(\tau) = 0$ for $\tau \in [\tau_1, \tau_2]$ with $\tau_2 > \tau_1$. During these dead-zones, which correspond to a sticking of the mass to the wall, $\text{sgn}(0) := 0$ in (1.1) would give $x(\tau_1) = \gamma \sin(\eta\tau)$ for $\tau \in [\tau_1, \tau_2]$, but this is impossible.

So a common device is to replace (1.1) by its multi-valued extension

$$\ddot{x}(\tau) + x(\tau) \in \gamma \sin(\eta\tau) - \text{Sgn}\,\dot{x}(\tau) \quad \text{a.e.}, \tag{1.2}$$

with

$$\text{Sgn}\,v = \begin{cases} \{v/|v|\} & : \quad v \neq 0 \\ [-1, 1] & : \quad v = 0 \end{cases}, \tag{1.3}$$

i.e., every $\text{Sgn}\,v$ is now a subset of $\mathbb{R}$ and (1.1) has become what is called a differential inclusion (or multi-valued differential equation). Note that by changing (1.1) to (1.2) in particular all $(x_0, 0)$ with $x_0 \in [-1, 1]$ have turned equilibria of (1.2) with $\gamma = 0$. Moreover, (1.2) is the same as (1.1) for $\dot{x}(\tau) \neq 0$.

By standard theory of differential inclusions, (1.2), considered as the autonomous system $\dot{y}(t) \in F(y(t))$ a.e., that is

$$\left. \begin{array}{l} \dfrac{d}{dt} \begin{pmatrix} x \\ v \\ \tau \end{pmatrix} \in F(x, v, \tau) \ \text{a.e.,} \quad y = (x, v, \tau) \in \mathbb{R}^3, \quad \text{where} \\[2em] F(x, v, \tau) = \{v\} \times \{\gamma \sin(\eta\tau) - x - w : w \in \text{Sgn}\,v\} \times \{1\} \end{array} \right\} \tag{1.4}$$

generates a global and continuous semiflow $\varphi : [0, \infty[\times \mathbb{R}^3 \to \mathbb{R}^3$, i.e., $\varphi(\cdot, y_0) : [0, \infty[\to \mathbb{R}^3$ is the unique solution of (1.4) on $[0, \infty[$ with initial value $\varphi(0, y_0) = y_0 = (x_0, v_0, \tau_0) \in \mathbb{R}^3$. Thus (1.4) gives rise to a (semi)-dynamical system.

There is yet another interesting class of non-smooth problems, which, contrary to the "dissipative" ones like (1.2), may be addressed as "conservative". To illustrate the difference, we consider the example

$$\ddot{x} + x + a \,\text{sgn}\,x = p(\tau) \tag{1.5}$$

with periodic forcing p and a parameter $a > 0$. Equations of this type are models for oscillators with state-dependent kicks, as shown in Fig. 1.3 below. The phase portrait of (1.5) without forcing $(p = 0)$ is given in Fig. 1.4.

In contrast to Fig. 1.2, all solutions are periodic and move on closed curves which are not differentiable at points on the axis $\{x = 0\}$. Since $\{\tau : x(\tau) = 0\}$ has measure zero, the value of $\text{sgn}(0)$ is not important here, and hence there is no need to change "sgn" to "Sgn".

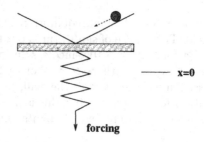

Fig. 1.3. A reflection pendulum

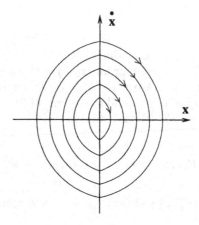

Fig. 1.4. Phase portrait of (1.5) with $p = 0$

After having introduced the two example models (1.2) and (1.5), for "dissipative" and "conservative" non-smooth systems, respectively (although this conceptual distinction will remain somewhat vague), we now turn to discuss some basic issues concerning these models, or concerning non-smooth dynamical systems in general. In an analysis of such non-smooth systems, the goal is the same as for their smooth counterparts: namely, to develop tools for a qualitative study of the problems. However, it has to be expected that it will be precisely the lack of smoothness that at some point or another will not allow standard concepts of dynamical system theory to be utilized for models like (1.2) or (1.5). It is the purpose of this book to show how several methods, successfully applied to smooth systems, can nevertheless be generalized; mostly these generalization are nontrivial, since at many instances classical dynamical system theory heavily relies on differentiability assumptions.

A brief outline of the organization of the book follows. The first chapter introduces some terminology and background material related to differential inclusions in general. In particular we will prove in some detail a theorem on the existence of solutions to differential inclusions that applies to (1.2) resp. (1.4).

Afterwards in Chapter 3 we turn to investigating special classes of solutions to non-smooth problems, like bounded, unbounded, periodic, or almost periodic solutions. The results that we obtain are sharp for the pendulum with dry friction (1.2), even when the forcing $\gamma \sin(\eta \tau)$ is replaced by a general (almost) periodic function p.

Especially when it comes to predicting "chaos", often Lyapunov exponents are a useful tool. For $\dot{y} = Ay$ with a matrix A, those are just the real parts of the eigenvalues of A, and hence closely related to the stability of the system. For general equations $\dot{y} = f(y)$ with smooth f, usually the multiplicative ergodic theorem of Oseledets is applied to ensure the existence of the Lyapunov exponents. As will be explained in detail in Chapter 4, it is not possible to proceed analogously for non-smooth systems like (1.2), because this requires the underlying semiflow to be C^1 w.r. to initial values in order to define what is called a cocycle. We thus have to develop first, in Section 4.1, a kind of "formal system" to describe equations like (1.2) appropriately. Then we construct a canonical cocycle for systems which fit this description. Afterwards it is motivated, by considering in Section 4.2 the example from (1.2), that in fact many dissipative non-smooth systems can be treated this way. In particular, we are able to show, besides other things, that there is no positive Lyapunov exponent for (1.2), and hence we conclude that the system does not show "chaotic behaviour", no matter what are γ and η.

Our next objective is then in Chapter 5 to generalize classical Conley index theory to non-smooth systems. Roughly speaking, the Conley index somehow measures the complexity of the set of such points (on the boundary of a neighborhood of an invariant set) which are going to leave this neighborhood immediately under the flow, either in forward or in backward time. Since a change of the Conley index indicates that the qualitative dynamical behaviour in this neighborhood has changed, Conley index theory may be used to prove bifurcation results. It is this point in which we are interested also for non-smooth systems like (1.2), in the presence of a suitable bifurcation parameter.

In Chapter 6 we turn to conservative systems like (1.5), and we show how classical Kolmogorov–Arnold–Moser (KAM) theory can be applied in order to prove the boundedness of all solutions in the $(x, \dot{x})$-phase plane, or in order to show the existence of periodic and quasiperiodic solutions of (1.5), under the assumption that $a > 0$ be large enough. On first sight it may seem surprising that KAM theory can be made work for systems like (1.5), because its application usually requires high regularity assumptions. We nevertheless can reduce the problem to fit the framework of (a variant of) Moser's twist theorem by means of a series of coordinate transformations, somehow passing the lack of regularity to the forcing p, which for that reason has to be smooth enough.

Since both Conley index theory and KAM theory are quite nontrivial issues, we included in Section 5.1 and Section 6.1, respectively, a short introduction to these subjects in order to explain, at least on a very basic level, briefly some ideas and notation.

In Chapter 7 we deal with planar non-smooth dynamical systems, mainly from the viewpoint of bifurcation of small periodic solutions (studied by means of Lyapunov constants and classical Hopf bifurcation methods), and we also mention a few results on piecewise linear planar systems.

Chapter 8 is devoted to describing some work on the generalization of Melnikov's method to non-smooth systems. This method can be used to analyze perturbations of planar systems with a homoclinic orbit. If sliding motion is excluded, we will see that the "non-smooth" Melnikov function contains two additional terms compared to the usual smooth one; both do result from impulses when the trajectory crosses the discontinuity line.

Finally, we conclude with Chapter 9, giving several related remarks and references, also on (part of) what has been omitted in the main text.

2. Some general theory of differential inclusions

In the introduction we have seen that the investigation of problems involving dry friction phenomena naturally leads to differential inclusions, also called multi-valued differential equations, a concept that generalizes the usual ordinary differential equations. This chapter is devoted to presenting some necessary background material for readers who are not familiar with those differential inclusions. Since the results that will be needed in later chapters are more or less on a basic level, we will often try to avoid the most general statements which may be found in standard references on the subject, as are DEIMLING [65], AUBIN/CELLINA [17], or FILIPPOV [82].

2.1 Preliminaries

To start with, we will need to introduce some notation. Let $D \subset \mathbb{R}^n$ be a nonempty set. Then a multi-valued mapping $F : D \to 2^{\mathbb{R}^n} \setminus \{\emptyset\}$ assigns to every $y \in D$ as value a set $F(y) \subset \mathbb{R}^n$, and $F(y) \neq \emptyset$. Here $2^{\mathbb{R}^n}$ stands for the power set of $\mathbb{R}^n$, i.e., the collection of all subsets of $\mathbb{R}^n$. Thus the values of F are the sets $F(y)$, and we say F has closed (or compact or convex) values, if each set $F(y) \subset \mathbb{R}^n$ is closed (or compact or convex). Finally, a selection f of F is a mapping $f : D \to \mathbb{R}^n$ such that $f(y) \in F(y)$ for all $y \in D$.

Example 2.1.1. We already have encountered in (1.3) a multi-valued mapping that will be of great importance to us later, namely

$$\text{Sgn}\, y = \begin{cases} \{y/|y|\} & : \quad y \neq 0 \\ [-1, 1] & : \quad y = 0 \end{cases}.$$

Thus $D = \mathbb{R}$, and $\text{Sgn}\, y = \{+1\}$ for $y > 0$ as well as $\text{Sgn}\, y = \{-1\}$ for $y < 0$ show that Sgn just agrees with the usual sgn-function for $y \neq 0$. The mapping $F = \text{Sgn}$ has compact convex values, and the selections of Sgn are given by $f(y) = \text{sgn}\, y$ for $y \neq 0$ and $f(y) =$ any value $\in [-1, 1]$ for $y = 0$. ◇

With such multi-valued mappings one can then try to solve the differential inclusions

$$\dot{y}(t) \in F(y(t)), \quad t \in I, \quad y(0) = y_0, \tag{2.1}$$

on some time interval I containing $t = 0$, but it should be clear that to do so at least some regularity of F will be required. Moreover, it will have to be made precise in which sense (2.1) should be understood.

Definition 2.1.1. *Let $F : D \to 2^{\mathbb{R}^n} \setminus \{\emptyset\}$ be a multi-valued mapping. Then F is said to be ε-δ upper semicontinuous (ε-δ-usc for short) if for every $y_0 \in D$ and $\varepsilon > 0$ there exists $\delta > 0$ such that $F(y) \subset F(y_0) + B_\varepsilon(0)$ for $y \in B_\delta(y_0) \cap D$.*

Here $F(y_0) + B_\varepsilon(0) = \{z + r : z \in F(y_0), |r| < \varepsilon\}$ denotes the "ball" of radius $\varepsilon > 0$ around the set $F(y_0) \subset \mathbb{R}^n$. Intuitively, an ε-δ-usc mapping always jumps upwards w.r. to inclusion.

Example 2.1.2. We re-consider Ex. 2.1.1 and argue that Sgn is ε-δ-usc. If e.g. $y_0 > 0$, then $F(y_0) + B_\varepsilon(0) = [1 - \varepsilon, 1 + \varepsilon]$, whence the required condition is satisfied for all $y > 0$, since then also $F(y) = \{+1\}$. However, if $y_0 = 0$, then $F(y_0) + B_\varepsilon(0) = [-1 - \varepsilon, 1 + \varepsilon]$. Thus $\delta > 0$ may even be chosen arbitrary, as all three possible values $\{+1\}$, $\{-1\}$, or $[-1, 1]$ of $F(y)$ are subsets of $[-1 - \varepsilon, 1 + \varepsilon]$. $\Diamond$

Remark 2.1.1. One may also introduce the concept of an upper semicontinuous (usc) multi-valued mapping, i.e., for every closed $A \subset \mathbb{R}^n$ the pre-image $F^{-1}(A) = \{y \in D : F(y) \cap A \neq \emptyset\}$ as well should be closed. In general, usc implies ε-δ-usc, but the converse may fail. However, if F has compact values, both notions of upper semicontinuity do agree. $\Diamond$

Remark 2.1.2. If, in Definition 2.1.1, $D_0 \subset D$ is a compact subset and F has compact values, then F is even uniformly ε-δ-usc on D_0, i.e., for $\varepsilon > 0$ there exists $\delta > 0$ with $F(y) \subset F(y_0) + B_\varepsilon(0)$ for $y_0 \in D_0$ and $y \in B_\delta(y_0) \cap D_0$. Indeed, otherwise there were an $\varepsilon_0 > 0$ and sequences $\delta_j \to 0^+$, $(y_{0,j}) \subset D_0$, $(y_j) \subset D_0$, and $(z_j) \subset \mathbb{R}^n$ such that $|y_j - y_{0,j}| < \delta_j$, $z_j \in F(y_j)$, as well as $\mathrm{dist}(z_j, F(y_{0,j})) \geq 2\varepsilon_0$. Due to compactness of D_0 we thus may assume $y_j \to y_0$ and $y_{0,j} \to y_0$ as $j \to \infty$ for some $y_0 \in D_0$. For fixed $\varepsilon > 0$ we have $z_j \in F(y_j) \subset F(y_0) + B_\varepsilon(0)$ for j large enough, since F is ε-δ-usc at y_0. By compactness of $F(y_0)$ we hence may as well assume that $z_j \to z_0 \in F(y_0)$ for $j \to \infty$. As also $F(y_{0,j}) \subset F(y_0) + B_{\varepsilon_0}(0)$ for j large enough, it follows that

$$2\varepsilon_0 \leq \mathrm{dist}(z_j, F(y_{0,j})) \leq \mathrm{dist}(z_j, F(y_0)) + \varepsilon_0 \leq |z_j - z_0| + \varepsilon_0$$

for those j, a contradiction. $\Diamond$

Already the example $F(y) = \mathrm{Sgn}\, y$ shows that an ε-δ-usc mapping need not have a continuous selection. However, at least a continuous "almost" selection can be found. This fact often allows the extension of concepts that are well-known for usual mappings (as the degree of mapping or Conley index theory) to multi-valued mappings.

Lemma 2.1.1. *Let $D \subset \mathbb{R}^n$ be compact and $F : D \to 2^{\mathbb{R}^n} \setminus \{\emptyset\}$ an ε-δ-usc mapping with convex values. Then for $\varepsilon > 0$ there exists an open neighborhood D_ε of D and a continuous $g : D_\varepsilon \to \mathbb{R}^n$ such that*

$$g(y) \in F(B_\varepsilon(y) \cap D) + B_\varepsilon(0), \quad y \in D_\varepsilon.$$

Proof: We follow DEIMLING [65, p. 7]. By Definition 2.1.1, for each $y \in D$ we find $\delta_y \in]0, \varepsilon]$ such that

$$F(\bar{y}) \subset F(y) + B_\varepsilon(0), \quad \bar{y} \in B_{3\delta_y}(y) \cap D. \tag{2.2}$$

Since D is compact it may be covered by finitely many such balls, i.e., we have $D \subset \bigcup_{i=1}^m B_{\delta_i}(y_i) =: D_\varepsilon$ for some $y_i \in D$ and $\delta_i = \delta_{y_i} \in]0, \varepsilon]$, $1 \leq i \leq m$. Correspondingly we can choose continuous functions $\varphi_i : \mathbb{R}^n \to [0, 1]$ satisfying $\varphi_i(y) = 0$ for $|y - y_i| \geq \delta_i$, and $\sum_{i=1}^m \varphi_i(y) = 1$ for all $y \in D_\varepsilon$; the φ_i constitute what is called a partition of unity. We select any $z_i \in F(y_i)$ and set $g(y) = \sum_{i=1}^m \varphi_i(y)z_i$ for $y \in D_\varepsilon$. Then fix $y_0 \in D_\varepsilon$ and introduce $\mathcal{I} = \{i : 1 \leq i \leq m, |y_0 - y_i| < \delta_i\}$; whence $\sum_{i \in \mathcal{I}} \varphi_i(y_0) = 1$. Define $p \in \mathcal{I}$ by requiring $\delta_p = \max\{\delta_i : i \in \mathcal{I}\}$. Then for $i \in \mathcal{I}$ we have $|y_i - y_p| \leq |y_i - y_0| + |y_0 - y_p| < \delta_i + \delta_p \leq 2\delta_p < 3\delta_p$, and thus by (2.2)

$$z_i \in F(y_i) \subset F(y_p) + B_\varepsilon(0),$$

since $y_i \in B_{3\delta_p}(y_p) \cap D$. Hence we may write $z_i = w_i + r_i$ for $i \in \mathcal{I}$, where $w_i \in F(y_p)$ and $|r_i| < \varepsilon$. In particular, $|\sum_{i \in \mathcal{I}} \varphi_i(y_0)r_i| < \varepsilon$, and consequently by convexity of $F(y_p)$

$$g(y_0) = \left(\sum_{i \in \mathcal{I}} \varphi_i(y_0)w_i \right) + \left(\sum_{i \in \mathcal{I}} \varphi_i(y_0)r_i \right) \in F(y_p) + B_\varepsilon(0).$$

Finally observe that $y_p \in B_\varepsilon(y_0) \cap D$ due to $|y_p - y_0| < \delta_p \leq \varepsilon$. $\square$

In case that the multi-valued mapping F in (2.1) is only defined on some subset D of $\mathbb{R}^n$, for a solution y to (2.1) we need to ensure that $y(I) \subset D$, so that $F(y(t))$ will make sense. This can be achieved by imposing a suitable boundary condition, formulated in terms of tangent cones, that is designed to guarantee that a solution starting in $y_0 \in D$ at time $t = 0$ will not be able to leave D at a later time.

Definition 2.1.2. *Let $D \subset \mathbb{R}^n$ be closed and $y \in D$. The tangent cone to D at y is*

$$T_D(y) = \left\{ z \in \mathbb{R}^n : \liminf_{\lambda \to 0+} \lambda^{-1}\mathrm{dist}(y + \lambda z, D) = 0 \right\}.$$

Sometimes $T_D(y)$ is called the Bouligand contingent cone; see for instance AUBIN/CELLINA [17]. Note that $0 \in T_D(y)$, $\mu T_D(y) \subset T_D(y)$ for $\mu \in [0, \infty[$, and

$$T_{\mathbb{R}^n}(y) = \mathbb{R}^n, \quad y \in \mathbb{R}^n, \quad \text{as well as} \quad T_D(y) = \mathbb{R}^n, \quad y \in D^0, \qquad (2.3)$$

whence $T_D(y)$ is interesting only for $y \in \partial D$. If D is not only closed but also convex, an easy characterization of $T_D(y)$ is available, cf. DEIMLING [65, Prop. 4.1].

Lemma 2.1.2. *Assume $D \subset \mathbb{R}^n$ is closed and convex. Then*

$$T_D(y) = \overline{\{\lambda(z - y) : \lambda \in [0, \infty[, z \in D\}}, \quad y \in D. \qquad (2.4)$$

Proof: Temporarily denote $S(y)$ the set on the right-hand side of (2.4). If $w \in T_D(y)$, then we find sequences $\lambda_j \to 0^+$ and $(z_j) \subset D$ such that $\lambda_j^{-1}|y - \lambda_j w - z_j| \to 0$ as $j \to \infty$. Whence $w = \lim_{j\to\infty} \lambda_j^{-1}(z_j - y)$ implies $w \in S(y)$. Conversely, suppose $w = \lim_{j\to\infty} \lambda_j(z_j - y) \in S(y)$ for some sequences $(\lambda_j) \subset [0, \infty[$ and $(z_j) \subset D$. Choose $\mu_j \in]0, \infty[$ such that $\mu_j \lambda_j \in [0, 1]$ and $\mu_j \to 0^+$. Thus by convexity of D,

$$y + \mu_j \lambda_j (z_j - y) = (1 - \mu_j \lambda_j)y + \mu_j \lambda_j z_j \in D,$$

thus

$$\mu_j^{-1}\text{dist}(y + \mu_j w, D) \leq \mu_j^{-1}\Big|[y + \mu_j w] - [y + \mu_j \lambda_j(z_j - y)]\Big|$$
$$= |w - \lambda_j(z_j - y)| \to 0, \quad j \to \infty,$$

and therefore $w \in T_D(y)$.

To explain the relation of the tangent cone to solutions not leaving D let us consider instead of (2.1) the simpler example of a usual ordinary differential equation $\dot{y} = f(y)$, where $f : D \to \mathbb{R}^n$ is continuous, and a solution $y : [0, a] \to D$ with $y(0) = y_0 \in D$ is given. Then for $t > 0$ small

$$t^{-1}\text{dist}(y_0 + tf(y_0), D) \leq t^{-1}|y_0 + tf(y_0) - y(t)|$$
$$\leq t^{-1} \int_0^t |f(y(s)) - f(y_0)| \, ds \to 0, \quad t \to 0^+,$$

and thus $f(y_0) \in T_D(y_0)$. As this argument applies to any point of D lying on a solution trajectory, it follows that

$$f(y) \in T_D(y), \quad y \in D, \qquad (2.5)$$

is a necessary condition for solvability of the initial-value problems. For ordinary differential equations it is well-know that this condition is also sufficient, cf. DEIMLING [62, Lemma 18.3], and it will turn out in the subsequent section that the appropriate generalization of (2.5) to multi-valued right-hand sides F is

$$F(y) \cap T_D(y) \neq \emptyset, \quad y \in D;$$

see Thm. 2.2.1.

Finally we state without proof a theorem on the existence of fixed points for multi-valued mappings $F : D \to 2^Y \setminus \{\emptyset\}$, defined on some closed, bounded, and convex subset D of a general Banach space Y. Here $y \in D$ is called a fixed point, if $y \in F(y)$.

Theorem 2.1.1. *Let Y be a Banach space, $\emptyset \neq D \subset Y$ closed, bounded, and convex, and $F : D \to 2^Y \setminus \{\emptyset\}$ an ε-δ-usc multi-valued mapping such that $F(D) \subset D$ and $\overline{F(D)} \subset Y$ is compact. Then F has a fixed point.*

Note that Thm. 2.1.1 generalizes to multi-valued mappings the classical Schauder's fixed point theorem which states that a compact self-map of a closed, bounded, and convex subset of a Banach space does have a fixed point. Thm. 2.1.1 in fact arises as a special case of a much more general result, see DEIMLING [63] and DEIMLING [65, Thm. 11.5/Cor. 11.3].

2.2 Existence and uniqueness of solutions

In this section we turn to solving differential inclusions of the type (2.1). Typically we cannot expect to find solutions that are differentiable everywhere.

Example 2.2.1. Consider $F(y) = -\text{Sgn}\, y$ for $y \in \mathbb{R}$ with initial condition $y(0) = 1$, and suppose $y : I = [0, 2] \to \mathbb{R}$ is a solution of (2.1). Then $y(t) = 1 + \int_0^t \dot{y}(s)\, ds \geq 1 - t$ for $t \in [0, 1]$, since $\dot{y}(s) \in F(y(s)) \subset [-1, 1]$, whence $\dot{y}(s) \geq -1$. In particular, $y(t) > 0$ for $t \in [0, 1[$, thus $\dot{y}(t) \in F(y(t)) = \{-1\}$ and continuity of y show that necessarily $y(t) = 1 - t$ for $t \in [0, 1]$. To determine the solution for $t \in [1, 2]$ denote $\varphi(\tau) = y^2(\tau+1)$. Then $\varphi(0) = 0$ as well as $\dot{\varphi}(\tau) = 2y(\tau+1)\dot{y}(\tau+1) = -2|y(\tau+1)| = -2\varphi^{1/2}(\tau)$, as $y > 0$ implies $F(y) = \{-1\}$ and $y < 0$ implies $F(y) = \{+1\}$. The differential equation for φ has the unique solution $\varphi(\tau) = 0$, and therefore $y(t) = 0$ for $t \in [1, 2]$. In summary, we have found that

$$y(t) = \begin{cases} 1 - t & : \quad t \in [0, 1] \\ 0 & : \quad t \in [1, 2] \end{cases} \tag{2.6}$$

is the unique solution of (2.1), which, however, is not of class C^1. $\Diamond$

Therefore it is necessary to relax the concept of a solution.

Definition 2.2.1. *Let $I \subset \mathbb{R}$ be an interval with $0 \in I$, $D \subset \mathbb{R}^n$, $y_0 \in D$, and $F : D \to 2^{\mathbb{R}^n} \setminus \{\emptyset\}$ a multi-valued mapping. A function y such that*

(a) $y : I \to D$ is absolutely continuous on I,
(b) $y(0) = y_0$, and
(c) $\dot{y}(t) \in F(y(t))$ for a.e. $t \in I$

is called a solution of the differential inclusion $\dot{y} \in F(y)$ a.e., $y(0) = y_0$.

The next remark recalls the notion of an absolutely continuous function.

Remark 2.2.1. Let $I \subset \mathbb{R}$ be a (bounded or unbounded) interval. A function $y : I \to \mathbb{R}^n$ is called absolutely continuous, if for every $\varepsilon > 0$ there exists $\delta > 0$ such that

$$\sum_{i=1}^{m} (\beta_i - \alpha_i) \le \delta \quad \text{implies} \quad \sum_{i=1}^{m} |y(\beta_i) - y(\alpha_i)| \le \varepsilon$$

whenever $m \in \mathbb{N}$ and $]\alpha_1, \beta_1[, \ldots,]\alpha_m, \beta_m[$ are mutually disjoint subintervals of I. Choosing $m = 1$ we find in particular that an absolutely continuous function is uniformly continuous. Equivalently, $y : I \to \mathbb{R}^n$ is absolutely continuous, if there exists $w \in L^1_{\text{loc}}(I; \mathbb{R}^n)$ such that

$$y(t) = y(s) + \int_s^t w(\tau) \, d\tau, \quad s, t \in I.$$

Then w is called the weak derivative of y. Moreover, an absolutely continuous function y admits a classical derivative $\dot{y}(t)$ for a.e. $t \in I$, and $\dot{y}(t) = w(t)$ for a.e. $t \in I$; for this reason, one simply writes $\dot{y}$ instead of w. The weak derivative w is unique only up to modification on sets of zero measure. See e.g. BUTTAZZO/GIAQUINTA/HILDEBRANDT [42, p. 80 ff.] for more information. $\diamondsuit$

Note in particular that, as $\dot{y}(t)$ is well-defined only for a.e. $t \in I$, the differential inclusion in (2.1) had to be replaced by "$\dot{y}(t) \in F(y(t))$ for a.e. $t \in I$" in Definition 2.2.1(c). Since the function y from (2.6) is absolutely continuous with weak derivative

$$w(t) = \dot{y}(t) = \begin{cases} -1 & : \quad t \in [0, 1[\\ 0 & : \quad t \in [1, 2] \end{cases},$$

it follows that y is a solution of $\dot{y} \in -\text{Sgn}\, y$ a.e., $y(0) = 1$, in the sense of Definition 2.2.1.

Next we turn to a standard existence theorem for solutions of the differential inclusion

$$\dot{y}(t) \in F(y(t)) \text{ for a.e. } t \in I, \quad y(0) = y_0. \tag{2.7}$$

Theorem 2.2.1. *Let $D \subset \mathbb{R}^n$ be closed and $F : D \to 2^{\mathbb{R}^n} \setminus \{\emptyset\}$ ε-δ-usc with closed convex values. Assume further that*

(a) $\|F(y)\| = \sup\{|z| : z \in F(y)\} \leq c_1(1 + |y|)$ for some constant $c_1 > 0$ and all $y \in D$, and
(b) $F(y) \cap T_D(y) \neq \emptyset$ for $y \in D$.

Then (2.7) has a solution on $I = [0, \infty[$ for every $y_0 \in D$.

Proof: We follow DEIMLING [65, Lemma 5.1]. Note that it is sufficient to prove the existence of a solution on any $I = [0, a]$, $a > 0$, since the problem is autonomous, whence solutions can be pieced together.

First we will argue that w.l.o.g. we may suppose that

$$\|F(y)\| \leq c_2, \quad y \in D. \tag{2.8}$$

Indeed, if y is a solution, then $|\dot{y}(t)| \leq c_1(1 + |y(t)|)$ for a.e. $t \in [0, a]$ by (2.7) and assumption (a). Therefore

$$|y(t)| \leq |y_0| + c_1 a + c_1 \int_0^t |y(s)| \, ds, \quad t \in [0, a],$$

implies, using Gronwall's lemma, that

$$|y(t)| \leq (|y_0| + c_1 a)e^{c_1 a} =: R - 1, \quad t \in [0, a]. \tag{2.9}$$

This a priori bound is verified for any solution. Assume now that we already can solve (2.7) under the hypothesis (2.8), and consider a general F that satisfies only the linear growth estimate from (a). Choose a continuous function $\varphi : [0, \infty[\to [0, 1]$ such that $\varphi(\xi) = 1$ for $\xi \in [0, R - 1]$ and $\varphi(\xi) = 0$ for $\xi \in [R, \infty[$. Then the multi-valued mapping $\tilde{F}(y) = \varphi(|y|)F(y)$, $y \in D$, is ε-δ-usc and has closed convex values. Moreover, $\tilde{F}(y) \cap T_D(y) \neq \emptyset$ for $y \in D$, since $F(y) \cap T_D(y) \neq \emptyset$ and $\mu T_D(y) \subset T_D(y)$ for $\mu \in [0, \infty[$, whence $\varphi(|y|)z \in T_D(y)$ provided that $z \in T_D(y)$. In addition,

$$\|\tilde{F}(y)\| \leq c_1\varphi(|y|)(1 + |y|) \leq c_1(1 + R) =: c_2.$$

Hence we find a solution of $\dot{y}(t) \in \tilde{F}(y(t))$ for a.e. $t \in [0, a]$, $y(0) = y_0$. But also $\|\tilde{F}(y)\| \leq c_1(1 + |y|)$, and therefore the above estimate applies and yields $|y(t)| \leq R - 1$ for $t \in [0, a]$. Thus $\varphi(|y(t)|) = 1$, and consequently y is a solution to (2.7).

The second step is to show the existence of suitable approximate solutions, where we now suppose (2.8). More precisely, we will verify that given $\varepsilon > 0$ there exists an absolutely continuous function $v = v_\varepsilon : [0, a] \to \mathbb{R}^n$ which satisfies

$$\left. \begin{array}{l} D \cap \overline{B}_{2\varepsilon}(v(t)) \neq \emptyset, \ t \in [0, a], \quad v(0) = y_0, \quad v(a) \in D, \quad \text{and} \\[2mm] \dot{v}(t) \in F\left(D \cap \overline{B}_{2\varepsilon}(v(t))\right) + \overline{B}_\varepsilon(0) \text{ for a.e. } t \in [0, a]. \end{array} \right\} \tag{2.10}$$

To see this, we choose $z_0 \in F(y_0) \cap T_D(y_0)$ and fix $\delta_0 \in]0, \varepsilon/c_2 \wedge a]$ such that $\operatorname{dist}(y_0 + \delta z_0, D) \le \delta\varepsilon/2$. Then we find $y_1 \in D$ with $|y_0 + \delta_0 z_0 - y_1| \le \delta_0\varepsilon$. Set $t_1 = \delta_0 \le a$ and

$$v(t) = y_0 + \frac{y_1 - y_0}{\delta_0} t, \quad t \in [0, t_1].$$

Then $v(0) = y_0$, $v(t_1) = y_1 \in D$, and $|v(t) - y_0| \le |y_1 - y_0| \le \delta_0\varepsilon + \delta_0|z_0| \le 2\varepsilon$, since $z_0 \in F(y_0)$ implies $|z_0| \le c_2$ by (2.8), and since we may suppose w.l.o.g. that $\varepsilon/c_2 \le 1$. Thus $y_0 \in D \cap \overline{B}_{2\varepsilon}(v(t))$ for $t \in [0, t_1]$. In addition, $|\dot{v}(t) - z_0| \le \varepsilon$ by choice of y_1, whence $z_0 \in F(y_0)$ shows

$$\dot{v}(t) \in F(y_0) + \overline{B}_\varepsilon(0) \subset F\Big(D \cap \overline{B}_{2\varepsilon}(v(t))\Big) + \overline{B}_\varepsilon(0), \quad t \in]0, t_1[.$$

In summary, v satisfies (2.10) on $[0, t_1]$ in place of $[0, a]$. Let

$$S = \Big\{(v, \delta) : \delta \in]0, a], \ v \text{ has the properties (2.10) on } [0, \delta]$$
$$\text{instead of } [0, a]\Big\},$$

and introduce $(v_1, \delta_1) \prec (v_2, \delta_2)$ iff $\delta_1 \le \delta_2$ and $v_1(t) = v_2(t)$ for $t \in [0, \delta_1]$. By the foregoing reasoning we see that $S \ne \emptyset$, and $\prec$ is a partial order on S. If $(v_\lambda, \delta_\lambda)_{\lambda \in \Lambda}$ denotes a linearly ordered subset of S, we let $\delta = \sup_{\lambda \in \Lambda} \delta_\lambda \le a$ and $v(t) = v_\lambda(t)$ for $t \in [0, \delta[\cap [0, \delta_\lambda]$. Then $v : [0, \delta[\to \mathbb{R}^n$ is a well-defined function that enjoys the properties listed in (2.10), besides that we need to ensure v has a continuous extension to $t = \delta$, with $v(\delta) \in D$. To verify the latter, note that $|\dot{v}(t)| \le c_2 + \varepsilon$ for a.e. $t \in [0, \delta[$ by (2.8). In particular, v is Lipschitz continuous on $[0, \delta[$, with Lipschitz constant $(c_2 + \varepsilon)$. This shows $v(\delta) := \lim_{t \to \delta-} v(t)$ does exist. But also $v(\delta) \in D$, since $v(\delta_\lambda) = v_\lambda(\delta_\lambda) \in D$, and $v(\delta_{\lambda_j}) \to v(\delta)$, if $\delta_{\lambda_j} \to \delta$. Therefore $(v, \delta) \in S$, and clearly (v, δ) is an upper bound for $(v_\lambda, \delta_\lambda)_{\lambda \in \Lambda}$. By Zorn's lemma, cf. YOSIDA [226, Ch. 0.1], S has a maximal element $(v_*, \delta_*) \in S$. In case that $\delta_* < a$, we could use $v_*(\delta_*) \in D$ to extend v_* to some interval strictly larger than $[0, \delta_*]$, by the argument employed above to construct v on $[0, t_1]$. As this would contradict the fact that (v_*, δ_*) is maximal, we must have $\delta_* = a$, and v_* satisfies (2.10) on $[0, a]$.

The third step consists in passing to the limit $\varepsilon \to 0^+$. Fix $\varepsilon_j \to 0^+$, $\varepsilon_j \le 1$, and denote $v_j = v_{\varepsilon_j}$ the corresponding approximate solutions from the previous step. For a.e. $t \in [0, a]$ and $j \in \mathbb{N}$ then

$$|\dot{v}_j(t)| \le c_2 + 1, \tag{2.11}$$

due to (2.8). Whence $v_j(0) = y_0$ implies

$$|v_j(t)| \le |y_0| + (c_2 + 1)a, \quad j \in \mathbb{N}, \quad t \in [0, a].$$

Consequently, $(v_j) \subset C([0, a]; \mathbb{R}^n)$ is bounded and uniformly Lipschitz continuous with Lipschitz constant $(c_2 + 1)$, in particular the sequence is equicontinuous. By the Arzelà–Ascoli theorem, cf. DEIMLING [62, Prop. 7.3], we thus may assume that $v_j \to y$ uniformly in $[0, a]$, for some continuous function $y : [0, a] \to \mathbb{R}^n$. Since $v_j(0) = y_0$, we note $y(0) = y_0$. In addition, from (2.11) we also infer $(\dot{v}_j) \subset L^2([0, a]; \mathbb{R}^n)$ is bounded. As this space is reflexive, upon passing to a further subsequence if necessary, we may as well suppose that $\dot{v}_j \to w$ weakly in $L^2([0, a]; \mathbb{R}^n)$, for some $w \in L^2([0, a]; \mathbb{R}^n)$. Fix any $\xi \in \mathbb{R}^n$ and $t \in [0, a]$. Then $z \in L^2([0, a]; \mathbb{R}^n)$ for $z(s) = \mathbf{1}_{[0,t]}(s)\xi$, $s \in [0, a]$, with $\mathbf{1}_{[0,t]}$ denoting the characteristic function of $[0, t]$. So as $j \to \infty$

$$
\begin{aligned}
\langle y(t), \xi \rangle \longleftarrow \langle v_j(t), \xi \rangle &= \langle y_0, \xi \rangle + \left\langle \int_0^t \dot{v}_j(s)\, ds, \xi \right\rangle \\
&= \langle y_0, \xi \rangle + \langle \dot{v}_j, z \rangle_{L^2([0,a];\mathbb{R}^n)} \\
&\longrightarrow \langle y_0, \xi \rangle + \langle w, z \rangle_{L^2([0,a];\mathbb{R}^n)} \\
&= \langle y_0, \xi \rangle + \left\langle \int_0^t w(s)\, ds, \xi \right\rangle.
\end{aligned}
$$

Therefore $y(t) = y_0 + \int_0^t w(s)\, ds$ for all $t \in [0, a]$. Since $w \in L^2([0, a]; \mathbb{R}^n)$, this proves $y : [0, a] \to \mathbb{R}^n$ is absolutely continuous, with $\dot{y}(t) = w(t)$ for a.e. $t \in [0, a]$. We need to show $y(t) \in D$ for all $t \in [0, a]$, and $w(t) \in F(y(t))$ for a.e. $t \in [0, a]$. Concerning the first assertion, by the properties of v_j we see that for $t \in [0, a]$

$$
\text{dist}(y(t), D) \le |y(t) - v_j(t)| + \text{dist}(v_j(t), D) \le |y(t) - v_j(t)| + 2\varepsilon_j \to 0
$$

as $j \to \infty$, whence $y(t) \in D$, as D is closed. To verify $w(t) \in F(y(t))$ for a.e. $t \in [0, a]$, note first that we find $I_0 \subset I$ of full measure such that for all $t \in I_0$ and $j \in \mathbb{N}$ there exist points $\zeta_j(t) \in D$ satisfying $|\zeta_j(t) - v_j(t)| \le 2\varepsilon_j$ and $\dot{v}_j(t) \in F(\zeta_j(t)) + \overline{B}_{\varepsilon_j}(0)$. Thus $|\zeta_j(t) - y(t)| \le 2\varepsilon_j + |v_j - y|_{C([0,a];\mathbb{R}^n)} =: \bar{\varepsilon}_j$, and $\bar{\varepsilon}_j \to 0$ as $j \to \infty$. Let

$$
D_0 = \{\zeta_j(t) : j \in \mathbb{N}, t \in [0, a]\} \cup \{y(t) : t \in [0, a]\} \subset D.
$$

From the uniform convergence of v_j it follows that $D_0 \subset \mathbb{R}^n$ is compact. Fix $\eta > 0$. Since F is uniformly ε-δ-usc on D_0, due to Rem. 2.1.2 we may choose $\delta > 0$ such that

$$
F(\xi) \subset F(\xi_0) + B_\eta(0) \quad \text{for} \quad \xi_0 \in D_0, \quad \xi \in B_\delta(\xi_0) \cap D_0.
$$

Fix j_0 so large that $\varepsilon_j < \eta$ and $\bar{\varepsilon}_j < \delta$ for $j \ge j_0$. Then

$$
\dot{v}_j(t) \in F(y(t)) + \overline{B}_{2\eta}(0), \quad j \ge j_0, \quad t \in I_0. \tag{2.12}
$$

Define $\mathcal{M} \subset L^2([0, a]; \mathbb{R}^n)$ by

$$\mathcal{M} = \left\{ \phi \in L^2([0,a]; \mathbb{R}^n) : \phi(t) \in F(y(t)) + \overline{B}_{2\eta}(0) \text{ for a.e. } t \in [0,a] \right\}.$$

Since F has convex values, $\mathcal{M}$ is convex, and $\mathcal{M}$ is also closed, as each $F(y(t)) + \overline{B}_{2\eta}(0)$ is closed and $\phi_j \to \phi$ in $L^2([0,a]; \mathbb{R}^n)$ implies $\phi_j(t) \to \phi(t)$ for a.e. $t \in [0,a]$, at least on a subsequence. Being convex and closed, $\mathcal{M}$ is weakly closed; cf. RUDIN [195, Thm. 3.12]. By (2.12) we have $\dot{v}_j \in \mathcal{M}$ for $j \geq j_0$, and consequently $w \in \mathcal{M}$. This means that for any $\eta > 0$

$$w(t) \in F(y(t)) + \overline{B}_{2\eta}(0), \qquad t \in I_\eta,$$

for some $I_\eta \subset I = [0,a]$ of full measure. Choosing a sequence $\eta_j \to 0^+$, we obtain $w(t) \in F(y(t))$ for a.e. $t \in [0,a]$, as was to be shown.

A further important issue is to prove uniqueness of solutions to (2.7). This usually is obtained as a consequence of some kind of dissipativity property of the right-hand side F.

Theorem 2.2.2. *Let $D \subset \mathbb{R}^n$ be closed and $F : D \to 2^{\mathbb{R}^n} \setminus \{\emptyset\}$ a multivalued mapping such that*

$$\langle y - \bar{y}, z - \bar{z} \rangle \leq c_2 |y - \bar{y}|^2, \quad y, \bar{y} \in D, \quad z \in F(y), \quad \bar{z} \in F(\bar{y}), \qquad (2.13)$$

for some constant $c_2 \in \mathbb{R}$. Then there is at most one solution of (2.7) on $I = [0,a]$.

Proof: Suppose $y : I \to D$ and $\bar{y} : I \to D$ are two such solutions. Defining $\varphi(t) = |y(t) - \bar{y}(t)|^2$ for $t \in I$, it follows that

$$\dot{\varphi}(t) = 2 \langle y(t) - \bar{y}(t), \dot{y}(t) - \dot{\bar{y}}(t) \rangle \quad \text{for a.e.} \quad t \in I.$$

Since $\dot{y}(t) \in F(y(t))$ and $\dot{\bar{y}}(t) \in F(\bar{y}(t))$ for a.e. $t \in I$, (2.13) implies that $\dot{\varphi}(t) \leq 2c_2\varphi(t)$ for a.e. $t \in I$. Due to $\varphi(0) = 0$ the differential inequality for φ has the unique solution $\varphi = 0$, whence $y(t) = \bar{y}(t)$ for $t \in I$.

Example 2.2.2. (a) Of course in general solutions to (2.7) are non-unique. Consider e.g. $F(y) = [-1,1]$ for $y \in \mathbb{R} = D$, and $y_0 = 0$. Then the only restrictions for an absolutely continuous function $y : [0,\infty[\to \mathbb{R}$ to be a solution to (2.7) are $y(0) = 0$ and $|\dot{y}(t)| \leq 1$ for a.e. $t \in [0,\infty[$.

(b) Even in case that solutions are unique on every $[0,a]$, it may happen that they are non-unique for negative times. Consider (1.2) with $\gamma = 0$, i.e., $\dot{y} \in F(y)$ a.e., where $y = (x,v)$ and

$$F(y) = F(x,v) = \{v\} \times \{-x - w : w \in \text{Sgn}\, v\}, \quad y \in \mathbb{R}^2.$$

Then (2.13) holds with $c_2 = 0$, cf. (2.15) below, whence solutions are unique to the right. Clearly, $x(\tau) \equiv x_0 \in [-1,1]$ are solutions to $\ddot{x} + x \in -\text{Sgn}\, \dot{x}$ a.e., but there are several other solutions. Following DEIMLING/HETZER/SHEN [67, Example 1] we define

$$\bar{x}(\tau) = \begin{cases} \sqrt{2} - 1 & : \ \tau \in [0, \infty[\\ \sin(\tau + \frac{\pi}{4}) + \cos(\tau + \frac{\pi}{4}) - 1 & : \ \tau \in [-\pi, 0[\\ (-1)^k(\sqrt{2} + 2k)\cos(\tau + k\pi) - (-1)^k & : \ k \in \mathbb{N}, \quad \text{and} \\ & \quad \tau \in [-(k+1)\pi, -k\pi[\end{cases}$$

to obtain that this $\bar{x}$ as well is a solution; note that we have $\dot{\bar{x}}(\tau) > 0$ in $\bigcup_{k \in \mathbb{N}_0}] - (2k+1)\pi, -2k\pi[$ and also $\dot{\bar{x}}(\tau) < 0$ in $\bigcup_{k \in \mathbb{N}_0}] - 2(k+1)\pi, -(2k+1)\pi[$. In particular, both $\bar{x}$ and $x(\tau) \equiv \sqrt{2} - 1$ are solutions with $x(0) = \sqrt{2} - 1$ and $\dot{x}(0) = 0$, they do agree for $\tau \geq 0$, but not for $\tau < 0$. Finally, we also remark that $\bar{x}$ is even unbounded, since $\bar{x}(-k\pi) = (-1)^k(\sqrt{2} - 1 + 2k)$. ◇

We are going to illustrate the application of Thms. 2.2.1 and 2.2.2 by means of an example of a dry friction problem that we will analyze in much greater detail later in Sect. 4.2.

Example 2.2.3. Let the multi-valued mapping $F : \mathbb{R}^3 \to 2^{\mathbb{R}^3} \setminus \{\emptyset\}$ be defined as in (1.4), i.e.,

$$F(x, v, \tau) = \{v\} \times \left\{ \gamma \sin(\eta\tau) - x - w : w \in \mathrm{Sgn}\, v \right\} \times \{1\} \subset \mathbb{R}^3$$

for $y = (x, v, \tau) \in \mathbb{R}^3$.

We first check the assumptions of Thm. 2.2.1 with $D = \mathbb{R}^3$. Similarly to Ex. 2.1.2 it can be verified that F is ε-δ-usc, and F has closed convex values, as has Sgn. If $z = (v, \gamma \sin(\eta\tau) - x - w, 1) \in F(y)$, then due to $|w| \leq 1$,

$$|z| \leq |v| + \gamma + |x| + 2 \leq \gamma + 2 + \sqrt{2}\left(|x|^2 + |v|^2\right)^{1/2} \leq c_1(1 + |y|), \quad (2.14)$$

with $c_1 = \gamma + 2$. Thus (a) holds, and according to (2.3) also (b) is satisfied; whence Thm. 2.2.1 applies.

To prove the uniqueness of the solutions so obtained we have to check (2.13). For this, observe that $(v - \bar{v})(w - \bar{w}) \geq 0$ for $v, \bar{v} \in \mathbb{R}$ and $w \in \mathrm{Sgn}\, v$, $\bar{w} \in \mathrm{Sgn}\, \bar{y}$. Let $z = (v, \gamma \sin(\eta\tau) - x - w, 1) \in F(y)$ and $\bar{z} = (\bar{v}, \gamma \sin(\eta\bar{\tau}) - \bar{x} - \bar{w}, 1) \in F(\bar{y})$. Then

$$\begin{aligned} \langle y - \bar{y}, z - \bar{z} \rangle &= \gamma \left[\sin(\eta\tau) - \sin(\eta\bar{\tau})\right](v - \bar{v}) + (v - \bar{v})(\bar{w} - w) \\ &\leq \gamma \left[\sin(\eta\tau) - \sin(\eta\bar{\tau})\right](v - \bar{v}) \\ &\leq \frac{\gamma\eta}{2}\left(|\tau - \bar{\tau}|^2 + |v - \bar{v}|^2\right) \leq \frac{\gamma\eta}{2}|y - \bar{y}|^2, \end{aligned} \quad (2.15)$$

and hence we may choose $c_2 = \frac{\gamma\eta}{2}$ and apply Thm. 2.2.2.

Consequently we have shown that for all $y_0 = (x_0, v_0, \tau_0) \in \mathbb{R}^3$ there exists a unique solution $\varphi(\cdot, y_0) : [0, \infty[\to \mathbb{R}^3$ of (1.4) with data $\varphi(0, y_0) =$

y_0, or equivalently, a unique solution $x(\cdot; y_0) : [\tau_0, \infty[\to \mathbb{R}$ of (1.2) with $x(\tau_0; y_0) = x_0$ and $\dot{x}(\tau_0; y_0) = v_0$. Note that, by definition of a solution, $\varphi(\cdot, y_0)$ is absolutely continuous on $[0, \infty[$. Whence $\dot{x}(\tau; y_0) = v(\tau; y_0)$ for a.e. $\tau \in [\tau_0, \infty[$ yields $x(\tau; y_0) = x_0 + \int_{\tau_0}^{\tau} v(\sigma; y_0)\, d\sigma$ for $\tau \in [\tau_0, \infty[$, and thus $x(\cdot; y_0) \in C^1([\tau_0, \infty[)$ as well as $\ddot{x}(\cdot; y_0) \in L^1_{\text{loc}}([\tau_0, \infty[)$. Therefore (1.4) induces a semiflow $\varphi : [0, \infty[\times \mathbb{R}^3 \to \mathbb{R}^3$, where

$$\varphi^t(y_0) = \big(x(\tau_0 + t; y_0), \dot{x}(\tau_0 + t; y_0), \tau_0 + t\big) \quad \text{for} \quad t \in [0, \infty[.$$

In addition, $\varphi^t(x_0, v_0, \tau_0 + k(2\pi/\eta)) = \varphi^t(y_0) + (0, 0, k(2\pi/\eta))$ for $k \in \mathbb{N}_0$ by periodicity of $\sin(\cdot)$ and uniqueness of solutions.

Finally we remark that the semiflow is locally Lipschitz continuous on $[0, \infty[\times \mathbb{R}^3$. To see this, we fix $a > 0$ and $y_0 \in \mathbb{R}^3$. We then obtain from (2.9) that

$$|\varphi(t, y_0)| \leq (|y_0| + c_1 a)e^{c_1 a}, \quad t \in [0, a], \quad y_0 \in \mathbb{R}^3.$$

This in turn implies by (2.14) that

$$|\partial_t \varphi(t, y_0)| \leq c_1 \left(1 + (|y_0| + c_1 a)e^{c_1 a}\right) \quad \text{a.e.} \tag{2.16}$$

Let $y(t) = \varphi(t, y_0)$ and $\bar{y}(t) = \varphi(t, \bar{y}_0)$ denote two solutions. By differentiating $|y(t) - \bar{y}(t)|^2$ and application of (2.15) we see that

$$|\varphi(t, y_0) - \varphi(t, \bar{y}_0)| \leq |y_0 - \bar{y}_0| e^{\gamma \eta a/2} \quad \text{for} \quad t \in [0, a]. \tag{2.17}$$

Thus (2.16) and (2.17) yield that indeed φ is locally Lipschitz continuous on $[0, \infty[\times \mathbb{R}^3$. $\diamond$

3. Bounded, unbounded, periodic, and almost periodic solutions

In concrete mechanical systems it is often desirable to have a priori knowledge about some special features of the motion, e.g., of the behaviour of several tools being part of some machine. For the underlying non-smooth dynamical system this means that one is particularly interested in special classes of solutions, like bounded, periodic, or almost periodic ones, since the existence of such solutions can prevent the machine from breaking down in finite time. On the other hand, it might also happen that e.g. periodic solutions have to be ruled out in order to exclude the possibility of wear.

Section 3.1 is devoted to boundedness questions. First we derive in Sect. 3.1.1 a criterion for all solutions of a general inclusion $\ddot{x} + x + F(x, \dot{x}) \ni p(t)$ to be unbounded in the $(x, \dot{x})$-plane. Then we study the dissipative problems $\ddot{x} + x + F(\dot{x}) \ni p(t)$ with "increasing" mappings F in Sect. 3.1.2, and it turns out that all solutions are bounded if and only if there is one bounded solution. More precise information is obtained in Sect. 3.1.3 for the special case $\ddot{x} + x + \operatorname{Sgn} \dot{x} \ni p(t)$ of the pendulum with dry friction, where we even can allow for almost periodic forcing functions p.

Afterwards in Sect. 3.2 we deal with the existence of periodic solutions, first for the pendulum with dry friction with an ω-periodic forcing, cf. Sect. 3.2.1, where the resonant case $\omega = 2k\pi$ for some $k \in \mathbb{N}$ is most interesting. For $\omega = 2\pi$, also a complete picture concerning the stability of periodic solutions is presented. Sect. 3.2.2 then contains similar (existence) results for the conservative problems $\ddot{x} + x + a \operatorname{Sgn} x \ni p(t)$.

In Sect. 3.3 we turn back to the pendulum with dry friction, with an almost periodic forcing. By deriving necessary and sufficient conditions, we clarify under which circumstances almost periodic solutions do exist and what are their stability properties. In case that there is more than one almost periodic solution, we also obtain a full characterization of this set of solutions.

The chapter is completed with an appendix on almost periodic functions in Sect. 3.4.

3.1 Bounded and unbounded solutions

3.1.1 Unbounded solutions at resonance

We follow KUNZE [112] and consider the general inclusion

$$\ddot{x} + x + F(x, \dot{x}) \ni p(t) \quad \text{a.e.} \tag{3.1}$$

at resonance, i.e., with a continuous and 2π-periodic p. Our aim is to provide a criterion for all solutions of (3.1) to be "uniformly" unbounded in the $(x, \dot{x})$-phase plane, cf. Thm. 3.1.1 below. This result will complement our findings later in Chap. 6, where we will investigate under which assumptions all solutions are bounded.

For simplicity we assume throughout that for every $\xi_0 = (x_0, v_0) \in \mathbb{R}^2$ and $t_0 \in [0, 2\pi]$ there exists a unique global solution $x(\cdot; \xi_0, t_0)$ of (3.1) on $[t_0, \infty[$ with initial values $x(t_0; \xi_0, t_0) = x_0$ and $\dot{x}(t_0; \xi_0, t_0) = v_0$.

We will need to introduce a suitable Lyapunov function, and in order to do so, we first let

$$\mathcal{E}_p(t; \xi_0, t_0) = x^2(t; \xi_0, t_0) + \dot{x}^2(t; \xi_0, t_0) \quad \text{for} \quad t \in [t_0, \infty[, \tag{3.2}$$

where the subscript p refers to the forcing p in (3.1). Moreover, we define

$$\sup(F) = \sup\{z : z \in F(x, v), \; x, v \in \mathbb{R}\}, \quad \text{and}$$
$$\inf(F) = \inf\{z : z \in F(x, v), \; x, v \in \mathbb{R}\}.$$

We start with a preliminary lemma.

Lemma 3.1.1. *In the setting described above,*

(a) $\mathcal{E}_p(t + t_0; \xi_0, t_0) = \mathcal{E}_{p(\cdot + t_0)}(t; \xi_0, 0)$ for $t \in [0, \infty[$, $t_0 \in [0, 2\pi]$, and $\xi_0 \in \mathbb{R}^2$.
(b) $\mathcal{E}_{p(\cdot + t_0)}(2n\pi + t; \xi_0, 0) = \mathcal{E}_{p(\cdot + t_0)}\big(t; (x(2n\pi; \xi_0, 0), \dot{x}(2n\pi; \xi_0, 0)), 0\big)$ for all $n \in \mathbb{N}_0$, $t \in [0, 2\pi]$, $t_0 \in [0, 2\pi]$, and $\xi_0 \in \mathbb{R}^2$.
(c) There exist constants $c_1, c_2 > 0$ such that for $t \in [0, 2\pi]$, $t_0 \in [0, 2\pi]$, and $\xi_0 = (x_0, v_0) \in \mathbb{R}^2$

$$\mathcal{E}_{p(\cdot + t_0)}(t; \xi_0, 0) \geq c_1 (x_0^2 + v_0^2) - c_2.$$

Proof: Ad (a): This follows from the assumed uniqueness of solutions to the right, since if x solves (3.1) on $[t_0, \infty[$ and has $x(t_0) = x_0$ and $\dot{x}(t_0) = v_0$, then $x(\cdot + t_0)$ is a solution of (3.1) on $[0, \infty[$ with the forcing p being replaced by $p(\cdot + t_0)$. Ad (b): Again this is implied by uniqueness of solutions to the right. Ad (c): The statement is the same as ALONSO/ORTEGA [7, Lemma 3.3] and may be shown as follows. By (3.1) we have $\ddot{x} + x + w = p(t + t_0)$ a.e. in $[0, \infty[$ for some function w satisfying $w(t) \in F(x(t), \dot{x}(t))$ a.e., whence

$$x(t) = x_0 \cos t + v_0 \sin t + S(t, t_0) \quad \text{and} \quad \dot{x}(t) = -x_0 \sin t + v_0 \cos t + C(t, t_0)$$

for $t \in [0, \infty[$, with

$$S(t, t_0) = \int_0^t [p(\sigma + t_0) - w(\sigma)] \sin(t - \sigma) d\sigma, \quad \text{and}$$

$$C(t, t_0) = \int_0^t [p(\sigma + t_0) - w(\sigma)] \cos(t - \sigma) d\sigma.$$

In particular, $|S(t, t_0)|, |C(t, t_0)| \leq 2\pi(|p|_\infty + \|F\|)$ for $t, t_0 \in [0, 2\pi]$. Moreover,

$$\begin{aligned}
\mathcal{E}_{p(\cdot + t_0)}(t; \xi_0, 0) &= x_0^2 + v_0^2 + S^2(t, t_0) + C^2(t, t_0) \\
&\quad + 2x_0 \left(S(t, t_0) \cos t - C(t, t_0) \sin t\right) \\
&\quad + 2v_0 \left(S(t, t_0) \sin t + C(t, t_0) \cos t\right) \\
&\geq x_0^2 + v_0^2 + S^2(t, t_0) + C^2(t, t_0) \\
&\quad - 2(|x_0| + |v_0|) \left(|S(t, t_0)| + |C(t, t_0)|\right) \\
&\geq \frac{1}{2}(x_0^2 + v_0^2) - 31\left(S^2(t, t_0) + C^2(t, t_0)\right),
\end{aligned}$$

and hence we may choose $c_1 = 1/2$ and $c_2 > 0$ appropriately. $\square$

Now we can proceed to

Theorem 3.1.1. *Let* $F : \mathbb{R} \times \mathbb{R} \to 2^{\mathbb{R}} \setminus \{\emptyset\}$ *be a multi-valued mapping such that* $\sup(F) < \infty$ *and* $\inf(F) > -\infty$, *let* p *be continuous and periodic with period* 2π, *and suppose*

$$\left| \int_0^{2\pi} p(t) e^{it} \, dt \right| > 2 \left(\sup(F) - \inf(F)\right). \tag{3.3}$$

Then for every $R > 0$

$$\inf_{|\xi_0| \leq R, t_0 \in [0, 2\pi]} \mathcal{E}_p(t; \xi_0, t_0) \to \infty \quad \text{as} \quad t \to \infty. \tag{3.4}$$

In particular, every solution of (3.1) is unbounded in the $(x, \dot{x})$-*phase plane.*

Proof: By Lemma 3.1.1(a) it is sufficient to show that

$$\inf_{|\xi_0| \leq R, t_0 \in [0, 2\pi]} \mathcal{E}_{p(\cdot + t_0)}(t; \xi_0, 0) \to \infty \quad \text{as} \quad t \to \infty,$$

and Lemma 3.1.1(b) & (c) imply that whence it is enough to prove

$$\inf_{|\xi_0| \leq R, t_0 \in [0, 2\pi]} \mathcal{E}_{p(\cdot + t_0)}(2n\pi; \xi_0, 0) \to \infty \quad \text{as} \quad n \to \infty. \tag{3.5}$$

For this, we fix $t_0 \in [0, 2\pi]$ and define $f_{t_0}(\xi_0) = (x(2\pi; \xi_0, 0), \dot{x}(2\pi; \xi_0, 0))$, with the solution x corresponding to the forcing $p(\cdot + t_0)$. Hence

$$\mathcal{E}_{p(\cdot+t_0)}(2n\pi;\xi_0,0) = |f_{t_0}^n(\xi_0)|_{\mathbb{R}^2}^2$$

by Lemma 3.1.1(b) and the uniqueness of solutions to the right. We choose $\phi \in [0, 2\pi[$ in such a way that

$$\int_0^{2\pi} p(t+t_0)\sin(t+\phi)\,dt = \left| \int_0^{2\pi} p(t+t_0)\,e^{it}\,dt \right| = \left| \int_0^{2\pi} p(t)\,e^{it}\,dt \right|.$$

Next we define the continuous $V_{t_0} : \mathbb{R}^2 \to \mathbb{R}$ through

$$V_{t_0}(x,v) = v\sin\phi - x\cos\phi, \quad (x,v) \in \mathbb{R}^2.$$

By (3.1) we have $\ddot{x} + x + w = p(t + t_0)$ a.e. on $[0, \infty[$ for some function w satisfying $w(t) \in F(x(t), \dot{x}(t))$ a.e. Therefore integration by parts gives

$$\int_0^{2\pi} \ddot{x}(t;\xi_0,0)\,\sin(t+\phi)dt = V_{t_0}(f_{t_0}(\xi_0)) - V_{t_0}(\xi_0)$$

$$- \int_0^{2\pi} x(t;\xi_0,0)\,\sin(t+\phi)dt,$$

and consequently by choice of ϕ, denoting by $\sin^+$ resp. $\sin^-$ the positive resp. negative part of sin, for every $\xi_0 \in \mathbb{R}^2$

$$V_{t_0}(f_{t_0}(\xi_0))$$

$$\geq V_{t_0}(\xi_0) + \left| \int_0^{2\pi} p(t)\,e^{it}\,dt \right|$$

$$+ \inf(F) \int_0^{2\pi} \sin^-(t+\phi)dt - \sup(F) \int_0^{2\pi} \sin^+(t+\phi)dt$$

$$= V_{t_0}(\xi_0) + \left| \int_0^{2\pi} p(t)\,e^{it}\,dt \right| - 2\,(\sup(F) - \inf(F)) = V_{t_0}(\xi_0) + \delta_0, \quad (3.6)$$

with $\delta_0 > 0$ being independent of t_0 by assumption (3.3). We find a constant $c(R) > 0$ such that $|V_{t_0}(\xi_0)| \leq c(R)$ for all $t_0 \in [0, 2\pi]$ and $|\xi_0| \leq R$. Hence iteration of (3.6) implies $V_{t_0}(f_{t_0}^n(\xi_0)) \geq V_{t_0}(\xi_0) + n\delta_0 \geq -c(R) + n\delta_0$ for all $n \in \mathbb{N}$, $t_0 \in [0, 2\pi]$, and $|\xi_0| \leq R$. Thus

$$\inf_{|\xi_0| \leq R, t_0 \in [0,2\pi]} |f_{t_0}^n(\xi_0)|_{\mathbb{R}^2}^2 \to \infty \quad \text{as} \quad n \to \infty$$

as was to be verified in order to prove (3.5). $\square$

Therefore we have shown that every solution of (3.1) is unbounded provided that the size of the first Fourier coefficient of p is large when compared to $\sup(F) - \inf(F)$. This will be illustrated by means of some examples.

Example 3.1.1. We first consider the pendulum with dry friction at resonance, i.e.,

$$\ddot{x} + x + \operatorname{Sgn}\dot{x} \ni \gamma \sin t \quad \text{a.e.,} \tag{3.7}$$

c.f. (1.2) in Chap. 1, with $\eta = 1$ and $\gamma \in [0, \infty[$. Since $F(\dot{x}) = \operatorname{Sgn}\dot{x}$ has $\sup(F) = 1$ and $\inf(F) = -1$, and $|\int_0^{2\pi} \sin t\, e^{it}\, dt| = \pi$, condition (3.3) reads as $\gamma \in]4/\pi, \infty[$. Hence for those γ, Thm. 3.1.1 applies to yield (3.4), where $p = \gamma \sin(\cdot)$. In particular, for $\gamma > 4/\pi$ the solutions of (3.7) are uniformly unbounded in the $(x, \dot{x})$-phase plane.

We also note that the example enlightens a difference between the differential equations considered in ALONSO/ORTEGA [7] and the inclusions studied above. While in condition (3.2) of ALONSO/ORTEGA [7] the equality corresponding to equality in (3.3) enforces the solutions to be unbounded, the same does not hold for (3.7), and thus as well not for (3.1) in general: in case that $\gamma = 4/\pi$ there exist (infinitely many) periodic solutions of (3.7), cf. Thm. 3.2.7(b3) below. Consequently, Ex. 3.1.4 to follow (see also Thm. 4.2.3(a) on p. 111) will imply that for $\gamma = 4/\pi$ every solution of (3.7) is globally bounded, i.e., there are no unbounded solutions at all.   ◇

Example 3.1.2. In Thm. 6.0.4 we will prove by means of KAM theory that for

$$\ddot{x} + x + a\operatorname{sgn}x = p(t), \tag{3.8}$$

with $p \in C^6(\mathbb{R})$ being 2π-periodic, every solution is bounded in the $(x, \dot{x})$-phase plane for parameters $a > 0$ sufficiently large. Now Thm. 3.1.1 shows that some condition on the size of a is necessary, since for $|\int_0^{2\pi} p(t)\, e^{it}\, dt| > 4a$ all solutions of

$$\ddot{x} + x + a\operatorname{Sgn}x \ni p(t) \quad \text{a.e.} \tag{3.9}$$

will be uniformly unbounded, and $\operatorname{sgn}x \in \operatorname{Sgn}x$ implies that solutions of (3.8) are solutions of (3.9). This also fits to Thm. 3.2.8 below, where the existence of periodic solutions of (3.9) is proved in case that $|\int_0^{2\pi} p(t)\, e^{it}\, dt| < 4a$, and, under an additional assumption, also when "=" holds.   ◇

3.1.2 A criterion for boundedness in dissipative systems

Here we want to devise a criterion to prove the boundedness of all solutions for certain dissipative differential inclusions. We consider

$$\ddot{x} + x + F(\dot{x}) \ni p(t) \quad \text{a.e.,} \tag{3.10}$$

first with a forcing $p \in C(\mathbb{R})$ not necessarily being periodic. Roughly speaking, we are going to show in Thm. 3.1.2 that all solutions to (3.10) will be bounded in case that there does exist only one bounded solution; this observation is particularly useful if we already know (3.10) has a periodic solution.

Again we assume for simplicity the global unique solvability of every initial value problem corresponding to (3.10), cf. the notation at the beginning of Sect. 3.1.1.

Lemma 3.1.2. *Let* $F : \mathbb{R} \to 2^{\mathbb{R}} \setminus \{\emptyset\}$ *be an increasing multi-valued mapping, i.e.,*

$$(v - \bar{v})(z - \bar{z}) \geq 0, \quad v, \bar{v} \in \mathbb{R}, \quad z \in F(v), \; \bar{z} \in F(\bar{v}). \tag{3.11}$$

Let $p \in C(\mathbb{R})$, *and fix* $\xi_0 = (x_0, v_0) \in \mathbb{R}^2$, $\bar{\xi}_0 = (\bar{x}_0, \bar{v}_0) \in \mathbb{R}^2$, *and* $t_0, \bar{t}_0 \in \mathbb{R}$ *with* $t_0 \leq \bar{t}_0$. *For* $t \in [\bar{t}_0, \infty[$ *define*

$$\Delta_{\mathcal{E}_p}(t; \xi_0, t_0, \bar{\xi}_0, \bar{t}_0) = \Big(x(t; \xi_0, t_0) - x(t; \bar{\xi}_0, \bar{t}_0) \Big)^2 + \Big(\dot{x}(t; \xi_0, t_0) - \dot{x}(t; \bar{\xi}_0, \bar{t}_0) \Big)^2. \tag{3.12}$$

Then ——

(a) for $t \in [t_0, \infty[$,

$$\Delta_{\mathcal{E}_p}(t; \xi_0, t_0, \bar{\xi}_0, t_0) \leq \Delta_{\mathcal{E}_p}(t_0; \xi_0, t_0, \bar{\xi}_0, t_0)$$
$$= (x_0 - \bar{x}_0)^2 + (v_0 - \bar{v}_0)^2 = |\xi_0 - \bar{\xi}_0|^2. \tag{3.13}$$

(b) for $t \geq \sigma \geq \bar{t}_0$,

$$\Delta_{\mathcal{E}_p}(t; \xi_0, t_0, \xi_0, \bar{t}_0) \leq \Delta_{\mathcal{E}_p}(\sigma; \xi_0, t_0, \xi_0, \bar{t}_0). \tag{3.14}$$

Proof: Ad (a): Let $x(t) = x(t; \xi_0, t_0)$, $\bar{x}(t) = x(t; \bar{\xi}_0, t_0)$, and let $\Delta_{\mathcal{E}_p}(t)$ denote the left-hand side of (3.13). Choose functions w resp. $\bar{w}$ such that $\ddot{x} + x + w = p(t)$, $\ddot{\bar{x}} + \bar{x} + \bar{w} = p(t)$, $w(t) \in F(\dot{x}(t))$, and $\bar{w}(t) \in F(\dot{\bar{x}}(t))$, all a.e. in $[t_0, \infty[$. Then differentiation gives a.e. in $[t_0, \infty[$

$$\frac{1}{2} \dot{\Delta}_{\mathcal{E}_p}(t) = [x(t) - \bar{x}(t)][\dot{x}(t) - \dot{\bar{x}}(t)] + [\dot{x}(t) - \dot{\bar{x}}(t)][\ddot{x}(t) - \ddot{\bar{x}}(t)]$$
$$= [p(t) - w(t) - p(t) + \bar{w}(t)][\dot{x}(t) - \dot{\bar{x}}(t)]$$
$$= -[w(t) - \bar{w}(t)][\dot{x}(t) - \dot{\bar{x}}(t)] \leq 0,$$

by assumption (3.11). Therefore $\Delta_{\mathcal{E}_p}(t) \leq \Delta_{\mathcal{E}_p}(t_0)$ for $t \in [t_0, \infty[$.

Ad (b): With $x(t) = x(t; \xi_0, t_0)$ and $\bar{x}(t) = x(t; \xi_0, \bar{t}_0)$, the calculation from (a) can be repeated. $\qquad \square$

Example 3.1.3. The conclusion of Lemma 3.1.2 is valid for the inclusions

$$\ddot{x} + x + \alpha \dot{x} + \beta \operatorname{Sgn} \dot{x} \ni p(t) \quad \text{a.e.}$$

with $\alpha, \beta \geq 0$, since in this case (3.11) holds with $F(v) = \alpha v + \beta \operatorname{Sgn} v$. In particular, Lemma 3.1.2 applies to the pendulum with dry friction (1.2). $\diamond$

By means of Lemma 3.1.2 we obtain a criterion for the boundedness of solutions of (3.10).

Theorem 3.1.2. *Let $F : \mathbb{R} \to 2^{\mathbb{R}} \setminus \{\emptyset\}$ satisfy (3.11), and assume $p \in C(\mathbb{R})$. Suppose that there exists an initial value $\xi_* \in \mathbb{R}^2$ and $c_1 > 0$ such that, with $\mathcal{E}_p$ from (3.2),*

$$\sup_{t \in [0,\infty[} \mathcal{E}_p(t; \xi_*, 0) \leq c_1. \tag{3.15}$$

Then for every $R > 0$ there exists a constant $c_2 > 0$ such that

$$\sup \left\{ \mathcal{E}_p(t; \xi_0, t_0) : |\xi_0| \leq R, t_0 \in [0, \infty[, t \in [t_0, \infty[\right\} \leq c_2. \tag{3.16}$$

Proof: For $\xi_0 = (x_0, v_0) \in \mathbb{R}^2$ with $|\xi_0| \leq R$ and $t \in [0, \infty[$ it follows from Hölder's inequality and Lemma 3.1.2(a) that

$$\mathcal{E}_p^{1/2}(t; \xi_0, 0) \leq \Delta_{\mathcal{E}_p}^{1/2}(t; \xi_0, 0, \xi_*, 0) + \mathcal{E}_p^{1/2}(t; \xi_*, 0) \leq |\xi_0 - \xi_*| + c_1^{1/2}$$

$$\leq R + |\xi_*| + c_1^{1/2} = c_*.$$

Thus for $t_0 \in [0, \infty[$ and $t \in [t_0, \infty[$ by application of Lemma 3.1.2(b) with $0, t_0$ in place of $t_0, \bar{t}_0$ there,

$$\mathcal{E}_p^{1/2}(t; \xi_0, t_0) \leq \Delta_{\mathcal{E}_p}^{1/2}(t; \xi_0, t_0, \xi_0, 0) + \mathcal{E}_p^{1/2}(t; \xi_0, 0)$$

$$\leq \Delta_{\mathcal{E}_p}^{1/2}(t_0; \xi_0, t_0, \xi_0, 0) + c_*$$

$$= \left((x_0 - x(t_0; \xi_0, 0))^2 + (v_0 - \dot{x}(t_0; \xi_0, 0))^2 \right)^{1/2} + c_*$$

$$\leq |\xi_0| + \mathcal{E}_p^{1/2}(t_0; \xi_0, 0) + c_* \leq 3R + 2|\xi_*| + 2c_1^{1/2}.$$

Hence $c_2 = \left(3R + 2|\xi_*| + 2c_1^{1/2} \right)^2$ is a suitable choice.

Example 3.1.4. In the setting of Thm. 3.1.2, suppose that (3.10) has an ω-periodic solution $x : [0, \omega] \to \mathbb{R}$ for some $\omega > 0$. Then (3.15) holds with

$$\xi_* = (x(0), \dot{x}(0)) \quad \text{and} \quad c_1 = \max_{t \in [0, \omega]} x^2(t) + \max_{t \in [0, \omega]} \dot{x}^2(t).$$

Hence Thm. 3.1.2 applies if F satisfies (3.11), in particular in case that $F(v) = \text{Sgn } v$. $\diamond$

3.1.3 Boundedness of solutions for the pendulum with dry friction

Contrary to the previous Sects. 3.1.1 and 3.1.2, where boundedness properties of solutions to the general differential inclusions (3.1) and (3.10) have been investigated, here we address the same question for the more specific

$$\ddot{x} + x + \operatorname{Sgn} \dot{x} \ni p(t) \quad \text{a.e.}, \tag{3.17}$$

with some forcing p.

The most general result in this direction deals with functions p that are almost periodic; see Sect. 3.4 for more information concerning this class of functions, and also for the relevant notation. Most of this section follows DEIMLING/HETZER/SHEN [67].

Theorem 3.1.3. *Suppose $p \in C_b(\mathbb{R})$ is almost periodic and such that*

$$|\langle p, e^{it} \rangle_{\mathrm{ap}}| < \frac{2}{\pi}. \tag{3.18}$$

Then every solution of (3.17) belongs to $C_b^1([0, \infty[)$.

Proof: It will be convenient to represent a solution $x(t)$ of (3.17) in polar coordinates
$$x(t) = r(t) \cos \theta(t), \quad \dot{x}(t) = r(t) \sin \theta(t),$$
which requires the compatibility condition $\dot{r} \cos \theta = r(\dot{\theta} + 1) \sin \theta$ to be satisfied. As long as $r(t) > 0$, we find that (3.17) is equivalent to

$$\dot{r} \sin \theta + r \dot{\theta} \cos \theta + \operatorname{Sgn}(\sin \theta) + r \cos \theta \ni p \quad \text{a.e.} \tag{3.19}$$

From the compatibility condition we obtain

$$\dot{r} \cos^2 \theta = r(\dot{\theta} + 1) \sin \theta \, \cos \theta, \quad \dot{r} \sin \theta \, \cos \theta = r(\dot{\theta} + 1) \sin^2 \theta,$$

and hence (3.19) yields upon multiplication with $\sin \theta$ and $\cos \theta$ that

$$\dot{r} + |\sin \theta| = p \sin \theta \quad \text{a.e.}, \quad r(\dot{\theta} + 1) + \operatorname{Sgn}(\sin \theta) \cos \theta \ni p \cos \theta \quad \text{a.e.} \tag{3.20}$$

In particular,

$$|\dot{r}| \le C = 1 + |p|_\infty \quad \text{a.e.}, \quad |\dot{\theta} + 1| \le \frac{C}{r} \quad \text{a.e.} \tag{3.21}$$

These estimates are verified on every time interval where $r(t) > 0$ holds.

Now we assume that $r^2(t_j) = |x(t_j)|^2 + |\dot{x}(t_j)|^2 \to \infty$ for some sequence $t_j \to \infty$, and we define $\delta > 0$ through $\frac{2}{\pi} - |\langle p, e^{it} \rangle_{\mathrm{ap}}| = 4\delta$. According to Lemma 3.4.3 we find $T_\delta > 0$ such that for $t \ge T_\delta$

$$\sup_{\tau \in [0,\infty[} \left| \frac{1}{t} \int_{\tau}^{\tau+t} p(s) e^{-is} \, ds \right| \leq |\langle p, e^{is} \rangle_{\mathrm{ap}}| + \delta = \frac{2}{\pi} - 3\delta, \quad \text{and} \quad (3.22a)$$

$$\sup_{\tau \in \mathbb{R}} \left(-\frac{1}{t} \int_{-\tau}^{t-\tau} |\sin s| \, ds \right) \leq -\langle |\sin s|, 1 \rangle_{\mathrm{ap}} + \delta = -\frac{2}{\pi} + \delta; \quad (3.22b)$$

see also Ex. 3.4.2. We claim that for $j \in \mathbb{N}$ large enough and $\tau \in [0,\infty[$

$$r(\tau) = r(t_j) \quad \Longrightarrow \quad r(\tau + t) > 0, \quad t \in [0, T_j], \quad \text{with} \quad T_j = \frac{r(t_j)}{2C}. \quad (3.23)$$

To see this, denote $T^* > 0$ the largest time such that $r(\tau + t) > 0$ for $t \in [0, T^*[$. If $T_j \geq T^*$, then by (3.21), $0 = r(\tau + T^*) \geq r(t_j) - CT^* \geq r(t_j) - CT_j = \frac{1}{2} r(t_j)$, a contradiction. Hence $T_j < T^*$, and (3.23) holds. Using (3.20), we calculate for $\tau \in [0,\infty[$ such that $r(\tau) = r(t_j)$ and for $t \in [0, T_j]$

$$\frac{1}{t} \Big(r(\tau + t) - r(\tau) \Big)$$

$$= \frac{1}{t} \int_{\tau}^{\tau+t} \Big(p(s) \sin \theta(s) - |\sin \theta(s)| \Big) \, ds$$

$$= \frac{1}{t} \int_{\tau}^{\tau+t} \Big(p(s) \sin[\tau + \theta(\tau) - s] - |\sin[\tau + \theta(\tau) - s]| \Big) \, ds$$

$$+ \frac{1}{t} \int_{\tau}^{\tau+t} \Big(|\sin[\tau + \theta(\tau) - s]| - |\sin \theta(s)| \Big) \, ds$$

$$+ \frac{1}{t} \int_{\tau}^{\tau+t} p(s) \Big(\sin \theta(s) - \sin[\tau + \theta(\tau) - s] \Big) \, ds. \quad (3.24)$$

Now (3.21) implies

$$\frac{1}{t} \int_{\tau}^{\tau+t} \Big| |\sin[\tau + \theta(\tau) - s]| - |\sin \theta(s)| \Big| \, ds$$

$$\leq \frac{1}{t} \int_{\tau}^{\tau+t} \Big| [\tau + \theta(\tau)] - [s + \theta(s)] \Big| \, ds$$

$$\leq \frac{C}{t} \int_{\tau}^{\tau+t} ds \int_{\tau}^{s} \frac{d\sigma}{r(\sigma)} = \frac{C}{t} \int_{0}^{t} ds \, \frac{(t-s)}{r(\tau+s)} \leq \frac{Ct}{r(\tau) - Ct}.$$

Recalling $C = 1 + |p|_\infty$, we hence obtain from (3.24), by means of Lemma 3.1.3 below, and with (3.22a) and (3.22b) the estimate

$$\frac{1}{t} \Big(r(\tau + t) - r(\tau) \Big) \leq \left| \frac{1}{t} \int_{\tau}^{\tau+t} p(s) e^{-is} \, ds \right| - \frac{1}{t} \int_{-\theta(\tau)}^{t-\theta(\tau)} |\sin s| \, ds$$

$$+ \frac{C^2 t}{r(\tau) - Ct}$$

$$\leq -2\delta + \frac{C^2 t}{r(\tau) - Ct}, \quad t \in [T_\delta, T_j];$$

note that $T_j \to \infty$, as $r(t_j) \to \infty$ for $j \to \infty$. Since we may assume $\delta \le c$, we have $T_j \ge \frac{\delta r(t_j)}{C(C+\delta)} = \hat{T}_j$, and thus finally

$$r(\tau) = r(t_j) \quad \Longrightarrow \quad r(\tau + t) - r(t_j) \le -\delta t < 0, \quad t \in [T_\delta, \hat{T}_j]. \qquad (3.25)$$

However, if $t \in [0, T_\delta]$, then $r(\tau + t) - r(\tau) \le Ct < 2CT_\delta$ by (3.20), whence

$$r(\tau) = r(t_j) \quad \Longrightarrow \quad r(\tau + t) - r(t_j) < 2CT_\delta, \quad t \in [0, \hat{T}_j]. \qquad (3.26)$$

We choose $\tau = t_j$ and denote $\hat{T}^*$ the largest time such that $r(t_j + t) - r(t_j) < 2CT_\delta$ for $t \in [0, \hat{T}^*[$. Hence $\hat{T}^* > \hat{T}_j$. At time $t = \hat{T}_j$ we even have $r(t_j + \hat{T}_j) - r(t_j) < 0$ by (3.25). If $r(t_j + t) - r(t_j) < 0$ for all $t \ge \hat{T}_j$, then $\hat{T}^* = \infty$. Otherwise we must have $r(\tau_1) = r(t_j)$ for a first $\tau_1 \in]t_j + \hat{T}_j, \hat{T}^*]$. Then $r(\tau_1 + s) - r(t_j) < 2CT_\delta$ for $s \in [0, \hat{T}_j]$ due to (3.26), whence $\hat{T}^* \ge \tau_1 + \hat{T}_j \ge t_j + 2\hat{T}_j$. This argument may be iterated to show that either $\hat{T}^* = \infty$ or $\hat{T}^* \ge t_j + k\hat{T}_j$ for all $k \in \mathbb{N}$. Consequently, $\hat{T}^* = \infty$ in any case, i.e., $r(t_j + t) < 2CT_\delta + 2r(t_j)$ for all $t \in [0, \infty[$ shows r were bounded, a contradiction. $\qquad \square$

The following fact has been used in the proof of Thm. 3.1.3.

Lemma 3.1.3. *Let p be real-valued. Then for every $\tau \in \mathbb{R}$, $t > 0$, and $\vartheta \in \mathbb{R}$*

$$\left| \int_\tau^{\tau+t} p(s) \sin(\vartheta - s) \, ds \right| \le \left| \int_\tau^{\tau+t} p(s) e^{-is} \, ds \right|.$$

Proof: The left-hand side equals $\left| \mathrm{Im}\left(e^{i\vartheta} \int_\tau^{\tau+t} p(s) e^{-is} \, ds \right) \right|$. $\qquad \square$

Remark 3.1.1. (a) Recall that Ex. 2.2.2(b) contains a solution of (3.17), even with $p = 0$, that is unbounded as $t \to -\infty$.

(b) In Ex. 3.4.2 we will see that

$$\langle \sin(\eta t), e^{it} \rangle_{ap} = \begin{cases} -\frac{i}{2} & : \ \eta = 1 \\ 0 & : \ \eta \ne 1 \end{cases}.$$

Hence we can apply Thm. 3.1.3 with $p(t) = \gamma \sin(\eta t)$ to find that the solutions x of (1.2) belong to $C_b^1([0, \infty[)$, either in case that $\eta \ne 1$, or if $\eta = 1$ and $\gamma \in [0, \frac{4}{\pi}[$. A much stronger result will be obtained later in Thm. 4.2.3(a). $\Diamond$

Although there might be solutions that are unbounded on $\mathbb{R}$, the next theorem will give evidence that there is at least one globally bounded solution. Let

$$m = \inf_{t \in \mathbb{R}} p(t) \quad \text{and} \quad M = \sup_{t \in \mathbb{R}} p(t). \qquad (3.27)$$

Theorem 3.1.4. *Assume (3.18) holds for an almost periodic function* $p \in C_b(\mathbb{R})$. *Then (3.17) has a solution in* $C_b^1(\mathbb{R})$. *Moreover, if*

$$M - m > 2, \qquad (3.28)$$

and if the set $\mathcal{B}$ *of solutions to (3.17) that lie in* $C_b^1(\mathbb{R})$ *has more than one element, then there exist* $x, \bar{x} \in \mathcal{B}$ *such that* $z = x - \bar{x} \neq 0$ *satisfies* $\ddot{z} + z = 0$ *a.e. in* $\mathbb{R}$.

Proof: To verify the first assertion, we consider (3.17) as the system

$$\dot{y} \in F(t,y) \quad \text{a.e.}, \quad \text{where} \quad F(t,y) = \{y_2\} \times \{p(t) - y_1 - w : w \in \operatorname{Sgn} y_2\}, \qquad (3.29)$$

for $y = (y_1, y_2) = (x, \dot{x})$. Let $y : [0, \infty[\to \mathbb{R}^2$ be any solution to (3.29); such a solution does exist according to Thm. 2.2.1, see also Ex. 2.2.3. Now observe in light of Thm. 3.1.3 that $y = (x, \dot{x})$ belongs to $C_b([0, \infty[; \mathbb{R}^2)$. Since p is almost periodic, Lemma 3.4.2 implies that $p_j \to p$ uniformly in $\mathbb{R}$ for some sequence $\tau_j \to \infty$, where $p_j(t) = p(t + \tau_j)$. For $j \in \mathbb{N}$ define $y_j : [-\tau_j, \infty[\to \mathbb{R}^2$ through $y_j(t) = y(t + \tau_j)$. Then y_j solves $\dot{y} \in F_j(y, t)$ a.e. in $[-\tau_j, \infty[$, where F_j is like F, only with p replaced by p_j. In particular, $|\dot{y}_j(t)| \leq \|F_j(y_j(t), t)\| \leq c_1(1 + |y_j(t)|) \leq c_2$ a.e. in $[-\tau_j, \infty[$, for some $c_2 > 0$ depending only upon $|p|_\infty$ and $|y|_\infty$. Hence y_j is Lipschitz continuous of constant c_2 on $[-\tau_j, \infty[$. Using the Arzelà–Ascoli theorem and a diagonal argument, we thus find a subsequence (still labeled $j \in \mathbb{N}$) and a continuous function $\bar{y} : \mathbb{R} \to \mathbb{R}^2$ such that $y_j \to \bar{y}$ as $j \to \infty$, uniformly on any compact subinterval of $\mathbb{R}$. Since $p_j \to p$ uniformly in $\mathbb{R}$ and $F(t, \cdot)$ is ε-δ-usc with compact convex values, the argument from the proof of Thm. 2.2.1 shows that $\bar{y}$ solves $\dot{\bar{y}} \in F(t, \bar{y})$ a.e. in $\mathbb{R}$, and $\bar{y} \in C_b(\mathbb{R}; \mathbb{R}^2)$ due to $|\bar{y}|_{C_b(\mathbb{R}; \mathbb{R}^2)} \leq |y|_{C_b([0, \infty[; \mathbb{R}^2)}$. Thus the first component of $\bar{y}$ is a solution to (3.17) that belongs to $C_b^1(\mathbb{R})$.

To validate the second claim, suppose that in addition (3.28) holds, and let $x, \bar{x}$ be any two elements of $\mathcal{B}$ such that $x \neq \bar{x}$. We choose measurable functions $w, \bar{w} : \mathbb{R} \to \mathbb{R}$ satisfying

$$w(t) \in \operatorname{Sgn} \dot{x}(t) \quad \text{a.e.}, \quad \ddot{x}(t) + x(t) + w(t) = p(t) \quad \text{a.e.}, \qquad (3.30a)$$

$$\bar{w}(t) \in \operatorname{Sgn} \dot{\bar{x}}(t) \quad \text{a.e.}, \quad \text{and} \quad \ddot{\bar{x}}(t) + \bar{x}(t) + \bar{w}(t) = p(t) \quad \text{a.e.} \qquad (3.30b)$$

Let $z = x - \bar{x}$ and define $\Delta_\varepsilon(t) = z^2(t) + \dot{z}^2(t)$. Then $\dot{\Delta}_\varepsilon = 2\dot{z}[\ddot{z} + z] = -2(w - \bar{w})(\dot{x} - \dot{\bar{x}}) \leq 0$ a.e., whence $\Delta_\varepsilon(t) \to e_{-\infty}$ as $t \to -\infty$, for some $e_{-\infty} \geq 0$. If $e_{-\infty} = 0$ we get $\Delta_\varepsilon(t) = 0$ for all $t \in \mathbb{R}$, since $\Delta_\varepsilon(\cdot)$ is decreasing. But then $0 = z = x - \bar{x}$, a contradiction. Thus $e_{-\infty} > 0$. Due to Lemma 3.4.2 there exists a sequence $\tau_j \to -\infty$ such that $p_{\tau_j} \to p$ as $j \to \infty$, uniformly in $\mathbb{R}$. Define $x_j(t) = x(t + \tau_j)$ and $\bar{x}_j(t) = \bar{x}(t + \tau_j)$ for $t \in \mathbb{R}$. By the argument just given for the first claim, we find solutions $x_\infty, \bar{x}_\infty : \mathbb{R} \to \mathbb{R}$ of (3.17) such that, at least on a subsequence, $x_j \to x_\infty$, $\bar{x} \to \bar{x}_\infty$, $\dot{x}_j \to \dot{x}_\infty$, and $\dot{\bar{x}} \to \dot{\bar{x}}_\infty$ as $j \to \infty$, uniformly on any compact subinterval of $\mathbb{R}$. In particular,

$$|x_\infty|_\infty \le |x|_\infty, \quad |\bar{x}_\infty|_\infty \le |\bar{x}|_\infty, \quad |\dot{x}_\infty|_\infty \le |\dot{x}|_\infty, \quad \text{and} \quad |\dot{\bar{x}}_\infty|_\infty \le |\dot{\bar{x}}|_\infty.$$
(3.31)

Denote $z_\infty = x_\infty - \bar{x}_\infty$ as well as $\Delta_{\mathcal{E},\infty}(t) = z_\infty^2(t) + \dot{z}_\infty^2(t)$. For fixed $t \in \mathbb{R}$ we have

$$\Delta_{\mathcal{E},\infty}(t) = \lim_{j \to \infty} \left[\left(x(t+\tau_j) - \bar{x}(t+\tau_j) \right)^2 + \left(\dot{x}(t+\tau_j) - \dot{\bar{x}}(t+\tau_j) \right)^2 \right]$$
$$= \lim_{j \to \infty} \Delta_{\mathcal{E}}(t+\tau_j) = e_{-\infty},$$

i.e., $\Delta_{\mathcal{E},\infty}(t) \equiv e_{-\infty} > 0$ is constant, so that $x_\infty \ne \bar{x}_\infty$. For simplicity we denote x_∞ and $\bar{x}_\infty$ again as x and $\bar{x}$. To summarize, we have shown that there exist solutions $x, \bar{x} \in B$ such that $z = x - \bar{x} \ne 0$,

$$(|\dot{x}| + |\dot{\bar{x}}|)(1 - w\bar{w}) = 0 \quad \text{a.e.,} \quad \text{and} \quad \dot{z}[\ddot{z} + z] = 0 \quad \text{a.e.,} \tag{3.32}$$

with some measurable functions $w, \bar{w} : \mathbb{R} \to \mathbb{R}$ satisfying (3.30a), (3.30b). Indeed, for the first relation in (3.32) observe that differentiating $\Delta_{\mathcal{E}}(t) \equiv e_{-\infty}$ yields $\dot{z}[\ddot{z} + z] = 0 = (w - \bar{w})(\dot{x} - \dot{\bar{x}})$ a.e. If both $\dot{x}(t) \ne 0$ and $\dot{\bar{x}}(t) \ne 0$, we have $w(t) = \operatorname{sgn} \dot{x}(t)$ and $\bar{w}(t) = \operatorname{sgn} \dot{\bar{x}}(t)$, and therefore $\operatorname{sgn} \dot{x}(t)\dot{\bar{x}}(t) + \operatorname{sgn} \dot{\bar{x}}(t)\dot{x}(t) = (|\dot{x}(t)| + |\dot{\bar{x}}(t)|)\operatorname{sgn} \dot{x}(t)\operatorname{sgn} \dot{\bar{x}}(t)$ and $(w - \bar{w})(\dot{x} - \dot{\bar{x}}) = 0$ show that (3.32) holds. On the other hand, if e.g. $\dot{x}(t) \ne 0$ and $\dot{\bar{x}}(t) = 0$, then

$$|\dot{x}(t)|(1 - w(t)\bar{w}(t)) = |\dot{x}(t)| - |\dot{x}(t)|w(t)\bar{w}(t) = \bar{w}(t)\left(\dot{x}(t) - |\dot{x}(t)|w(t) \right) = 0$$

as desired. Hence (3.32) holds in any case. Note that up to this point we did not make use of (3.28).

To invoke this assumption, we see that then (3.17) does not have any constant solution. Indeed, if $x(t) \equiv x_0$ in $\mathbb{R}$, then $x_0 + w(t) = p(t)$ a.e. From $|w(t)| \le 1$ a.e. it follows that $x_0 - 1 \le p(t) \le x_0 + 1$ for $t \in \mathbb{R}$, and consequently $x_0 - 1 \le m$ as well as $M \le x_0 + 1$, with m and M from (3.27). Addition then gives the contradiction $M - m \le 2$. In particular, our special solutions x and $\bar{x}$ are non-constant, and therefore $|\dot{x}(t)| + |\dot{\bar{x}}(t)| > 0$ for all t in a neighborhood of some $t_0 \in \mathbb{R}$. From (3.32) we deduce that $w(t)\bar{w}(t) = 1$ a.e. in this neighborhood, and therefore necessarily $w(t) = \bar{w}(t)$ there. Hence $z = x - \bar{x}$ is non-constant, since otherwise $z(t) \equiv z_0$ would yield, due to (3.30a) and (3.30b), that $z_0 = \ddot{z}(t) + z(t) = \bar{w}(t) - w(t) = 0$, the latter a.e. around t_0. However, $z(t) \equiv z_0 = 0$ is impossible, as $e_{-\infty} > 0$. Therefore we find an open interval $]a, b[\subset \mathbb{R}$ such that $\dot{z}(t) \ne 0$ in $]a, b[$. By (3.32) we thus have $\ddot{z} + z = 0$ a.e. in $]a, b[$, and hence $z(t) = A\cos(t+\theta)$ for $t \in]a, b[$, with suitable $A, \theta \in \mathbb{R}$, and $A \ne 0$. Denote $[a_1, b_1]$ the largest interval containing $]a, b[$ such that $z(t) = A\cos(t+\theta)$ for $t \in [a_1, b_1]$. Then b_1 and a_1 are at least of distance π apart, since otherwise $\dot{z}(a_1) \ne 0$ or $\dot{z}(b_1) \ne 0$, and in both cases the continuity of $\dot{z}$ together with (3.32) would yield that $\ddot{z} + z = 0$ a.e. in an interval strictly larger than $[a_1, b_1]$. Therefore we must indeed have $b_1 - a_1 \ge \pi$. We note that as well $\dot{z}(a_1) = 0 = \dot{z}(a_1 + \pi)$ and $\dot{z}(t) \ne 0$ for $t \in]a_1, a_1 + \pi[$: if $\dot{z}(a_1) \ne 0$, then the previous reasoning showed that $[a_1, b_1]$ were not maximal. Next we deduce that $\dot{z}(t) \ne 0$ for some $t \in]a_1 + \pi, \infty[$. To verify this, assume that $z(t) \equiv z_0$

for $t > a_1 + \pi$. If $|\dot{x}(t_0)| + |\dot{\bar{x}}(t_0)| > 0$ for a $t_0 > a_1 + \pi$, then the argument just given above once more implies $z_0 = 0$, contradicting $\Delta_{\varepsilon}(t) \equiv e_{-\infty} > 0$. Hence both $x(t) \equiv x_0$ and $\bar{x} \equiv \bar{x}_0$ will be constant in $[a_1 + \pi, \infty[$. Then again $x_0 - 1 \le p(t) \le x_0 + 1$ for $t \in [a_1 + \pi, \infty[$, cf. the foregoing argument. Choosing now $\tau_j \to \infty$ such that $p_{\tau_j} \to p$ uniformly on $\mathbb{R}$, see Lemma 3.4.2, it follows that in fact $x_0 - 1 \le p(t) \le x_0 + 1$ for all $t \in \mathbb{R}$, in contradiction to (3.28). Thus we have shown that $\dot{z}(t_1) \ne 0$ for some $t_1 \in]a_1 + \pi, \infty[$. As before hence $\dot{z}(t) \ne 0$ at least on some interval $]a_1', b_1'[$ of length π that contains t_1. Since $\dot{z}(a_1 + \pi) = 0$ we must have $a_1' \ge a_1 + \pi$, therefore

$$a_2 = \min\left\{a_1' \ge a_1 + \pi : \dot{z}(t) \ne 0 \text{ in }]a_1', a_1' + \pi[,\ \dot{z}(a_1') = 0\right\}$$

is well-defined, and $a_2 \ge a_1 + \pi$. Suppose that $a_2 > a_1 + \pi$ and $\dot{z}(t_2) \ne 0$ for a $t_2 \in]a_1 + \pi, a_2[$. Again $\dot{z}(t) \ne 0$ on some interval $]a_1'', b_1''[$ of length π that contains t_2. Then $a_1 + \pi \le a_1'' < b_1'' \le a_2$ is impossible due to the definition of a_2. But $a_1'' < a_1 + \pi$ is also impossible, as $\dot{z}(a_1 + \pi) = 0$, and similarly $b_1'' > a_2$ is impossible, since $\dot{z}(a_2) = 0$. This contradiction enforces $\dot{z}(t) = 0$ for $t \in [a_1 + \pi, a_2]$, i.e., $z(t) \equiv z_0 \ne 0$ there. W.l.o.g. we may assume that $z_0 > 0$. Employing the arguments used above, we see that both $x(t) \equiv x_0$ as well as $\bar{x}(t) \equiv \bar{x}_0$ are constant on $[a_1 + \pi, a_2]$. From $\dot{z}(t) \ne 0$ for $t \in]a_2, a_2 + \pi[$ it follows that $|\dot{x}(t)| + |\dot{\bar{x}}(t)| > 0$ for $t \in]a_2, a_2 + \pi[$, whence $1 = w(t)\bar{w}(t)$ a.e. in $]a_2, a_2 + \pi[$ according to (3.32). This in turn yields

$$w(t) = \bar{w}(t) \quad \text{a.e. in }]a_2, a_2 + \pi[. \tag{3.33}$$

Suppose now that there exists a sequence $t_j \searrow a_2$ such that $\dot{x}(t_j) \ne 0$ for $j \in \mathbb{N}$. If $\dot{x}(t_j) > 0$ for infinitely many (w.l.o.g. all) $j \in \mathbb{N}$, we may choose $s_j \in [a_2, t_j[$ such that $\dot{x}(s_j) = 0$ and also $\dot{x}(t) > 0$ for $t \in]s_j, t_j]$. Then $w(t) = 1$ a.e. in $]s_j, t_j]$, and therefore (3.30a) implies through integration

$$\int_{s_j}^{t_j} x(t)\,dt + (t_j - s_j) \le \int_{s_j}^{t_j} [\ddot{x}(t) + x(t) + w(t)]\,dt = \int_{s_j}^{t_j} p(t)\,dt.$$

As both x and p are continuous we may divide by $(t_j - s_j)$ and take the limit $j \to \infty$ to infer that $x_0 + 1 = x(a_2) + 1 \le p(a_2)$. But we also have

$$x_0 + w(t) = p(t) = \bar{x}_0 + \bar{w}(t) \quad \text{a.e. in } [a_1 + \pi, a_2], \tag{3.34}$$

due to (3.30a) and (3.30b). This shows that in particular $p(t) \le \bar{x}_0 + 1$ for all $t \in [a_1 + \pi, a_2]$, thus $x_0 + 1 \le p(a_2) \le \bar{x}_0 + 1$ yields a contradiction to $z_0 = x_0 - \bar{x}_0 > 0$. Hence provided that there is a sequence $t_j \searrow a_2$ with $\dot{x}(t_j) \ne 0$, then necessarily $\dot{x}(t_j) < 0$ for all j large enough (w.l.o.g. for $j \ge 1$). The same argument as above then can be employed to give $x_0 - 1 \ge p(a_2)$. However, (3.34) implies that also $x_0 - 1 \le p(a_2)$, whence $x_0 - 1 = p(a_2)$. Since $\dot{x}(t_j) < 0$ and $\dot{x}(a_2) = 0$, we may choose $s_j \in [a_2, t_j[$ such that $\dot{x}(s_j) = 0$ and $\dot{x}(t) < 0$ for $t \in]s_j, t_j]$. From (3.30a) we obtain $w(t) = -1$ a.e. in $]s_j, t_j]$, and thus also $\bar{w}(t) = -1$ a.e. in $]s_j, t_j]$ by (3.33). Hence (3.30b) shows $\dot{\bar{x}}(t) \leq 0$ in $[s_j, t_j]$. Assume now that for infinitely many $j' \in \mathbb{N}$ we have $\dot{\bar{x}}(\tau_{j'}) < 0$ for some $\tau_{j'} \in [s_{j'}, t_{j'}]$. Then the same argument as for x yields $\bar{x}_0 - 1 = p(a_2)$, and therefore $z_0 = x_0 - \bar{x}_0 = [1 + p(a_2)] - [1 + p(a_2)] = 0$, a contradiction. Consequently we must have $\dot{\bar{x}}(t) = 0$ for $t \in [s_j, t_j]$ and j large enough, w.l.o.g. for $j \geq 1$. Since $\bar{w}(t) = -1$ a.e. in $]s_j, t_j]$, we see from the differential equation in (3.30b) that $\bar{x}(t) - 1 = p(t)$ for $t \in]s_j, t_j]$, in particular $\bar{x}(t_j) - 1 = p(t_j)$. As $j \to \infty$ we obtain the contradiction $\bar{x}_0 - 1 = \bar{x}(a_2) - 1 = p(a_2)$. In summary, we have found that there cannot exist a sequence $t_j \searrow a_2$ with $\dot{x}(t_j) \neq 0$, i.e., $x(t) = x_0$ as well for $t \in [a_2, a_2 + \delta]$, with some $\delta > 0$. But it is impossible that also $\bar{x}(t) = \bar{x}_0$ for $t \in [a_2, a_2 + \bar{\delta}]$ with some $\bar{\delta} > 0$, as $\dot{z}(t) = \dot{x}(t) - \dot{\bar{x}}(t) \neq 0$ for $t \in]a_2, a_2 + \pi[$. If $\dot{\bar{x}}(t_j) < 0$ for a sequence $t_j \searrow a_2$, then the foregoing reasoning shows that $\bar{x}_0 - 1 \geq p(a_2)$ in contradiction to $z_0 > 0$ and $x_0 - 1 \leq p(a_2)$, the latter by (3.34). Thus finally $\dot{\bar{x}}(t_j) > 0$ for a sequence $t_j \searrow a_2$. Then again $\bar{x}_0 + 1 \leq p(a_2)$, thus $\bar{x}_0 + 1 = p(a_2)$ due to (3.33). Moreover, $\dot{\bar{x}}(s_j) = 0$ and $\dot{\bar{x}}(t) > 0$ for $t \in]s_j, t_j]$ with an appropriate $s_j \in [a_2, t_j[$. Whence $w(t) = \bar{w}(t) = 1$ a.e. in $]s_j, t_j]$ by (3.33). If $t_j \leq a_2 + \delta$, this yields $x_0 + 1 = p(t)$ for $t \in [s_j, t_j]$, and hence $x_0 + 1 = p(a_2)$ in the limit $j \to \infty$, in contradiction to $\bar{x}_0 + 1 = p(a_2)$ and $z_0 > 0$.

We thus have shown that in fact $a_2 = a_1 + \pi$, whence $\dot{z}(a_1) = \dot{z}(a_1 + \pi) = \dot{z}(a_1 + 2\pi) = 0$ as well as $\dot{z}(t) \neq 0$ for $t \in]a_1, a_1 + \pi[\cup]a_1 + \pi, a_1 + 2\pi[$; in particular, $w(t) = \bar{w}(t)$ a.e. in $]a_1, a_1 + \pi[\cup]a_1 + \pi, a_1 + 2\pi[$ by (3.32). This argument may be iterated, also to the left instead of to the right, to finally yield $\dot{z}(t) \neq 0$ for $t \in]a_1 + j\pi, a_1 + (j+1)\pi[$ for all $j \in \mathbb{Z}$, hence $\ddot{z} + z = 0$ a.e. in $\mathbb{R}$ due to (3.32). $\qquad \square$

Inspection of the preceding proof also gives the following

Corollary 3.1.1. *Assume $p \in C_b(\mathbb{R})$ is almost periodic, and $x, \bar{x} \in B$ are solutions of (3.17) such that $z = x - \bar{x} \neq 0$ satisfies (3.32), and moreover $|\dot{x}(t_0)| + |\dot{\bar{x}}(t_0)| > 0$ for some $t_0 \in \mathbb{R}$. Then $\ddot{z} + z = 0$ a.e. in $\mathbb{R}$ as well as $w(t) = \bar{w}(t)$ a.e. in $\mathbb{R}$, with w and $\bar{w}$ satisfying (3.30a) and (3.30b), respectively.*

Remark 3.1.2. Note that B from Thm. 3.1.4 indeed may have more than one element. To give an example, let $p(t) = -\lambda \cos(\frac{t}{2})$ with some $\lambda > 0$. Then $\langle p, e^{it} \rangle_{\mathrm{ap}} = 0$, whence (3.18) holds, and (3.28) requires $M - m = 2\lambda > 2$, i.e., $\lambda > 1$. Define

$$x(t) = \begin{cases} (x_0 + 1 + \frac{4}{3}\lambda)\cos(t) - \frac{4}{3}\lambda \cos(\frac{t}{2}) - 1 & : \quad t \in]0, 2\pi[\\ (x_0 - 1 + \frac{4}{3}\lambda)\cos(t) - \frac{4}{3}\lambda \cos(\frac{t}{2}) + 1 & : \quad t \in]2\pi, 4\pi[\end{cases}.$$

Then x, periodically continued to all of $\mathbb{R}$, is a C^1-function and globally bounded. One moreover calculates that

$$\ddot{x}(t) + x(t) = \begin{cases} p(t) - 1 & : \quad t \in]0, 2\pi[\\ p(t) + 1 & : \quad t \in]2\pi, 4\pi[\end{cases}.$$

Hence x will be a solution to (3.17) provided we can ensure, by suitable choice of x_0 and λ, that $w(t) \in \operatorname{Sgn} \dot{x}(t)$ for $t \in]0, 2\pi[\cup]2\pi, 4\pi[$, where

$$w(t) = \begin{cases} 1 & : \quad t \in]0, 2\pi[\\ -1 & : \quad t \in]2\pi, 4\pi[\end{cases}.$$

Thus we need to guarantee that $\dot{x}(t) \geq 0$ for $t \in]0, 2\pi[$ and $\dot{x}(t) \leq 0$ for $t \in]2\pi, 4\pi[$. Observing that $\sin(t) = 2\cos(\frac{t}{2})\sin(\frac{t}{2})$ as well as e.g. $\sin(\frac{t}{2}) \geq 0$ for $t \in [0, 2\pi]$, these both conditions read as

$$\left| x_0 + 1 + \frac{4}{3}\lambda \right| \leq \frac{\lambda}{3} \quad \text{and} \quad \left| x_0 - 1 + \frac{4}{3}\lambda \right| \leq \frac{\lambda}{3}. \tag{3.35}$$

Thus necessarily $\lambda \geq 3$, since $2 = (x_0 + 1 + \frac{4}{3}\lambda) + (-x_0 + 1 - \frac{4}{3}\lambda) \leq \frac{2}{3}\lambda$. For $\lambda = 3$, (3.35) enforces $x_0 = -4$, but in case that $\lambda > 3$ we have the choice of $x_0 = -1 - \lambda - A$, where $A \in [0, \frac{2}{3}(\lambda - 3)]$. Thus we have found several globally bounded (even 4π-periodic) solutions to (3.17). $\Diamond$

3.2 Periodic solutions

3.2.1 Periodic solutions for the pendulum with dry friction

Here we continue to study

$$\ddot{x} + x + \operatorname{Sgn} \dot{x} \ni p(t) \quad \text{a.e.}, \tag{3.17}$$

see Sect. 3.1.3, this time under the viewpoint of existence of periodic solutions, provided that the forcing p is periodic. We start with the non-resonant case.

Theorem 3.2.1 (Non-resonant case). *Suppose that $\omega > 0$ and $p \in C(\mathbb{R})$ is ω-periodic. If $\omega \neq 2k\pi$ for $k \in \mathbb{N}$, then there exists an ω-periodic solution of (3.17).*

Proof: Denoting $y = (y_1, y_2) = (x, \dot{x})$, we first rewrite (3.17) as

$$\dot{y} \in Ay + F(t, y) \quad \text{a.e.,} \quad \text{where}$$

$$A = \begin{pmatrix} 0 & 1 \\ -1 & 0 \end{pmatrix} \quad \text{and} \quad F(t, y) = \{0\} \times \{p(t) - w : w \in \operatorname{Sgn} y_2\}.$$

Define

$$g(t, s) = U(t)(\operatorname{id} - U(\omega))^{-1} U(s)^{-1}, \quad t, s \in \mathbb{R},$$

with the fundamental matrix $U(t) = \begin{pmatrix} \cos t & -\sin t \\ \sin t & \cos t \end{pmatrix}$ of $\dot{y} = Ay$, and let

$$\Gamma(t,s) = \begin{cases} g(t,s) & : \quad 0 \le s \le t \le \omega \\ g(t+\omega,s) & : \quad 0 \le t < s \le \omega \end{cases}.$$

Note that $(\mathrm{id} - U(\omega))^{-1}$ is well-defined due to our assumption $\omega \ne 2k\pi$ for $k \in \mathbb{N}$. For $y \in Y = C([0,\omega]; \mathbb{R}^2)$ we moreover introduce the multi-valued mapping

$$G(y) = \Big\{ z \in Y : z(t) = \int_0^\omega \Gamma(t,s) f(s)\, ds \text{ for some measurable function}$$

$$f : [0,\omega] \to \mathbb{R}^2 \text{ such that } f(t) \in F(t, y(t)) \text{ a.e. in } [0,\omega] \Big\}.$$

For each $y = (y_1, y_2) \in Y$ such a selection f of $F(\cdot, y(\cdot))$ does exist; this follows either from an abstract result like DEIMLING [65, Exerc. 3.10], or for this particular F we may explicitly set

$$f(t) = \mathbf{1}_A(t)\, (0, p(t) - 1) + \mathbf{1}_{\mathbb{R}\setminus A}(t)\, (0, p(t) + 1),$$

with the measurable $A = \{t \in \mathbb{R} : y_2(t) \ge 0\}$; whence $Gy \ne \emptyset$. Using Thm. 2.1.1, we are going to show that $G : \overline{B}_r(0) \subset Y \to 2^Y \setminus \{\emptyset\}$ has a fixed point, if $r > 0$ is chosen sufficiently large. Since $\|F(t,y)\| \le |p|_\infty + 1$, the boundedness of g implies that for any $y \in Y$ and $z \in G(y)$ we have

$$|z(t)| \le C \int_0^\omega |f(s)|\, ds \le C\omega(|p|_\infty + 1) =: r.$$

Hence in particular $G(\overline{B}_r(0)) \subset \overline{B}_r(0)$. In addition, $G(Y) \subset Y$ is relatively compact. To see this, let $y \in Y$ and $z \in G(y)$. For $0 \le t_1 < t_2 \le \omega$ we then find, using $|g(\tau_1, s) - g(\tau_2, s)| \le C|\tau_1 - \tau_2|$ and $|g(\tau, s)| \le C$, that

$$|z(t_1) - z(t_2)|$$
$$= \Big| \int_0^{t_1} [g(t_1, s) - g(t_2, s)] f(s)\, ds - \int_{t_1}^{t_2} g(t_2, s) f(s)\, ds$$
$$+ \int_{t_2}^{\omega} [g(t_1 + \omega, s) - g(t_2 + \omega, s)] f(s)\, ds + \int_{t_1}^{t_2} g(t_1 + \omega, s) f(s)\, ds \Big|$$
$$\le C|t_1 - t_2|.$$

Hence $G(Y) \subset Y$ is equicontinuous, and therefore relatively compact, by the Arzelà–Ascoli theorem. Thus $G(\overline{B}_r(0)) \subset Y$ is relatively compact, too. In order to show that G is ε-δ-usc, fix $\varepsilon > 0$ and $y^{(0)} = (y_1^{(0)}, y_2^{(0)}) \in Y$. Then $A_0 = \{\xi : |\xi - y_2^{(0)}(t)| \le 1 \text{ for some } t \in [0,\omega]\} \subset \mathbb{R}$ is compact. Since $\mathrm{Sgn}(\cdot)$ is ε-δ-usc on $\mathbb{R}$ and has compact values, Rem. 2.1.2 implies that we find $\delta > 0$ such that $\mathrm{Sgn}\, v \subset \mathrm{Sgn}\, v_0 + B_\varepsilon(0)$ for $v, v_0 \in A_0$ with $|v - v_0| \le \delta$. Let $y \in Y$ satisfy $|y - y^{(0)}| \le \min\{\delta, 1\}$, and let $z \in G(y)$, i.e., $z(t) = \int_0^\omega \Gamma(t, s) f(s)\, ds$ for some measurable $f : [0, \omega] \to \mathrm{I\!R}^2$ with $f(t) \in F(t, y(t))$ a.e. in $[0, \omega]$. Writing $f = (f_1, f_2)$, the latter means $f_1(t) = 0$ a.e. and $p(t) - f_2(t) \in \mathrm{Sgn}\, y_2(t) \subset \mathrm{Sgn}\, y_2^{(0)}(t) + B_\varepsilon(0)$ a.e. This may be seen to imply the existence of measurable functions $s, r : [0, \omega] \to \mathrm{I\!R}$ such that $p(t) - f_2(t) = s(t) + r(t)$, $s(t) \in \mathrm{Sgn}\, y_2^{(0)}(t)$, and $|r(t)| < \varepsilon$, all a.e. in $[0, \omega]$; one may e.g. use DEIMLING [65, Exerc. 12.1] to verify this. Then we define $f^{(0)} = (f_1^{(0)}, f_2^{(0)})$, where $f_1^{(0)}(t) = 0$ and $f_2^{(0)}(t) = f_2(t) + r(t)$ for $t \in [0, \omega]$. We deduce that $f^{(0)}(t) \in F(t, y^{(0)}(t))$ a.e., hence $z^{(0)} \in G(y^0)$ for $z^{(0)}(t) = \int_0^\omega \Gamma(t, s) f^{(0)}(s)\, ds$, $t \in [0, \omega]$. Moreover,

$$|z(t) - z^{(0)}(t)| = \left| \int_0^\omega \Gamma(t, s)[f(s) - f^{(0)}(s)]\, ds \right| \le \omega |\Gamma|_\infty |r|_\infty \le \omega |\Gamma|_\infty \varepsilon,$$

and thus G in fact is ε-δ-usc on Y.

Consequently, all assumptions of Thm. 2.1.1 are verified, and thus $y \in G(y)$ for some $y \in D$. By definition, we find that $y(t) = \int_0^\omega \Gamma(t, s) f(s)\, ds$ for some measurable function f with $f(t) \in F(t, y(t))$ a.e. Since $\Gamma(0, s) = g(\omega, s) = \Gamma(\omega, s)$, we obtain that y is ω-periodic. Observing

$$\frac{\partial g}{\partial t}(t, s) = A g(t, s) \quad \text{and} \quad g(t, t) - g(t + \omega, t) = \mathrm{id},$$

the latter due to the group property $U(t + s) = U(t) U(s)$, we also have

$$\dot{y}(t) = \frac{d}{dt} \left(\int_0^t g(t, s) f(s)\, ds + \int_t^\omega g(t + \omega, s) f(s)\, ds \right)$$
$$= A y(t) + [g(t, t) - g(t + \omega, t)] f(t) = A y(t) + f(t)$$
$$\in A y(t) + F(t, y(t)) \quad \text{a.e.,}$$

which completes the proof. $\qquad\qquad\qquad\qquad\qquad\qquad\qquad\qquad \square$

We remark that for the special $p(t) = \gamma \sin(\eta t + \phi)$, Thm. 3.2.1 was first proved in DEIMLING/SZILÁGYI [68, Thm. 1], using a degree theory argument to find a fixed point of G. Our approach follows DEIMLING [66, Sect. 2]; see also IRISOV/TONKOVA/TONKOV [102], as outlined in DEIMLING [65, Exerc. 13.4]. It moreover has to be mentioned that (contrary to the resonant case $\omega = 2\pi$ to follow) in the non-resonant case it is not known whether periodic solutions have to be unique or uniformly attracting, or whether they will stick to $\dot{x} = 0$ for a while.

Now we turn to the resonant case, where p is 2π-periodic; cf. Rem. 3.2.1 for $\omega = 2k\pi$ with some $k \ge 2$.

Lemma 3.2.1. *Suppose $p \in C(\mathbb{R})$ is 2π-periodic and (3.17) has an 2π-periodic solution. Then*

$$\left| \int_0^{2\pi} p(t)\,e^{-it}\,dt \right| \leq 4. \qquad (3.36)$$

Proof: We find a measurable function $w : \mathbb{R} \to \mathbb{R}$ such that $w(t) \in \operatorname{Sgn}\dot{x}(t)$ a.e. and $\ddot{x}(t) + x(t) + w(t) = p(t)$ a.e. Multiplication by e^{-it} and integration by parts twice yields

$$\int_0^{2\pi} p(t)e^{-it}\,dt = \int_0^{2\pi} w(t)e^{-it}\,dt + \int_0^{2\pi} [\ddot{x}(t) + x(t)]e^{-it}\,dt$$

$$= \int_0^{2\pi} w(t)e^{-it}\,dt + \left([\dot{x}(t) + ix(t)]e^{-it} \right)\Big|_{t=0}^{t=2\pi}$$

$$= \int_0^{2\pi} w(t)e^{-it}\,dt, \qquad (3.37)$$

due to periodicity of x. If we choose the unique $\tau \in [0, 2\pi[$ with

$$e^{i\tau} \left(\int_0^{2\pi} w(t)\,e^{-it}\,dt \right) = \left| \int_0^{2\pi} w(t)\,e^{-it}\,dt \right|,$$

then $\left| \int_0^{2\pi} w(t)\,e^{it}\,dt \right| = \int_0^{2\pi} w(t)\cos(t - \tau)\,dt$ by taking real parts, and this shows the necessity of (3.36), since we have $|w(t)| \leq 1$ a.e. and moreover $\int_0^{2\pi} |\cos(t - \tau)|\,dt = 4$. $\qquad \square$

Theorem 3.2.2. *Assume $p \in C(\mathbb{R})$ is 2π-periodic. If "$<$" holds in (3.36), then (3.17) has a 2π-periodic solution. In case that "$=$" in (3.36), then a 2π-periodic solution exists if and only if*

$$\int_0^{\pi} p(t + \tau - \pi/2)\cos(t)\,dt = 0, \qquad (3.38)$$

with $\tau \in [0, 2\pi[$ being the argument of $\int_0^{2\pi} p(t)\,e^{-it}\,dt \in \mathbb{C}$. In this case there are infinitely many (explicitly known) such solutions.

Proof: See DEIMLING [66, Sect. 3 & 4]. In Thm. 3.2.8(b) below we will give an analogous proof of the corresponding result for the conservative system (3.47). $\qquad \square$

Next we will investigate multiplicity and stability of periodic solutions to (3.17). For this we recall from (3.27) the notation

$$m = \inf_{t \in \mathbb{R}} p(t) \quad \text{and} \quad M = \sup_{t \in \mathbb{R}} p(t).$$

Lemma 3.2.2. *Let $\omega > 0$ and $0 < M - m \leq 2$. If x is an ω-periodic solution to (3.17) with $p \in C(\mathbb{R})$ being ω-periodic, then $x(t) \equiv c$, for some $c \in [M - 1, m + 1]$.*

Proof: This is a special case of Lemma 3.3.2 below.

In the non-resonant case we have

Theorem 3.2.3. *Assume that $p \in C(\mathbb{R})$ is ω-periodic, $\omega \neq 2k\pi$ for $k \in \mathbb{N}$, and $M - m > 2$. Then there is exactly one ω-periodic solution of (3.17).*

Proof: Due to Thm. 3.2.1 we have to verify uniqueness of the periodic solution. Suppose $x, \bar{x}$ are two such solutions with $z = x - \bar{x} \neq 0$. Then both x and $\bar{x}$ are globally bounded. Choosing $w, \bar{w}$ as in (3.30a), (3.30b), and defining $\Delta_{\mathcal{E}}(t) = z^2(t) + \dot{z}^2(t)$, as in the proof of Thm. 3.1.4 it follows that $\dot{\Delta}_{\mathcal{E}} = 2\dot{z}[\ddot{z} + z] = -2(w - \bar{w})(\dot{x} - \dot{\bar{x}}) \leq 0$ a.e. Thus $\Delta_{\mathcal{E}}$ is decreasing and ω-periodic, hence constant, i.e., (3.32) is satisfied. Therefore $\ddot{z} + z = 0$ a.e. in $\mathbb{R}$ according to Cor. 3.1.1, as neither x nor $\bar{x}$ can be constant. This in turn implies $z(t) = A\cos(t + \theta)$ for suitable $A, \theta \in \mathbb{R}$, but since $\omega \neq 2k\pi$ for $k \in \mathbb{N}$, and z is ω-periodic, we see that $A = 0$, a contradiction. $\square$

A similar approach can be employed in the resonant case to yields uniqueness of periodic solutions.

Theorem 3.2.4. *Suppose $p \in C(\mathbb{R})$ is 2π-periodic, "$<$" holds in (3.36), and $M - m > 2$. Then there is exactly one 2π-periodic solution of (3.17).*

Proof: If $x, \bar{x}$ both are 2π-periodic solutions, and $z = x - \bar{x} \neq 0$, then once more Cor. 3.1.1 applies. Whence $x(t) = \bar{x}(t) + A\cos(t + \theta)$ for some $A, \theta \in \mathbb{R}$ with $A \neq 0$ as well as $w(t) = \bar{w}(t)$ a.e. in $\mathbb{R}$, with w and $\bar{w}$ satisfying (3.30a) and (3.30b), respectively. We therefore obtain

$$\begin{cases} \dot{x}(t) > 0 & \implies \dot{x}(t) - \dot{z}(t) \geq 0 \\ \dot{\bar{x}}(t) > 0 & \implies \dot{\bar{x}}(t) + \dot{z}(t) \geq 0 \end{cases}. \tag{3.39}$$

To see e.g. the first assertion, if $\dot{x}(s) > 0$ for s in a neighborhood of some $t \in \mathbb{R}$, then $1 = w(s) = \bar{w}(s) \in \mathrm{Sgn}\,\bar{x}(s) = \mathrm{Sgn}\,(\dot{x}(s) - \dot{z}(s))$ a.e. in this neighborhood implies $\dot{x}(s) - \dot{z}(s) \geq 0$ for s close to t. Thus (3.39) holds. Next we choose $t_0 \in \mathbb{R}$ such that $\dot{z}(t) = \dot{x}(t) - \dot{\bar{x}}(t) = -A\sin(t + \theta) > 0$ for $t \in]t_0, t_0 + \pi[$ as well as $\dot{z}(t) < 0$ for $t \in]t_0 + \pi, t_0 + 2\pi[$. As $\dot{x} \not\equiv 0$, we moreover find, w.l.o.g., $t_* \in \mathbb{R}$ such that $\dot{x}(t_*) > 0$. Shifting t_0 by a multiple of 2π and changing t_* a little, we may suppose that $t_* \in]t_0, t_0 + 2\pi[$ and $t_* \neq t_0 + \pi$. We are going to show that $t_* \in]t_0, t_0 + \pi[$ is impossible, the case $t_* \in]t_0 + \pi, t_0 + 2\pi[$ being dealt with analogously. So we assume $t_* \in]t_0, t_0 + \pi[$. Then $\dot{x}(t) > 0$ in a neighborhood of t_* implies $\dot{x}(t) \geq \dot{z}(t) > 0$ for t in this neighborhood by (3.39). Since $\dot{z}(t) > 0$ in $]t_0, t_0 + \pi[$, by considering the largest interval on which the estimate $\dot{x}(t) > 0$ is verified, we see that in fact $\dot{x}(t) > 0$ for all $t \in]t_0, t_0 + \pi[$. Assume next that $\dot{x}(t_1) \geq 0$ for some $t_1 \in]t_0 + \pi, t_0 + 2\pi[$. Then $\dot{\tilde{x}}(t_1) = \dot{x}(t_1) - \dot{z}(t_1) > 0$, whence locally $\dot{\tilde{x}}(t) > 0$ close to t_1. Utilizing the foregoing argument and (3.39), it follows that then $\dot{\tilde{x}}(t) > 0$ for all $t \in]t_0 + \pi, t_0 + 2\pi[$. Due to $w(t) = \bar{w}(t)$ a.e. in $\mathbb{R}$, this in turn implies $\dot{x}(t) \geq 0$ for $t \in]t_0 + \pi, t_0 + 2\pi[$. Hence x is strictly increasing over $]t_0, t_0 + \pi[$ and non-decreasing over $[t_0 + \pi, t_0 + 2\pi]$, contradicting the fact that x is 2π-periodic. Consequently we must have $\dot{x}(t) < 0$ for all $t \in]t_0 + \pi, t_0 + 2\pi[$, and for w this means

$$w(t) = \begin{cases} 1 & \text{a.e. in }]t_0, t_0 + \pi[\\ -1 & \text{a.e. in }]t_0 + \pi, t_0 + 2\pi[\end{cases}. \tag{3.40}$$

Then due to periodicity of p and x, and as in (3.37), we calculate using (3.40)

$$\left| \int_0^{2\pi} p(t) e^{-it} dt \right| = \left| \int_{t_0}^{t_0+2\pi} p(t) e^{-it} dt \right| = \left| \int_{t_0}^{t_0+2\pi} w(t) e^{-it} dt \right| = 4,$$

but this is impossible according to "$<$" in (3.36).

Remark 3.2.1. If $p \in C(\mathbb{R})$ is $2k\pi$-periodic for some $k \geq 2$, then a necessary condition for the existence of a $2k\pi$-periodic solution to (3.17) is

$$\left| \int_0^{2k\pi} p(t) e^{-it} dt \right| \leq 4k. \tag{3.41}$$

Conversely, if even "$<$" holds in (3.41), then (3.17) has a $2k\pi$-periodic solution. This may be verified in the same manner as for $k = 1$. Note, however, that in Rem. 3.1.2 we had $p(t) = -\lambda \cos(\frac{t}{2})$, i.e., $\omega = 4\pi$ and $k = 2$, but a whole family of 4π-periodic solutions for $\lambda > 3$. ◊

Next we consider the problem of stability of the periodic solutions to (3.17). If it is unique, the periodic solution turns out to be asymptotically stable.

Theorem 3.2.5. *Assume $M - m > 2$, and $\omega \neq 2k\pi$ for $k \in \mathbb{N}$, or $\omega = 2\pi$ and "$<$" holds in (3.36). Let x_* denote the corresponding unique ω-periodic solution from Thm. 3.2.3 or Thm. 3.2.4, respectively. Then*

$$\left(x(t) - x_*(t) \right)^2 + \left(\dot{x}(t) - \dot{x}_*(t) \right)^2 \to 0 \quad \text{as} \quad t \to \infty \tag{3.42}$$

for any solution x of (3.17).

Proof: Defining $\Delta_{\mathcal{E}}(t) = (x(t) - x_*(t))^2 + (\dot{x}(t) - \dot{x}_*(t))^2$, we obtain that $\dot{\Delta}_{\mathcal{E}}(t) \leq 0$ a.e., cf. the proof of Thm. 3.2.3. Thus $\Delta_{\mathcal{E}}$ is decreasing, whence $\Delta_{\mathcal{E}}(t) \to e_\infty$ as $t \to \infty$ for some $e_\infty \geq 0$, and we need to show that in fact $e_\infty = 0$. Now we may proceed similar to the first part of the proof of

Thm. 3.1.4: for $j \in \mathbb{N}$ we let $x_j(t) = x(j\omega + t)$. Then $\ddot{x}_j(t) + x_j(t) + \mathrm{Sgn}\,\dot{x}_j(t) \ni p(t)$ a.e. due to the periodicity of p. Therefore we may select a subsequence (w.l.o.g. labeled $j \in \mathbb{N}$) and a C^1-function x_∞ such that $x_j \to x_\infty$ and $\dot{x}_j \to \dot{x}_\infty$ as $j \to \infty$, uniformly on any compact subinterval of $\mathbb{R}$, and x_∞ solves (3.17). Now observe that for every $t \in \mathbb{R}$ by periodicity of x_*

$$
\begin{aligned}
\Big(x_\infty(t) - x_*(t)\Big)^2 &+ \Big(\dot{x}_\infty(t) - \dot{x}_*(t)\Big)^2 \\
&= \lim_{j \to \infty} \left[\Big(x(j\omega + t) - x_*(j\omega + t)\Big)^2 + \Big(\dot{x}(j\omega + t) - \dot{x}_*(j\omega + t)\Big)^2 \right] \\
&= \lim_{j \to \infty} \Delta_\varepsilon(j\omega + t) = e_\infty.
\end{aligned}
\tag{3.43}
$$

On the other hand, $x_\infty(t+\omega) = \lim_{j\to\infty} x_j(t+\omega) = \lim_{j\to\infty} x([j+1]\omega+t) = \lim_{j\to\infty} x_{j+1}(t) = x_\infty(t)$, i.e., x_∞ is ω-periodic as well. Thus $x_\infty = x_*$, since the ω-periodic solution is unique, and this implies $e_\infty = 0$ by (3.43). $\square$

So it remains to deal with the case $0 < M - m \le 2$. Then we will find that the set $[M-1, m+1]$ of constant solutions is asymptotically stable.

Theorem 3.2.6. *Let $\omega > 0$, let $p \in C(\mathbb{R})$ be ω-periodic, and assume $0 < M - m \le 2$. If x is any solution of (3.17), then there is $c = c_x \in [M-1, m+1]$ such that*

$$
(x(t) - c)^2 + \dot{x}^2(t) \to 0 \quad as \quad t \to \infty.
$$

Proof: In Thm. 3.3.3 below we will verify a more general statement. $\square$

Most results of this section are taken from DEIMLING [64, 66], DEIM-LING/SZILÁGYI [68], and DEIMLING/HETZER/SHEN [67]. In BOTHE [28] the more general $\ddot{x} + x + \mu(x)\,\mathrm{Sgn}\,\dot{x} \ni p(t)$ is investigated, with an ω-periodic forcing p. This means the coefficient of friction μ is allowed to depend on the state x. Under some assumptions relating p and μ, it is shown that the equation has an ω-periodic solution, and also the boundedness of solutions is studied. Moreover, a "principle of linearized stability" is provided for solutions that do not stick to $\dot{x} = 0$. Some related, mostly non-rigorous or semi-rigorous, results for more general friction laws and models are contained in e.g. CAUGHEY/MASRI [46], CAUGHEY/VIJAYARAGHAVAN [47, 48], CAPECCHI [43], DANKOWICZ [60], DANKOWICZ/NORDMARK [61], and IN-AUDI/KELLY [101]. The reference REITHMEIER [193, Sect. 3] deals with the existence of periodic solutions for non-resonant forcings, see also KLOTTER [109, Chap. 6.6], and in addition PARNES [168] for an application to a real-world problem concerning buried pipes.

Finally we turn to the special case

$$
\ddot{x} + x + \mathrm{Sgn}\,\dot{x} \ni \gamma \sin(\eta t) \quad \text{a.e.}
\tag{3.44}
$$

from (1.2), i.e., $p(t) = \gamma \sin(\eta t)$ with $\gamma, \eta \ge 0$.

Theorem 3.2.7. *(a) If $\eta \neq 1$, then (3.44) has a $(2\pi/\eta)$-periodic solution. This solution is unique for $\gamma > 1$, and denoting it x_*, then*

$$\left(x(t) - x_*(t)\right)^2 + \left(\dot{x}(t) - \dot{x}_*(t)\right)^2 \to 0 \quad as \quad t \to \infty \qquad (3.45)$$

for every other solution x of (3.44). For $\gamma \in [0,1]$, the $(2\pi/\eta)$-periodic solutions are exactly the equilibria $x_0 \in [\gamma - 1, -\gamma + 1]$. In this case, for any solution x of (3.44) there is $c = c_x \in [\gamma - 1, -\gamma + 1]$ satisfying

$$(x(t) - c)^2 + \dot{x}^2(t) \to 0 \quad as \quad t \to \infty. \qquad (3.46)$$

(b) For $\eta = 1$ the following holds.

  (b1) If $\gamma \in [0,1]$, then the equilibria $x_0 \in [\gamma - 1, -\gamma + 1]$ are the only 2π-periodic solutions of (3.44). Moreover, (3.46) holds.

  (b2) If $\gamma \in]1, 4/\pi[$, then there exists a unique 2π-periodic solution $x_\gamma : [0, 2\pi] \to \mathbb{R}$ of (3.44). This x_γ is globally asymptotically stable for (3.44), in the sense of (3.45) with x_γ in place of x_*. In addition, there are numbers $t_1^\gamma \in]0, \pi/2[$ and $t_2^\gamma \in]\pi/2, \pi[$ such that x_γ is given through

$$\begin{cases} x_\gamma(t) \equiv -1 + \gamma \sin t_1^\gamma & for \quad t \in [0, t_1^\gamma], \\ \dot{x}_\gamma(t) > 0 & for \quad t \in]t_1^\gamma, t_2^\gamma[, \\ x_\gamma(t) \equiv 1 - \gamma \sin t_1^\gamma & for \quad t \in [t_2^\gamma, \pi], \\ x_\gamma(t) = -x_\gamma(t + \pi) & for \quad t \in \mathbb{R} \end{cases}$$

Here t_1^γ and t_2^γ are the unique solutions of

$$\cot an t_2^\gamma = -\frac{t_2^\gamma - t_1^\gamma - \sin t_1^\gamma \cos t_1^\gamma}{\sin^2 t_1^\gamma}, \quad 2/\gamma = \frac{1}{2}\frac{\sin^2 t_1^\gamma}{\sin t_2^\gamma} + \frac{1}{2}\sin t_2^\gamma + \sin t_1^\gamma,$$

or equivalently

$$t_2^\gamma - t_1^\gamma = \frac{\sin t_1^\gamma}{\sin t_2^\gamma}\sin(t_2^\gamma - t_1^\gamma), \quad 4\sin t_2^\gamma = \gamma\left(\sin t_1^\gamma + \sin t_2^\gamma\right)^2.$$

  (b3) If $\gamma = 4/\pi$, then there are infinitely many 2π-periodic solutions of (3.44), and they may be calculated explicitly.

  (b4) If $\gamma \in]4/\pi, \infty[$, then (3.44) has no 2π-periodic solution.

Proof: Ad (a): From Thm. 3.2.1 we obtain existence of a periodic solution, and since $M - m = 2\gamma$, uniqueness in case that $\gamma > 1$ is implied by Thm. 3.2.3, whereas (3.45) follows from Thm. 3.2.5. Next, Lemma 3.2.2 says that the periodic solutions coincide with the constants $x_0 \in [M - 1, m + 1] = [\gamma - 1, -\gamma + 1]$ if $\gamma \in [0, 1]$ (for $\gamma = 0$ obviously satisfied), and the associated stability assertion is contained in Thm. 3.2.6. Accordingly, also (b1) is verified. Ad (b2): Since $\int_0^{2\pi} \sin(t)e^{it} dt = i\pi$, we first note that $\gamma \in]1, 4/\pi[$ corresponds to "<"

in (3.36) and $M - m > 2$. Hence Thm. 3.2.4 applies to yield a unique 2π-periodic solution, and (3.45) once more follows from Thm. 3.2.5. Concerning the special form of x_γ, this was shown in DEIMLING/SZILÁGYI [68, Prop. 3]. Ad (b3): We can apply Thm. 3.2.2, as $\int_0^{2\pi} \sin(t)e^{-it}\,dt = -i\pi$ has $\tau = 3\pi/2$, and $\int_0^\pi \sin(s + \pi)\cos(s)\,ds = 0$. Ad (b4): Here $|\int_0^{2\pi} p(t)e^{it}\,dt| = \gamma\pi > 4$, thus Lemma 3.2.1 ensures that there are no periodic solutions at all.

According to Ex. 3.1.4 and Thm. 3.2.7(a), (b1), (b2), (b3), we also obtain the following

Corollary 3.2.1. *If $\eta \neq 1$, or if $\eta = 1$ and $\gamma \in [0, \frac{4}{\pi}]$, then Thm. 3.1.2 applies to (3.44), i.e., to (1.2).*

3.2.2 Periodic solutions of conservative problems

In this section we consider

$$\ddot{x} + x + a\,\mathrm{Sgn}\,x \ni p(t) \quad \text{a.e.} \tag{3.47}$$

with a (piecewise) continuous forcing p and a parameter $a > 0$, and we want to decide whether or not periodic solutions do exist. The results are taken from KUNZE [113], and the approach is similar to DEIMLING [66]; in FEČKAN [78] the problems $\ddot{x} + x + g(x) = p(t)$ are investigated, with a bounded but not necessarily continuous g.

For (3.47) we obtain

Theorem 3.2.8. *(a) If $\omega \neq 2k\pi$ for all $k \in \mathbb{N}$, then (3.47) has an ω-periodic solution for every $a > 0$.*
(b) Consider the case $\omega = 2\pi$. Then a necessary condition for (3.47) to have a 2π-periodic solution is

$$\left| \int_0^{2\pi} p(t)\,e^{it}\,dt \right| \leq 4a. \tag{3.48}$$

On the other hand, if "$<$" holds in (3.48), then (3.47) has a 2π-periodic solution. Finally, if we have "$=$" in (3.48), then a 2π-periodic solution does exist if and only if

$$\int_0^\pi \sin(\pi - t)\,p(t - \tau - \pi/2)\,dt = 2a, \tag{3.49}$$

with $\tau \in [0, 2\pi[$ being the argument of $\int_0^{2\pi} p(t)\,e^{it}\,dt \in \mathbb{C}$. In this case there are infinitely many (explicitly known) of such solutions.

Proof: Ad (a): This may be shown analogously to Thm. 3.2.1, since with $y = (y_1, y_2) = (x, \dot{x})$ we see that (3.47) reads as

$$\dot{y} \in Ay + F(t, y) \quad \text{a.e.,} \quad \text{where}$$

$$A = \begin{pmatrix} 0 & 1 \\ -1 & 0 \end{pmatrix} \quad \text{and} \quad F(t, y) = \{0\} \times \{p(t) - aw : w \in \operatorname{Sgn} y_1\}.$$

Ad (b): The fact that (3.48) is a necessary condition is verified as in Lemma 3.2.1. Next we suppose that "=" holds in (3.48) and that there exists a 2π-periodic solution x. Choose a measurable function w such that

$$\ddot{x}(t) + x(t) + a\,w(t) = p(t) \quad \text{a.e. in } [0, 2\pi], \quad \text{where} \quad (3.50a)$$

$$w(t) \in \operatorname{Sgn} x(t) \quad \text{a.e. in } [0, 2\pi], \quad (3.50b)$$

and moreover let $\tau \in [0, 2\pi[$ satisfy $e^{i\tau} \left(\int_0^{2\pi} w(t)\,e^{it}\,dt \right) = \left| \int_0^{2\pi} w(t)\,e^{it}\,dt \right|$. Calculating as in (3.37) with e^{it} in place of e^{-it}, we conclude that

$$\int_0^{2\pi} w(t) \cos(t + \tau)\,dt = 4,$$

and hence $\int_0^{2\pi} [\operatorname{sgn} \cos(t + \tau) - w(t)] \cos(t + \tau)\,dt = 0$. Because the integrand is ≥ 0 a.e., we obtain $w(t) = \operatorname{sgn} \cos(t + \tau)$ a.e. Define $y, \psi : [0, 2\pi] \to \mathbb{R}$ by $y(t) = x(t - \tau - \pi/2)$ and $\psi(t) = p(t - \tau - \pi/2)$. Then (3.50a), (3.50b) give

$$\ddot{y}(t) + y(t) + a\,\bar{w}(t) = \psi(t) \quad \text{a.e. in } [0, 2\pi], \quad \text{with}$$

$$\bar{w}(t) = \operatorname{sgn} \sin t \in \operatorname{Sgn} y(t) \quad \text{a.e. in } [0, 2\pi]. \quad (3.51)$$

The latter and the continuity of y imply $y(t) \geq 0$ in $[0, \pi]$ resp. $y(t) \leq 0$ in $[\pi, 2\pi]$. Thus $y(0) = y(\pi) = 0$ and

$$y(t) = \beta \sin t + \int_0^t \sin(t - s) [\psi(s) - a\bar{w}(s)]\,ds \quad \text{for} \quad t \in [0, 2\pi] \quad (3.52)$$

with $\beta = \dot{y}(0)$. Since $\int_0^\pi \sin(\pi - s)\,\bar{w}(s)\,ds = 2$, condition (3.49) is necessary.

On the other hand, assume that "=" holds in (3.48) and that (3.49) is satisfied. We define ψ as above and first derive from (3.48) that

$$\int_0^{2\pi} \psi(t) \sin t\,dt = 4a \quad \text{and} \quad \int_0^{2\pi} \psi(t) \cos t\,dt = 0. \quad (3.53)$$

For that, the continuous 2π-periodic p can be written as a Fourier series

$$c_0 + \sum_{k=1}^\infty [c_k \cos kt + d_k \sin kt] \quad (3.54)$$

being $L^2([0, 2\pi])$-convergent to p, hence $|p|_2^2 = 2\pi c_0^2 + \pi \sum_{k=1}^\infty [c_k^2 + d_k^2]$, where

$$c_0 = \frac{1}{2\pi} \int_0^{2\pi} p(t)\, dt \quad \text{and} \quad \begin{pmatrix} c_k \\ d_k \end{pmatrix} = \frac{1}{\pi} \int_0^{2\pi} p(t) \begin{pmatrix} \cos kt \\ \sin kt \end{pmatrix} dt, \quad k \geq 1.$$
(3.55)

This yields $\left| \int_0^{2\pi} p(t)\, e^{it}\, dt \right| = \pi \sqrt{c_1^2 + d_1^2}$ and $\int_0^{2\pi} p(t) \cos(t + \tau)\, dt = \pi\, [c_1 \cos(\tau) - d_1 \sin(\tau)]$, and therefore

$$4a = \pi \sqrt{c_1^2 + d_1^2} = \pi\, [c_1 \cos\tau - d_1 \sin\tau]$$

from (3.48). In particular, $(c_1, d_1) \in \mathbb{R}^2$ and $(\cos\tau, -\sin\tau) \in \mathbb{R}^2$ are linearly dependent, hence $c_1 \sin\tau + d_1 \cos\tau = 0$. Consequently, (3.53) follows from

$$\int_0^{2\pi} \psi(t) \sin t\, dt = \pi\, [c_1 \cos\tau - d_1 \sin\tau], \quad \text{and}$$

$$\int_0^{2\pi} \psi(t) \cos t\, dt = -\pi\, [c_1 \sin\tau + d_1 \cos\tau].$$

Now let $\theta(t) = \int_0^t \sin(t - s)\, \psi(s)\, ds$, choose $\beta \in \mathbb{R}$ such that

$$\beta \geq \max_{t \in [0,\pi]} \left(\frac{a(1 - \cos t) - \theta(t)}{\sin t} \right) \quad \text{as well as}$$

$$\beta \geq \max_{t \in [\pi, 2\pi]} \left(\frac{-a(1 + 3\cos t) - \theta(t)}{\sin t} \right),$$
(3.56)

and define y by (3.52), with $\bar{w}(t) = \operatorname{sgn} \sin t$. Note that both maxima exist in (3.56), due to (3.49) and (3.53). We find

$$y(t) = \begin{cases} \beta \sin t + \theta(t) - a(1 - \cos t) & : \ t \in [0, \pi] \\ \beta \sin t + \theta(t) + a(1 + 3\cos t) & : \ t \in [\pi, 2\pi] \end{cases},$$

and y is C^1 and 2π-periodic by (3.53). Moreover, by (3.56), $y(t) \geq 0$ in $[0, \pi]$ and $y(t) \leq 0$ in $[\pi, 2\pi]$. Hence we clearly obtain $\ddot{y} + y + a \operatorname{Sgn} y \ni \psi(t)$ a.e., and transformation back by means of $x(t) = y(t + \tau + \pi/2)$ yields a 2π-periodic solution $x = x_\beta$ of (3.47) for every β satisfying (3.56).

Finally we have to show the existence of a 2π-periodic solution of (3.47) in the most interesting case that $\left| \int_0^{2\pi} p(t)\, e^{it}\, dt \right| = \pi \sqrt{c_1^2 + d_1^2} < 4a$, cf. (3.54), (3.55). For that, we consider the approximate problems

$$\ddot{x}_\varepsilon + (1 + \varepsilon)x_\varepsilon + a \operatorname{Sgn} x_\varepsilon \ni p(t) \quad \text{a.e.},$$
(3.57)

which have 2π-periodic solutions $x_\varepsilon \in W^{2,2}([0, 2\pi]) \subset C^1([0, 2\pi])$, as may be verified analogously to Thm. 3.2.1.

We first want to derive a uniform $L^2([0, 2\pi])$-bound on the solutions $\mathcal{X} = \{x_\varepsilon : \varepsilon \in]0, 1]\} \subset L^2([0, 2\pi])$, and for that it is sufficient to show that $\mathcal{X} \subset L^2([0, 2\pi])$ is weakly bounded. It will be convenient to expand every x_ε into a Fourier series, like in (3.54), (3.55), with coefficients $a_{k,\varepsilon}$ and $b_{k,\varepsilon}$. These series for x_ε resp. $\dot{x}_\varepsilon$ are uniformly resp. $L^2([0, 2\pi])$-convergent, and we thus have

$$|x_\varepsilon|_2^2 = 2\pi a_{0,\varepsilon}^2 + \pi \sum_{k=1}^{\infty} [a_{k,\varepsilon}^2 + b_{k,\varepsilon}^2] \quad \text{and} \quad |\dot{x}_\varepsilon|_2^2 = \pi \sum_{k=1}^{\infty} k^2 [a_{k,\varepsilon}^2 + b_{k,\varepsilon}^2].$$

(3.58)

We define a continuous linear projection $P : L^2([0, 2\pi]) \to L^2([0, 2\pi])$ through

$$(Py)(t) = \left(\frac{1}{\pi} \int_0^{2\pi} y(s) \cos s \, ds \right) \cos t + \left(\frac{1}{\pi} \int_0^{2\pi} y(s) \sin s \, ds \right) \sin t.$$

This P is also symmetric, i.e., $\langle Py, z \rangle = \langle y, Pz \rangle$ holds, and consequently $\langle (\mathrm{id} - P)y, z \rangle = \langle y, (\mathrm{id} - P)z \rangle$, with $\langle y, z \rangle = \int_0^{2\pi} y(s) z(s) \, ds$ for $y, z \in L^2([0, 2\pi])$. Next we choose measurable functions w_ε with

$$\ddot{x}_\varepsilon(t) + (1 + \varepsilon) x_\varepsilon(t) + a \, w_\varepsilon(t) = p(t) \quad \text{a.e. in } [0, 2\pi], \quad \text{where}$$
$$w_\varepsilon(t) \in \mathrm{Sgn}\, x_\varepsilon(t) \quad \text{a.e. in } [0, 2\pi], \tag{3.59}$$

hence in particular $|w_\varepsilon(t)| \leq 1$ a.e. Consequently, if we also let

$$A_{k,\varepsilon} = \frac{1}{\pi} \int_0^{2\pi} w_\varepsilon(s) \cos ks \, ds \quad \text{and} \quad B_{k,\varepsilon} = \frac{1}{\pi} \int_0^{2\pi} w_\varepsilon(s) \sin ks \, ds$$

for $k \in \mathbb{N}_0$, then we have $\sup\{|A_{k,\varepsilon}| : k \in \mathbb{N}_0, \varepsilon \in]0, 1]\} \leq 2$ and also $\sup\{|B_{k,\varepsilon}| : k \in \mathbb{N}, \varepsilon \in]0, 1]\} \leq 2$. Additionally, (3.59) implies $(1 + \varepsilon) a_{0,\varepsilon} + a A_{0,\varepsilon} = c_0$ and $(1 + \varepsilon - k^2) a_{k,\varepsilon} + a A_{k,\varepsilon} = c_k$ resp. $(1 + \varepsilon - k^2) b_{k,\varepsilon} + a B_{k,\varepsilon} = d_k$ for $k \geq 1$. For $k \geq 2$ and $\varepsilon \in]0, 1]$ we therefore obtain the estimates

$$|a_{k,\varepsilon}| = \left| \frac{c_k - a A_{k,\varepsilon}}{1 + \varepsilon - k^2} \right| \leq \frac{2(a + |p|_\infty)}{|k^2 - 1 - \varepsilon|} \leq \frac{4}{k^2}(a + |p|_\infty),$$
$$|b_{k,\varepsilon}| \leq \frac{4}{k^2}(a + |p|_\infty).$$

This implies for $\varepsilon \in]0, 1]$

$$|\dot{x}_\varepsilon|_2^2 - |x_\varepsilon|_2^2 = \pi \sum_{k=1}^{\infty} k^2 [a_{k,\varepsilon}^2 + b_{k,\varepsilon}^2] - 2\pi a_{0,\varepsilon}^2 - \pi \sum_{k=1}^{\infty} [a_{k,\varepsilon}^2 + b_{k,\varepsilon}^2]$$

$$\leq \pi \sum_{k=2}^{\infty} k^2 [a_{k,\varepsilon}^2 + b_{k,\varepsilon}^2] \leq c, \tag{3.60}$$

and

$$|(\mathrm{id} - P)(x_\varepsilon)|_2^2 = \left| a_{0,\varepsilon} + \sum_{k=2}^{\infty} [a_{k,\varepsilon} \cos kt + b_{k,\varepsilon} \sin kt] \right|_2^2$$

$$= 2\pi a_{0,\varepsilon}^2 + \pi \sum_{k=2}^{\infty} [a_{k,\varepsilon}^2 + b_{k,\varepsilon}^2] \leq c \tag{3.61}$$

for some $c > 0$ being independent of $\varepsilon \in]0, 1]$.

Next, observe that

$$\left\{ \int_0^{2\pi} w(s) \, e^{is} \, ds \; : \; w : [0, 2\pi] \to \mathbb{R} \text{ is measurable with } |w(s)| \leq 1 \text{ a.e.} \right\} \tag{3.62}$$

equals $\overline{B}_4(0) \subset \mathbb{C}$. To verify this, denote $W \subset \mathbb{C}$ the set in (3.62). Then we have $W \subset \overline{B}_4(0)$, as for $z = \int_0^{2\pi} w(s) \, e^{is} \, ds \in W$ we can choose $\tau \in [0, 2\pi]$ such that $e^{i\tau} z = |z|$, whence upon taking real parts we see that

$$\left| \int_0^{2\pi} w(s) \, e^{is} \, ds \right| = \int_0^{2\pi} w(s) \, \cos(s + \tau) \, ds \leq \int_0^{2\pi} |\cos(s + \tau)| \, ds = 4.$$

On the other hand, if $|z| = 4$, i.e., $z = 4e^{i\tau}$ for some $\tau \in [0, 2\pi]$, then we can define $w(s) = \operatorname{sgn} \cos(s - \tau)$ to find $|w(s)| \leq 1$ a.e. and

$$\int_0^{2\pi} w(s) \, e^{is} \, ds = e^{i\tau} \int_0^{2\pi} \operatorname{sgn} \cos(s - \tau) \Big[\cos(s - \tau) + i \sin(s - \tau) \Big] \, ds$$

$$= e^{i\tau} \int_0^{2\pi} |\cos(s - \tau)| \, ds = 4e^{i\tau} = z.$$

Thus $\partial \overline{B}_4(0) \subset W$, and since $0 \in W$ and W is convex, we indeed have $W = \overline{B}_4(0)$.

Since $\left| \int_0^{2\pi} p(s) \, e^{is} \, ds \right| < 4a$ by assumption, we find $\delta_0 > 0$ such that

$$\left| \int_0^{2\pi} [p(s) + h(s)] \, e^{is} \, ds \right| \leq 4a \quad \text{for} \quad h \in L^2([0, 2\pi]) \quad \text{with} \quad |h|_2 \leq \delta_0 .$$

We fix such $h \in L^2([0, 2\pi])$ with $|h|_2 \leq \delta_0$. Then $a^{-1} \int_0^{2\pi} [p(s) + h(s)] \, e^{is} \, ds \in \overline{B}_4(0)$. Thus we find a measurable $w_h : [0, 2\pi] \to \mathbb{R}$ with $|w_h(s)| \leq 1$ a.e. and $\int_0^{2\pi} [p(s) + h(s)] \, e^{is} \, ds = a \int_0^{2\pi} w_h(s) \, e^{is} \, ds$, in other words $P(p + h) = a \, P(w_h)$. By multiplication of (3.59) with x_ε and integration over $[0, 2\pi]$ we hence obtain

$$\langle h, x_\varepsilon \rangle = |\dot{x}_\varepsilon|_2^2 - (1 + \varepsilon) |x_\varepsilon|_2^2 + \langle P(p + h) - a \, w_\varepsilon, x_\varepsilon \rangle$$
$$+ \langle (\operatorname{id} - P)(p + h), x_\varepsilon \rangle$$
$$= |\dot{x}_\varepsilon|_2^2 - (1 + \varepsilon) |x_\varepsilon|_2^2 + a \, \langle P(w_h) - w_\varepsilon, x_\varepsilon \rangle + \langle p + h, (\operatorname{id} - P)(x_\varepsilon) \rangle$$
$$= |\dot{x}_\varepsilon|_2^2 - (1 + \varepsilon) |x_\varepsilon|_2^2 + a \, \langle w_h - w_\varepsilon, x_\varepsilon \rangle$$
$$+ \langle p + h - a \, w_h, (\operatorname{id} - P)(x_\varepsilon) \rangle$$
$$\leq |\dot{x}_\varepsilon|_2^2 - |x_\varepsilon|_2^2 + \langle p + h - a \, w_h, (\operatorname{id} - P)(x_\varepsilon) \rangle,$$

since Sgn is a monotonous multi-valued mapping, $w_h(s) \in \operatorname{Sgn}(0)$ a.e., thus $(w_\varepsilon(s) - w_h(s)) \, x_\varepsilon(s) \geq 0$ a.e. by (3.59). Hence (3.60) and (3.61) imply that there are $C > 0$ and $\delta_0 > 0$ such that

$\langle h, x_\varepsilon \rangle \leq C \left(1+|h|_2\right)$ for $h \in L^2([0, 2\pi])$ with $|h|_2 \leq \delta_0$ and $\varepsilon \in]0, 1]$.

Therefore $\mathcal{X} = \{x_\varepsilon : \varepsilon \in]0, 1]\} \subset L^2([0, 2\pi])$ is weakly bounded. By the uniform boundedness principle, see KATO [104, Thm. III 1.27], this implies that $\mathcal{X} \subset L^2([0, 2\pi])$ is bounded in norm. Together with (3.60) it follows that

$$\sup_{\varepsilon \in]0,1]} |x_\varepsilon|_2^2 + \sup_{\varepsilon \in]0,1]} |\dot{x}_\varepsilon|_2^2 < \infty.$$

Thus (3.59) shows $\mathcal{X} \subset W^{2,2}([0, 2\pi])$ is bounded in norm, with $W^{2,2}([0, 2\pi])$ being the usual Sobolev space; see ADAMS [2]. Because the embedding $W^{2,2}([0, 2\pi]) \subset C^1([0, 2\pi])$ is compact, we may assume that $x_{\varepsilon_j} \to x$ in $C^1([0, 2\pi])$ for some sequence $\varepsilon_j \to 0$. Finally, standard arguments for ε-δ-usc differential inclusions, cf. the proof of Thm. 2.2.1, together with (3.57) imply that this x solves (3.47). □

An important ingredient in the preceding proof was the uniform boundedness principle; a similar approach has been used before in DEIMLING [66], where it was reported that this trick originated in the paper BREZIS/NIRENBERG [31].

3.3 Almost periodic solutions for the pendulum with dry friction

This section studies, following DEIMLING/HETZER/SHEN [67], the existence of an almost periodic solution to the dry friction problem

$$\ddot{x} + x + \operatorname{Sgn} \dot{x} \ni p(t) \quad \text{a.e.} \tag{3.17}$$

from Sect. 3.1.3, provided that the forcing p is almost periodic. Section 3.4 contains some background material on almost periodic functions.

We start with a necessary condition. For simplicity, here and throughout, we will only deal with the case that both the forcing p and the (perspective) solution x are defined on all of $\mathbb{R}$, with obvious generalizations to functions only defined on the half-line $[0, \infty[$.

Lemma 3.3.1. *Suppose* $p \in C_b(\mathbb{R})$ *is almost periodic and (3.17) has an almost periodic solution* $x \in C_b^1(\mathbb{R})$ *on* $\mathbb{R}$. *Then*

$$\left| \langle p, e^{it} \rangle_{\mathrm{ap}} \right| \leq \frac{2}{\pi}.$$

Proof: We find a measurable function $w : \mathbb{R} \to \mathbb{R}$ such that $w(t) \in \operatorname{Sgn} \dot{x}(t)$ a.e. and $\ddot{x}(t) + x(t) + w(t) = p(t)$ a.e. Then multiplication by e^{-it} and integration by parts twice yields

$$\int_0^T p(t)e^{-it}\,dt = \int_0^T w(t)e^{-it}\,dt + \int_0^T [\ddot{x}(t) + x(t)]e^{-it}\,dt$$

$$= \int_0^T w(t)e^{-it}\,dt + \left([\dot{x}(t) + ix(t)]e^{-it}\right)\Big|_{t=0}^{t=T}.$$

Since both x and $\dot{x}$ are bounded, we deduce that

$$\lim_{T\to\infty} \frac{1}{T}\int_0^T w(t)e^{-it}\,dt = \langle p, e^{it}\rangle_{ap} \tag{3.63}$$

does exist. For every $T > 0$ we find $\tau = \tau(T) \in \mathbb{R}$ such that $|\int_0^T w(t)e^{-it}\,dt| = \int_0^T w(t)\cos(t+\tau)\,dt$, by suitably rotating $\zeta = \int_0^T w(t)e^{-it}\,dt \in \mathbb{C}$ to $e^{-i\tau}\zeta \in [0, \infty[$. We moreover choose $m = m(T)$ with $2m\pi \le T \le 2(m+1)\pi$. Then $m \to \infty$ as $T \to \infty$, hence $|w(t)| \le 1$ a.e. implies

$$\left|\frac{1}{T}\int_0^T w(t)e^{-it}\,dt\right| \le \frac{1}{T}\int_0^T |\cos(t+\tau)|\,dt \le \frac{1}{2m\pi}\int_\tau^{\tau+2(m+1)\pi} |\cos t|\,dt$$

$$= \frac{1}{2m\pi}\int_0^{2(m+1)\pi} |\cos t|\,dt = \frac{4(m+1)}{2m\pi} \to \frac{2}{\pi}$$

as $m \to \infty$. Therefore (3.63) shows that $|\langle p, e^{it}\rangle_{ap}| \le \frac{2}{\pi}$ holds. $\square$

Remark 3.3.1. (a) Since $|\langle p, e^{it}\rangle_{ap}| = \frac{1}{2\pi}|\int_0^{2\pi} p(t)e^{-it}\,dt|$ for a 2π-periodic function $p \in C(\mathbb{R})$, in particular Lemma 3.3.1 recovers Lemma 3.2.1.

(b) In the limiting case

$$|\langle p, e^{it}\rangle_{ap}| = \frac{2}{\pi} \tag{3.64}$$

there may or may not be almost periodic solutions to (3.17): if $p(t) = \frac{4}{\pi}\sin t$, then (3.64) holds, cf. Ex. 3.4.2. In this case (3.17) even does have infinitely many periodic solutions, recall Thm. 3.2.7(b3). To give an example of non-existence of an almost periodic solution under condition (3.64), we follow DEIMLING [66, Sect. 3], and for $\alpha \in]0, 2\pi[$ we introduce

$$p_\alpha(t) = \begin{cases} \frac{t}{\alpha} & : \ t \in [0, \alpha] \\ \frac{2\pi-t}{2\pi-\alpha} & : \ t \in [\alpha, 2\pi] \end{cases},$$

periodically continued to all of $\mathbb{R}$. It then may be calculated that

$$\int_0^{2\pi} p_\alpha(t)e^{-it}\,dt = \frac{2\pi}{\alpha(2\pi-\alpha)}(e^{-i\alpha} - 1), \tag{3.65}$$

thus

$$|\langle p_\alpha, e^{it}\rangle_{ap}| = \frac{1}{2\pi}\left|\int_0^{2\pi} p_\alpha(t)e^{-it}\,dt\right| = \frac{\sqrt{2(1-\cos\alpha)}}{\alpha(2\pi-\alpha)} =: \eta_\alpha.$$

It follows that (3.64) is satisfied for every $p = \frac{2}{\pi \eta_\alpha} p_\alpha$.

Next we will argue that with such p, (3.17) does have a periodic solution if and only if $\alpha = \pi$. To verify this, we employ Thm. 3.2.2, where we have to check the additional condition (3.38), as "$=$" holds in (3.36) due to (3.64). By (3.65), the argument $\tau \in [0, 2\pi[$ of $\int_0^{2\pi} p_\alpha(t)e^{-it}\, dt \in \mathbb{C}$ is $\tau = \frac{3\pi - \alpha}{2}$, whence one evaluates

$$
\int_0^\pi p_\alpha(t + \tau - \pi/2)\cos(t)\, dt
$$

$$
= \begin{cases}
\dfrac{2}{2\pi - \alpha} & : \ \alpha \in]0, \tfrac{2\pi}{3}[\\[2mm]
\dfrac{2}{\alpha(2\pi - \alpha)}\left(\alpha - \pi\left[1 + \cos(\tfrac{3\alpha}{2})\right]\right) & : \ \alpha \in [\tfrac{2\pi}{3}, \tfrac{4\pi}{3}] \\[2mm]
-\dfrac{2}{\alpha} & : \ \alpha \in]\tfrac{4\pi}{3}, 2\pi[
\end{cases}
$$

This integral may be seen to vanish if and only if $\alpha = \pi$, hence by Thm. 3.2.2 there is a periodic solution of (3.17) if and only if $\alpha = \pi$.

Finally we fix $\alpha \neq \pi$, choose $p = \frac{2}{\pi \eta_\alpha} p_\alpha$ as forcing, and we assume (3.17) has an almost periodic solution x. Let $x_j(t) = x(t + 2\pi j)$ for $j \in \mathbb{N}$ and $t \in \mathbb{R}$. Then x_j as well is a solution of (3.17), since p is 2π-periodic, and all x_j and $\dot{x}_j$ are uniformly bounded. By the arguments from the proof of Thm. 3.1.4, we find a solution $\bar{x}$ of (3.17) such that, as $j \to \infty$, $x_j \to \bar{x}$ and $\dot{x}_j \to \dot{\bar{x}}$ uniformly on every compact subinterval of $\mathbb{R}$. In particular, for $t \in \mathbb{R}$ we obtain $\bar{x}(t + 2\pi) = \lim_{j\to\infty} x_j(t + 2\pi) = \lim_{j\to\infty} x(t + 2\pi[j + 1]) = \lim_{j\to\infty} x_{j+1}(t) = \bar{x}(t)$, i.e., x is 2π-periodic. But this contradicts the previous step. $\qquad \diamond$

Thus we consider the most interesting range

$$
\left| \langle p, e^{it} \rangle_{\mathrm{ap}} \right| < \frac{2}{\pi}, \tag{3.18}
$$

recall also Thm. 3.1.3 from Sect. 3.1.3, where this condition was shown to be sufficient for any solution of (3.17) belonging to $C_b^1([0, \infty[)$. In the proof of Thm. 3.1.4 we have moreover seen that

$$
M - m > 2, \tag{3.28}
$$

with

$$
m = \inf_{t \in \mathbb{R}} p(t) \quad \text{and} \quad M = \sup_{t \in \mathbb{R}} p(t),
$$

excludes constant solutions of (3.17). Contrary to this we are going to show next that for $0 < M - m \leq 2$ the only almost periodic solutions are constants.

Lemma 3.3.2. *If $0 < M - m \leq 2$ and $x \in C_b^1(\mathbb{R})$ is an almost periodic solution to (3.17) with $p \in C_b(\mathbb{R})$ being almost periodic, then $x(t) \equiv c$, for some $c \in [M - 1, m + 1]$.*

Proof: Let x and $\bar{x}$ be any almost periodic solutions to (3.17). Differentiating $\Delta_\varepsilon(t) = (x(t) - \bar{x}(t))^2 + (\dot{x}(t) - \dot{\bar{x}}(t))^2$, we obtain as in the proof of Thm. 3.1.4 that

$$\dot{\Delta}_\varepsilon(t) = -2\left(|\dot{x}(t)| + |\dot{\bar{x}}(t)|\right)(1 - w(t)\bar{w}(t)) \leq 0 \tag{3.66}$$

for a.e. $t \in \mathbb{R}$, where w and $\bar{w}$ are like in (3.30a) and (3.30b), respectively. Since Δ_ε is decreasing and almost periodic, it must be constant, cf. Lemma 3.4.5, and hence equality holds in (3.66) a.e. in $\mathbb{R}$. Stated differently, this means that $\dot{z}[\ddot{z} + z] = 0$ a.e. in $\mathbb{R}$, for $z = x - \bar{x}$.

We first consider the case $0 < M - m < 2$ and choose $\bar{x}(t) \equiv c$, with $c \in]M - 1, m + 1[$; observe that this $\bar{x}$ is a solution to (3.17). If we assume that, w.l.o.g., $\dot{x}(t) > 0$ in some interval $[t_0 - \delta, t_0 + \delta]$, then $w(t) = 1$ a.e. in $[t_0 - \delta, t_0 + \delta]$ by (3.30a), and moreover $1 - w(t)\bar{w}(t) = 0$ a.e. in $[t_0 - \delta, t_0 + \delta]$ due to (3.66), i.e., $\bar{w}(t) = w(t) = 1$ a.e. in $[t_0 - \delta, t_0 + \delta]$. From (3.30b) we infer that $c + \bar{w}(t) = p(t)$ a.e. in $\mathbb{R}$, and hence $M - 1 < c = p(t) - 1 \leq M - 1$, a contradiction.

For $M - m = 2$, we again let $\bar{x}(t) \equiv c = M - 1 = m + 1$. Supposing that $x(t) \not\equiv c$, i.e., $z \neq 0$, the above observations show that (3.32) is satisfied. Hence Cor. 3.1.1 applies to yield $\ddot{z} + z = 0$ a.e. in $\mathbb{R}$; observe that if x is constant, then $x(t) \equiv c$, since this is the only constant solution of (3.17). Consequently, $z(t) = A\cos(t + \theta)$ for some $A \neq 0$ and $\theta \in \mathbb{R}$, and this means $x(t) = A\cos(t + \theta) + c$ for $t \in \mathbb{R}$. Since $\bar{x}$ is constant, (3.30b) reads as $c + \bar{w}(t) = p(t)$ a.e. in $\mathbb{R}$. Thus $\dot{x}(t) \neq 0$ a.e. in $\mathbb{R}$ in conjunction with (3.66) implies that $w(t) = \bar{w}(t) = p(t) - c$ a.e. in $\mathbb{R}$. Assuming w.l.o.g. that $A > 0$, we finally infer from $p(t) - c = w(t) \in \text{Sgn}\,\dot{x}(t) = \text{Sgn}(-A\sin(t+\theta)) = -\text{Sgn}(\sin(t+\theta)) = \{-\text{sgn}(\sin(t+\theta))\}$ a.e. in $\mathbb{R}$ that $p(t) = -\text{sgn}(\sin(t+\theta)) + c$ for $t \notin \{j\pi - \theta : j \in \mathbb{Z}\}$. However, this is impossible, as p is continuous. $\square$

Thus it became obvious that in order to find non-trivial almost periodic solutions of (3.17), we should assume (3.18) and (3.28). We next employ the information obtained in Thms. 3.1.3 and 3.1.4 to derive a result on the existence of almost periodic solutions to (3.17). This will rely much on the following minimization property, where we again consider (3.17) as the system (3.29) with $y = (y_1, y_2) = (x, \dot{x})$.

Theorem 3.3.1. *Suppose that (3.18) and (3.28) are satisfied for an almost periodic function $p \in C_b(\mathbb{R})$, and let*

$$S = \left\{ y \in C_b(\mathbb{R}; \mathbb{R}^2) : y \text{ is a solution of } (3.29) \right\}.$$

Then there exists a unique $y^* \in S$ such that

$$|y^*|_{C_b(\mathbb{R};\mathbb{R}^2)} = \min\left\{ |y|_{C_b(\mathbb{R};\mathbb{R}^2)} : y \in S \right\}.$$

Proof: First note that $\mathcal{S} \neq \emptyset$ by Thm. 3.1.4. To prove existence of a minimizer, consider a minimizing sequence $(y_j) \subset C_b(\mathbb{R}; \mathbb{R}^2)$, i.e., $|y_j|_{C_b(\mathbb{R};\mathbb{R}^2)} \to \min\{\ldots\} =: c_0 \geq 0$ as $j \to \infty$. Now we may use the argument from the beginning of the proof of Thm. 3.1.4 to find a minimizer: since p is almost periodic, we can choose a sequence $\tau_j \to \infty$ such that $p_{\tau_j} \to p$ uniformly on $\mathbb{R}$; see Lemma 3.4.2. Define $y_j^*(t) = y_j(t + \tau_j)$ for $t \in \mathbb{R}$. Then y_j^* is a solution to (3.29) with p replaced by p_{τ_j}, and $|y_j^*|_{C_b(\mathbb{R};\mathbb{R}^2)} = |y_j|_{C_b(\mathbb{R};\mathbb{R}^2)}$ shows that $(y_j^*) \subset C_b(\mathbb{R}; \mathbb{R}^2)$ is bounded. Since also $(\ddot{y}_j^*) \subset L^\infty(\mathbb{R})$ is bounded, as follows from the corresponding differential inclusions, we may select a subsequence (still labeled by $j \in \mathbb{N}$) and a function y^* such that $y_j^* \to y^*$ as $j \to \infty$ uniformly on any compact subinterval of $\mathbb{R}$. Hence in particular $|y^*|_{C_b(\mathbb{R};\mathbb{R}^2)} \leq \liminf_{j \to \infty} |y_j|_{C_b(\mathbb{R};\mathbb{R}^2)} = c_0$. However, y^* as well is a solution to (3.29), and therefore $y^* \in \mathcal{S}$ implies $c_0 \leq |y^*|_{C_b(\mathbb{R};\mathbb{R}^2)}$.

So it remains to prove uniqueness of a minimizer, and for that we assume that both $y = (x, \dot{x})$ and $\bar{y} = (\bar{x}, \dot{\bar{x}})$ are such minimizers with $x \neq \bar{x}$. Since $x, \bar{x} \in C_b^1(\mathbb{R})$, we can use Thm. 3.1.4 to find solutions $x_\infty, \bar{x}_\infty \in C_b^1(\mathbb{R})$ of (3.17) such that, denoting $y_\infty = (x_\infty, \dot{x}_\infty)$ and $\bar{y}_\infty = (\bar{x}_\infty, \dot{\bar{x}}_\infty)$, we have $|y_\infty|_{C_b(\mathbb{R};\mathbb{R}^2)} \leq |y|_{C_b(\mathbb{R};\mathbb{R}^2)} = c_0$ as well as $|\bar{y}_\infty|_{C_b(\mathbb{R};\mathbb{R}^2)} \leq |\bar{y}|_{C_b(\mathbb{R};\mathbb{R}^2)} = c_0$, by (3.31). Moreover, $z_\infty = x_\infty - \bar{x}_\infty \neq 0$ has $\ddot{z}_\infty + z_\infty = 0$ a.e. in $\mathbb{R}$. We see that y_∞ and $\bar{y}_\infty$ are minimizers, too, whence we continue to denote x_∞ and $\bar{x}_\infty$ as x and y, for simplicity. Thus $z = x - \bar{x} \neq 0$ and $\ddot{z} + z = 0$ a.e. in $\mathbb{R}$, i.e., $z(t) = A \cos(t + \theta)$ for $t \in \mathbb{R}$, with suitable $A \neq 0$ and $\theta \in \mathbb{R}$. Moreover,

$$w(t) = \bar{w}(t) \quad \text{a.e. in} \quad \mathbb{R} \tag{3.67}$$

for the corresponding selections of Sgn, as is obtained by subtracting (3.30b) from (3.30a).

Next we are going to verify that $\lambda x + (1 - \lambda)\bar{x} = \bar{x} + \lambda z$ is a solution to (3.17) for any $\lambda \in [0, 1]$, and w.l.o.g. we consider $\lambda \in]0, 1[$. To this end, note first that

$$t \in \mathbb{R}, \quad \dot{x}(t) > 0 \quad \text{or} \quad \dot{\bar{x}}(t) > 0 \quad \Longrightarrow \quad \begin{cases} \dot{x}(t) - \dot{z}(t) \geq 0 \\ \dot{\bar{x}}(t) + \dot{z}(t) \geq 0 \end{cases}. \tag{3.68}$$

To see this, assume e.g. that $\dot{x}(s) > 0$ for s close to some $t \in \mathbb{R}$. Then $1 = w(s) \in \mathrm{Sgn}\,\dot{x}(s) = \mathrm{Sgn}\,(\dot{\bar{x}}(s) + \dot{z}(s))$, and analogously, by (3.67), we obtain $1 = \bar{w}(s) \in \mathrm{Sgn}\,\dot{\bar{x}}(s) = \mathrm{Sgn}\,(\dot{x}(s) - \dot{z}(s))$ a.e. for s close to t. Whence $\dot{\bar{x}}(s) + \dot{z}(s) \geq 0$ as well as $\dot{x}(s) - \dot{z}(s) \geq 0$ for those s, and we see that (3.68) holds. Then we choose $t_0 \in \mathbb{R}$ such that $\dot{z}(t) = \dot{x}(t) - \dot{\bar{x}}(t) = -A \sin(t + \theta) > 0$ for $t \in]t_0, t_0 + \pi[$ and $\dot{z}(t) < 0$ for $t \in]t_0 + \pi, t_0 + 2\pi[$. Since $\dot{\bar{x}} \not\equiv 0$, we find $t_* \in \mathbb{R}$ with, w.l.o.g., $\dot{\bar{x}}(t_*) > 0$. Shifting t_0 by a multiple of 2π and changing t_* a little, we may assume that $t_* \in]t_0, t_0 + 2\pi[$ and $t_* \neq t_0 + \pi$. We only consider the case that $t_* \in]t_0, t_0 + \pi[$, if $t_* \in]t_0 + \pi, t_0 + 2\pi[$, a similar argument applies. According to (3.68) we thus have $\dot{x}(t_*) \geq \dot{z}(t_*) > 0$. Introducing the largest interval where $\dot{x}(t) > 0$ holds, it thus follows from $\dot{z}(t) > 0$ in $]t_0, t_0 + \pi[$ that

even $\dot{x}(t) > 0$ for all $t \in]t_0, t_0 + \pi[$. Using (3.67), this moreover yields $\ddot{\bar{x}}(t) \geq 0$
for $t \in]t_0, t_0 + \pi[$. Having now clarified the signs of $\dot{x}$ and $\ddot{\bar{x}}$ in $]t_0, t_0, \pi[$,
we consider the next interval $]t_0 + \pi, t_0 + 2\pi[$. As a first case, if $\dot{x}(t_1) \geq 0$
for some $t_1 \in]t_0 + \pi, t_0 + 2\pi[$, then $\dot{\bar{x}}(t_1) = \dot{x}(t_1) - \dot{z}(t_1) > 0$. Hence (3.68)
and the preceding argument enforces $\dot{\bar{x}}(t) > 0$ for all $t \in]t_0 + \pi, t_0 + 2\pi[$,
and then also $\dot{x}(t) \geq 0$ in $]t_0 + \pi, t_0 + 2\pi[$ by (3.67). On the other hand, if
$\dot{x}(t) < 0$ for all $t \in]t_0 + \pi, t_0 + 2\pi[$, then (3.67) additionally implies $\ddot{\bar{x}}(t) \leq 0$
in $]t_0 + \pi, t_0 + 2\pi[$. Using the analogue of (3.68) for the case that one of the
derivatives is negative at some point, we can hence continue this reasoning
to see that on every interval $]t_0 + j\pi, t_0 + (j+1)\pi[, j \in \mathbb{Z}$, we have

$$
(i): \quad \left(\dot{x} > 0 \text{ and } \ddot{\bar{x}} \geq 0\right), \quad \text{or} \quad (ii): \quad \left(\dot{x} \geq 0 \text{ and } \ddot{\bar{x}} > 0\right), \quad \text{or}
$$

$$
(iii): \quad \left(\dot{x} < 0 \text{ and } \ddot{\bar{x}} \leq 0\right), \quad \text{or} \quad (iv): \quad \left(\dot{x} \leq 0 \text{ and } \ddot{\bar{x}} < 0\right).
$$

In cases (i) or (ii) we obtain $w(t) = \bar{w}(t) = 1$ a.e. in $]t_0 + j\pi, t_0 + (j+1)\pi[$
due to (3.67), whence $w(t) \in \{1\} = \text{Sgn}(\lambda \dot{x}(t) + (1 - \lambda)\ddot{\bar{x}}(t))$ for a.e. $t \in$
$]t_0 + j\pi, t_0 + (j+1)\pi[$. If (iii) or (iv) happen to hold, then in the same manner
$w(t) \in \{-1\} = \text{Sgn}(\lambda \dot{x}(t) + (1 - \lambda)\ddot{\bar{x}}(t))$ for a.e. $t \in]t_0 + j\pi, t_0 + (j+1)\pi[$. To
summarize, we have shown that

$$
\bar{w}(t) = w(t) \in \text{Sgn}(\lambda \dot{x}(t) + (1 - \lambda)\ddot{\bar{x}}(t)) \quad \text{for a.e.} \quad t \in \mathbb{R}. \tag{3.69}
$$

This implies that indeed $\lambda x + (1 - \lambda)\bar{x} = \bar{x} + \lambda z$ is a solution of (3.17) for
every $\lambda \in]0, 1[$, since by (3.30b) and (3.69) a.e. in $\mathbb{R}$

$$
p(t) - [\ddot{\bar{x}}(t) + \lambda \ddot{z}(t)] - [\bar{x}(t) + \lambda z(t)] = p(t) - \ddot{\bar{x}}(t) - \bar{x}(t)
$$
$$
= \bar{w}(t) \in \text{Sgn}(\dot{\bar{x}}(t) + \lambda \dot{z}(t)).
$$

Having verified that $\bar{x} + \lambda z$ solves (3.17) for every $\lambda \in [0, 1]$, in particular
we find that $\bar{y}_\lambda = \bar{y} + \lambda \zeta \in S$ for $\lambda \in [0, 1]$, where $\zeta = y - \bar{y} = (z, \dot{z}) =$
$A\left(\cos(\cdot + \theta), -\sin(\cdot + \theta)\right)$. Noting $|y(t) - \bar{y}(t)| = A$, we obtain from the
parallelogram identity for the Euclidean norm in $\mathbb{R}^2$ that

$$
\frac{A^2}{4} + \frac{1}{4}|y(t) + \bar{y}(t)|^2 = \frac{1}{2}|y(t)|^2 + \frac{1}{2}|\bar{y}(t)|^2
$$
$$
\leq \frac{1}{2}|y|^2_{C_b(\mathbb{R};\mathbb{R}^2)} + \frac{1}{2}|\bar{y}|^2_{C_b(\mathbb{R};\mathbb{R}^2)} = c_0^2.
$$

Since $\frac{1}{2}(y + \bar{y}) = \bar{y}_{\frac{1}{2}} \in S$ for $\lambda = \frac{1}{2}$, this leads to

$$
c_0^2 \leq |\bar{y}_{\frac{1}{2}}|^2_{C_b(\mathbb{R};\mathbb{R}^2)} \leq c_0^2 - \frac{A^2}{4},
$$

a contradiction to $z \neq 0$.

From the preceding proof we additionally obtain

Corollary 3.3.1. *If (3.18) and (3.28) hold for an almost periodic function $p \in C_b(\mathbb{R})$, and if $x, \bar{x} \in C_b^1(\mathbb{R})$ are two solutions of (3.17) such that $\ddot{z} + z = 0$ a.e. in $\mathbb{R}$ for $z = x - \bar{x}$, then $\lambda x + (1 - \lambda)\bar{x}$ as well is a solution of (3.17), for all $\lambda \in [0, 1]$.*

We need a further auxiliary lemma.

Lemma 3.3.3. *Assume $p \in C_b(\mathbb{R})$, and $y^* \in C_b(\mathbb{R}; \mathbb{R}^2)$ is a solution to (3.29) that minimizes $|\cdot|_{C_b(\mathbb{R}; \mathbb{R}^2)}$ over the class of solutions to (3.29). If $p_{\tau_j} \to \bar{p}$ uniformly on $\mathbb{R}$ and $y^*_{\tau_j} \to \bar{y}$ uniformly on every compact subset of $\mathbb{R}$, then $\bar{y}$ is a minimizer of $|\cdot|_{C_b(\mathbb{R}; \mathbb{R}^2)}$ over the class of bounded solutions y to*

$$\dot{y} \in \bar{F}(y, t) \quad a.e., \quad with \quad \bar{F}(y, t) = \{y_2\} \times \{\bar{p}(t) - y_1 - w : w \in \operatorname{Sgn} y_2\},$$
$$(3.70)$$

for $y = (y_1, y_2) \in \mathbb{R}^2$.

Proof: To begin with, the argument from the proof of Thm. 2.2.1 shows that $\bar{y}$ indeed solves (3.70) under the hypotheses stated in the lemma; see also the proof of Thm. 3.1.4. Now, $|\bar{p}_{-\tau_j} - p|_\infty \to 0$ as $j \to \infty$, since for $t \in \mathbb{R}$ and denoting $s = t - \tau_j$ it follows that

$$|\bar{p}_{-\tau_j}(t) - p(t)| = |\bar{p}(t - \tau_j) - p(t)| = |\bar{p}(s) - p(s + \tau_j)| \le |\bar{p} - p_{\tau_j}|_\infty.$$

Let $\hat{y} \in C_b(\mathbb{R}; \mathbb{R}^2)$ be any solution of (3.70). To verify that $|\bar{y}|_{C_b(\mathbb{R}; \mathbb{R}^2)} \le |\hat{y}|_{C_b(\mathbb{R}; \mathbb{R}^2)}$, note that $\hat{y}_{-\tau_j}$ is a solution to the differential inclusion with forcing $\bar{p}_{-\tau_j}$. Hence, passing to a subsequence if necessary, we find a solution $y \in C_b(\mathbb{R}; \mathbb{R}^2)$ of (3.29) such that $\hat{y}_{-\tau_j} \to y$ uniformly on every compact subset of $\mathbb{R}$. In particular, $|y|_{C_b(\mathbb{R}; \mathbb{R}^2)} \le \liminf_{j \to \infty} |\hat{y}_{-\tau_j}|_{C_b(\mathbb{R}; \mathbb{R}^2)} = |\hat{y}|_{C_b(\mathbb{R}; \mathbb{R}^2)}$. But also $|\bar{y}|_{C_b(\mathbb{R}; \mathbb{R}^2)} \le |y^*|_{C_b(\mathbb{R}; \mathbb{R}^2)}$ according to the assumptions. Consequently it follows from y^* being a minimizer that

$$|\bar{y}|_{C_b(\mathbb{R}; \mathbb{R}^2)} \le |y^*|_{C_b(\mathbb{R}; \mathbb{R}^2)} \le |y|_{C_b(\mathbb{R}; \mathbb{R}^2)} \le |\hat{y}|_{C_b(\mathbb{R}; \mathbb{R}^2)},$$

as was to be shown.

Now we are ready for a main result of this section.

Theorem 3.3.2. *Assume (3.18) and (3.28) hold for an almost periodic function $p \in C_b(\mathbb{R})$. Then (3.17) has an almost periodic solution.*

Proof: Let y^* denote the unique minimizer from Thm. 3.3.1. We are going to show $y = (x, \dot{x})$ is almost periodic, then x will be an almost periodic solution to (3.17). To verify that $\{y^*_\tau : \tau \in \mathbb{R}\}$ is relatively compact w.r. to $|\cdot|_{C_b(\mathbb{R}; \mathbb{R}^2)}$, we consider a sequence $y^*_{\tau_j}$ with $(\tau_j) \subset \mathbb{R}$. Since p is almost periodic, we may assume w.l.o.g. that $p_{\tau_j} \to \bar{p}$ uniformly on $\mathbb{R}$; see Lemma 3.4.2. By passing to a further subsequence we may as well suppose that $y^*_{\tau_j} \to \bar{y}$ uniformly on every compact subset of $\mathbb{R}$, for some function $\bar{y} \in C_b(\mathbb{R}; \mathbb{R}^2)$ that is a solution to (3.70). According to Lemma 3.3.3, $\bar{y}$ minimizes $|\cdot|_{C_b(\mathbb{R};\mathbb{R}^2)}$ over the class of bounded solutions to (3.70).

We would be done if even $y^*_{\tau_j} \to \bar{y}$ uniformly on all of $\mathbb{R}$. If this were false, then for some $\varepsilon_0 > 0$, $j' \to \infty$, and some (unbounded) sequence $(t_{j'}) \subset \mathbb{R}$ we had

$$|y^*_{\tau_{j'}+t_{j'}}(0) - \bar{y}_{t_{j'}}(0)| = |y^*_{\tau_{j'}}(t_{j'}) - \bar{y}(t_{j'})| \geq \varepsilon_0. \tag{3.71}$$

For simplicity we henceforth omit the prime. Due to Lemma 3.4.2, w.l.o.g. also $p_{\tau_j+t_j} \to \psi$ as $j \to \infty$ uniformly on $\mathbb{R}$ for some function $\psi \in C_b(\mathbb{R})$. In addition, Lemma 3.4.4 shows that

$$\inf_{\mathbb{R}} \psi = \inf_{\mathbb{R}} p = m, \quad \sup_{\mathbb{R}} \psi = \sup_{\mathbb{R}} p = M, \quad \text{and} \quad |\langle \psi, e^{it} \rangle_{\mathrm{ap}}| = |\langle p, e^{it} \rangle_{\mathrm{ap}}|, \tag{3.72}$$

and ψ is almost periodic. Since

$$|p_{\tau_j+t_j} - \bar{p}_{t_j}|_\infty = |p_{\tau_j} - \bar{p}|_\infty \to 0, \quad j \to \infty,$$

we see that as well $\bar{p}_{t_j} \to \psi$ uniformly on $\mathbb{R}$. Next note that $y^*_{\tau_j+t_j}$ is a solution with forcing $p_{\tau_j+t_j}$, and $\bar{y}_{t_j}$ is a solution with forcing $\bar{p}_{t_j}$. Hence we may select further subsequences (w.l.o.g. the whole sequences) and functions $u^* \in C_b(\mathbb{R}; \mathbb{R}^2)$ and $\bar{u} \in C_b(\mathbb{R}; \mathbb{R}^2)$ such that $y^*_{\tau_j+t_j} \to u^*$ and $\bar{y}_{t_j} \to \bar{u}$ as $j \to \infty$, both uniformly on compact subsets of $\mathbb{R}$. Moreover, then u^* and $\bar{u}$ are solutions to

$$\dot{y} \in \Psi(y,t) \quad \text{a.e.,} \quad \text{where} \quad \Psi(y,t) = \{y_2\} \times \{\psi(t) - y_1 - w : w \in \mathrm{Sgn}\, y_2\}, \tag{3.73}$$

and (3.71) implies that

$$|u^*(0) - \bar{u}(0)| \geq \varepsilon_0. \tag{3.74}$$

Since, according to (3.72), ψ satisfies (3.18) and (3.28), we know from Thm. 3.3.1 that there is a unique minimizer of $|\cdot|_{C_b(\mathbb{R};\mathbb{R}^2)}$ among the bounded solutions of (3.73). However, since both y^* and $\bar{y}$ are minimizers of $|\cdot|_{C_b(\mathbb{R};\mathbb{R}^2)}$ among the bounded solutions of (3.29) and (3.70), respectively, we deduce from Lemma 3.3.3 that also the limits u^* and $\bar{u}$ do minimize $|\cdot|_{C_b(\mathbb{R};\mathbb{R}^2)}$ over the bounded solutions of the respective limiting problems, i.e., of (3.73). Thus uniqueness implies $u^* = \bar{u}$, in contradiction to (3.74). $\qquad \square$

According to Lemma 3.3.1, Rem. 3.3.1(b), Lemma 3.3.2, and Thm. 3.3.2 we thus have obtained a complete picture under what conditions on p almost periodic solutions to (3.17) will or will not exist.

Next we proceed to discussing the stability of almost periodic solutions.

Theorem 3.3.3. *Suppose $p \in C(\mathbb{R})$ is almost periodic, and assume that $0 < M - m \le 2$ holds. If x is any solution of (3.17), then there exists $c = c_x \in [M - 1, m + 1]$ such that*

$$(x(t) - c)^2 + \dot{x}^2(t) \to 0 \quad as \quad t \to \infty.$$

Proof: We start with a preliminary remark. Let x and $\bar{x}$ be any two solutions of (3.17), and assume $\bar{x}$ is almost periodic. As in the proof of Thm. 3.2.5 it follows that $\Delta_\varepsilon(t) \searrow e_\infty$ as $t \to \infty$ for some $e_\infty \ge 0$, where we let $\Delta_\varepsilon(t) = (x(t) - \bar{x}(t))^2 + (\dot{x}(t) - \dot{\bar{x}}(t))^2$. Then the function $t \mapsto (p(t), \bar{x}(t), \dot{\bar{x}}(t)) \in \mathbb{R}^3$ is almost periodic. Due to Lemma 3.4.2 we consequently find $\tau_j \to \infty$ such that $p_{\tau_j} \to p$, $\bar{x}_{\tau_j} \to \bar{x}$, and $\dot{\bar{x}}_{\tau_j} \to \dot{\bar{x}}$ as $j \to \infty$, all uniformly on $\mathbb{R}$. Let $x_j(t) = x(t + \tau_j)$ for $j \in \mathbb{N}$ and $t \in \mathbb{R}$. Then $\ddot{x}_j(t) + x_j(t) + \mathrm{Sgn}\,\dot{x}_j(t) \ni p_{\tau_j}(t)$ a.e. implies that w.l.o.g. $x_j \to x_\infty$ and $\dot{x}_j \to \dot{x}_\infty$ as $j \to \infty$, uniformly on any compact subinterval of $\mathbb{R}$, and x_∞ solves (3.17). Since $\Delta_\varepsilon(t + \tau_j) \to e_\infty$ as $j \to \infty$ for every fixed $t \in \mathbb{R}$, it follows that

$$(x_\infty(t) - \bar{x}(t))^2 + (\dot{x}_\infty(t) - \dot{\bar{x}}(t))^2 = e_\infty, \quad t \in \mathbb{R}. \tag{3.75}$$

To prove the assertions of the theorem, we first deal with the case $0 < M - m < 2$, i.e., $M - 1 < \frac{1}{2}(M + m) < m - 1$. Let x be any solution of (3.17). Choosing the constant solution $\bar{x}(t) \equiv \frac{1}{2}(M + m)$ in the foregoing argument, we then construct x_∞ accordingly. Assume that $\dot{x}_\infty(t_0) \ne 0$ for some $t_0 \in \mathbb{R}$. Then Cor. 3.1.1 applies with x_∞ and $\bar{x}$ to yields $w_\infty(t) = \bar{w}(t)$ a.e. in $\mathbb{R}$, where w_∞ and $\bar{w}$ are analogous to (3.30a) and (3.30b), respectively; note that $x_\infty \in C_b^1(\mathbb{R})$ due to (3.75). But $\bar{x}$ is a constant solution, hence $\bar{w}(t) = p(t) - \bar{x}(t)$ shows $|w_\infty(t)| = |\bar{w}(t)| = |p(t) - \frac{1}{2}(M+m)| \le \frac{1}{2}(M-m) < 1$ a.e. by definition of m and M. Since $w_\infty(t) \in \mathrm{Sgn}\,\dot{x}_\infty(t)$ a.e., we see that $\dot{x}_\infty(t) = 0$ for $t \in \mathbb{R}$, a contradiction. Hence x_∞ is constant, and as it is a solution to (3.17), we must have $x_\infty(t) \equiv c$ for some $c \in [M - 1, m - 1]$. By definition of x_∞ thus in particular $x(\tau_j) \to c$ and $\dot{x}(\tau_j) \to 0$ as $j \to \infty$ for some sequence $\tau_j \to \infty$. Since both x and $x_\infty(t) \equiv c$ are solutions to (3.17), we may as well define $\hat{\Delta}_\varepsilon(t) = (x(t) - c)^2 + \dot{x}^2(t)$ and deduce $\hat{\Delta}_\varepsilon(t) \searrow \hat{e}_\infty$ as $t \to \infty$, for some $\hat{e}_\infty \ge 0$. Setting $t = \tau_j$ we obtain $\hat{e}_\infty = 0$ as desired.

For $M - m = 2$, we again obtain (3.75), where $\bar{x}(t) \equiv \frac{1}{2}(M + m) = M - 1 = m + 1$, and x_∞ is derives from the given solution x of (3.17). Now we need to verify $e_\infty = 0$, as $\bar{x}$ is the only constant solution. To see $e_\infty = 0$, we first assume that $\dot{x}_\infty(t_0) \ne 0$ for some $t_0 \in \mathbb{R}$. Then application of Cor. 3.1.1 once more yields $w_\infty(t) = \bar{w}(t) = p(t) - \bar{x}(t)$ a.e. in $\mathbb{R}$ and $\ddot{z} + z = 0$ a.e. for $z = x_\infty - \bar{x} \ne 0$. This leads to a contradiction as in the proof of Lemma 3.3.2: we have $x_\infty(t) = \frac{1}{2}(M + m) + A\cos(t + \theta)$, for some $A, \theta \in \mathbb{R}$, w.l.o.g. $A > 0$. Hence $p(t) - \frac{1}{2}(M + m) = w_\infty(t) \in \mathrm{Sgn}\,\dot{x}_\infty(t) = \mathrm{Sgn}(-A\sin(t + \theta)) = -\mathrm{Sgn}(\sin(t + \theta)) = \{-\mathrm{sgn}(\sin(t + \theta))\}$ a.e. in $\mathbb{R}$. It follows that $p(t) = -\mathrm{sgn}(\sin(t + \theta)) + \frac{1}{2}(M + m)$ for $t \notin \{j\pi - \theta : j \in \mathbb{Z}\}$. However, this contradicts that fact that p is continuous, and hence x_∞ must be constant. Since it is a solution to (3.17), we find $x_\infty = \bar{x}$, thus $e_\infty = 0$ by (3.75). $\qquad\square$

With the same argument as in the preceding proof it may be shown that if (3.18) and (3.28) are satisfied for an almost periodic function $p \in C_b(\mathbb{R})$, and if (3.17) does have a unique almost periodic solution x_*, then this solution attracts all other solutions x of (3.17) as $t \to \infty$, in the sense of (3.42). Hence it only remains to clarify what happens if there is more than one almost periodic solution; the fact that in general almost periodic solutions are non-unique has already been noted in Rem. 3.1.2, where we gave an example with a continuum of periodic solutions.

Theorem 3.3.4. *Assume (3.18) and (3.28) are verified for an almost periodic function $p \in C_b(\mathbb{R})$, and denote S_{ap} the set of almost periodic solutions x to (3.17) such that x and $\dot{x}$ are globally bounded. If S_{ap} has more than one element, then*

$$S_{\mathrm{ap}} = \left\{ x_* + A\cos(\cdot + \theta_0) : A \in [0, A_*] \right\} \tag{3.76}$$

for some $x_* \in S_{\mathrm{ap}}$, $\theta_0 \in \mathbb{R}$, and $A_* > 0$. In addition, if x is any solution of (3.17), then there exists $A \in [0, A_*]$ satisfying

$$\left(x(t) - x_A(t) \right)^2 + \left(\dot{x}(t) - \dot{x}_A(t) \right)^2 \to 0 \quad \text{as} \quad t \to \infty,$$

where $x_A(t) = x_*(t) + A\cos(t + \theta_0)$.

Proof: Let $x_0, x \in S_{\mathrm{ap}}$ be different. Since both solutions are bounded and non-constant, the latter due to (3.28), Cor. 3.1.1 applies to yield $x(t) = x_0(t) + A\cos(t + \theta_0)$ for some $A, \theta_0 \in \mathbb{R}$, $A \neq 0$. Moreover, $w(t) = w_0(t)$ a.e. in $\mathbb{R}$ for the corresponding a.e. selections w of $\mathrm{Sgn}\,\dot{x}$ and w_0 of $\mathrm{Sgn}\,\dot{x}_0$. Upon multiplying (3.30a) by $\dot{x}$ and integrating over $[0, T]$ we find that

$$\int_0^T p(t)\dot{x}(t)\,dt = \frac{1}{2}(x^2(T) - x^2(0)) + \frac{1}{2}(\dot{x}^2(T) - \dot{x}^2(0)) + \int_0^T |\dot{x}(t)|\,dt.$$

We hence deduce

$$|A| \int_0^T |\sin(t + \theta_0)|\,dt \leq \int_0^T |\dot{x}_0(t)|\,dt + \int_0^T |\dot{x}(t)|\,dt$$

$$\leq \int_0^T |\dot{x}_0(t)|\,dt + \left| \int_0^T p(t)\dot{x}_0(t)\,dt \right|$$

$$+ \frac{1}{2}(x^2(0) - x^2(T)) + \frac{1}{2}(\dot{x}^2(0) - \dot{x}^2(T))$$

$$+ |A| \left| \int_0^T p(t)\sin(t + \theta_0)\,dt \right|$$

$$\leq (1 + |p|_\infty)T|\dot{x}_0|_\infty + \frac{1}{2}(x^2(0) - x^2(T))$$

$$+ \frac{1}{2}(\dot{x}^2(0) - \dot{x}^2(T)) + |A| \left| \int_0^T p(t)e^{-it}\,dt \right|,$$

where we have used Lemma 3.1.3 with $\tau = 0$, $t = T$, and $\vartheta = -\theta_0$. Dividing by T and taking the limit $T \to \infty$, it follows that

$$\frac{2}{\pi}|A| \leq (1 + |p|_\infty)|\dot{x}_0|_\infty + |A|\,|\langle p, e^{it}\rangle_{\mathrm{ap}}|,$$

cf. (3.80). Thus

$$|A| \leq \frac{(1 + |p|_\infty)|\dot{x}_0|_\infty}{\left(\frac{2}{\pi} - |\langle p, e^{it}\rangle_{\mathrm{ap}}|\right)} =: A_1 \tag{3.77}$$

is an a priori bound for the possible values of A; recall (3.18).

We claim that

$$\mathcal{S}_{\mathrm{ap}} \subset \left\{ x_0 + A\cos(\cdot + \theta_0) : |A| \leq A_1 \right\}. \tag{3.78}$$

To verify this, let $\bar{x} \in \mathcal{S}_{\mathrm{ap}}$ be a further almost periodic solution. Then, as before, $\bar{x}(t) = x_0(t) + \bar{A}\cos(t + \bar{\theta})$ for some $\bar{A}, \bar{\theta} \in \mathbb{R}$ with $\bar{A} \neq 0$, and $\bar{w}(t) = w_0(t)$ a.e. in $\mathbb{R}$ by Cor. 3.1.1. From (3.77) we then infer $|\bar{A}| \leq A_1$, hence it suffices to show that $\bar{\theta} \in \{\theta_0 + j\pi : j \in \mathbb{Z}\}$. If this were wrong, then the situation were as displayed in Fig. 3.1.

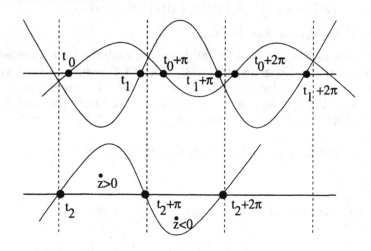

Fig. 3.1. $\bar{\theta} \notin \{\theta_0 + j\pi : j \in \mathbb{Z}\}$

Here t_0 is chosen in such a way that $\dot{x}(t) - \dot{x}_0(t) = -A\sin(t + \theta_0) > 0$ for $t \in\,]t_0, t_0 + \pi[$ and $\dot{x}(t) - \dot{x}_0(t) < 0$ for $t \in\,]t_0 + \pi, t_0 + 2\pi[$, whereas t_1 is characterized by $\dot{\bar{x}}(t) - \dot{x}_0(t) = -\bar{A}\sin(t + \bar{\theta}) > 0$ for $t \in\,]t_1, t_1 + \pi[$ and $\dot{\bar{x}}(t) - \dot{x}_0(t) < 0$ for $t \in\,]t_1 + \pi, t_1 + 2\pi[$. The intersections of these both curves determines t_2 satisfying $\dot{z}(t) = \dot{x}(t) - \dot{\bar{x}}(t) = -A\sin(t + \theta_0) + \bar{A}\sin(t + \bar{\theta}) > 0$ for $t \in\,]t_2, t_2 + \pi[$ and $\dot{z}(t) < 0$ for $t \in\,]t_2 + \pi, t_2 + 2\pi[$, where $z = x - \bar{x}$.

To see that this leads to a contradiction, we choose $t_* \in \mathbb{R}$ with $\dot{x}_0(t_*) \neq 0$, w.l.o.g. $\dot{x}_0(t_*) > 0$. By shifting the t_j by a multiple of 2π and changing t_* a little if necessary, we may assume that $t_* \in]t_0, t_0 + 2\pi[$ and t_* does not coincide with $t_1, t_0 + \pi$, or $t_1 + \pi$. We will show that $t_* \in]t_0, t_2 + \pi[$ is impossible, the other cases being handled similarly. So assume that $t_* \in]t_0, t_2 + \pi[$. Since $\dot{x}_0(t_*) > 0$, we have $\bar{w}(t) = w_0(t) \in \mathrm{Sgn}\, \dot{x}_0(t) = \{1\}$ a.e. close to t_*, whence $\bar{w}(t) \in \mathrm{Sgn}\, \dot{\bar{x}}(t)$ a.e. shows $\dot{x}(t) - \dot{z}(t) = \dot{\bar{x}}(t) \geq 0$, and thus $\dot{x}(t) \geq \dot{z}(t) > 0$ close to t_*. By considering the largest interval where this estimate is verified, and observing $\dot{z}(t) > 0$ in $]t_0, t_2 + \pi[$, is follows that in fact $\dot{x}(t) > 0$ for $t \in]t_0, t_2 + \pi[$. This in turn yields $\dot{x}_0(t) \geq 0$ and $\dot{\bar{x}}(t) \geq 0$ for $t \in]t_0, t_2 + \pi[$, since $w(t) = w_0(t) = \bar{w}(t)$ a.e. in $\mathbb{R}$. Hence in particular all three functions x, x_0, and $\bar{x}$ are non-decreasing over $t \in]t_0, t_2 + \pi[$. Suppose next that $\dot{x}(t_2 + \pi) = \dot{x}_0(t_2 + \pi) = \dot{\bar{x}}(t_2 + \pi) = 0$. Then $A \sin(t_2 + \pi + \theta_0) = \bar{A} \sin(t_2 + \pi + \bar{\theta}) = 0$, but this is impossible due to our assumption $\bar{\theta} \notin \{\theta_0 + j\pi : j \in \mathbb{Z}\}$. Hence at least one of the three derivatives must be strictly positive at $t = t_2 + \pi$. Then $w(t) = w_0(t) = \bar{w}(t)$ a.e. shows that $1 = w(t) \in \mathrm{Sgn}\, \dot{x}(t) = \mathrm{Sgn}\, (\dot{\bar{x}}(t) + \dot{z}(t))$ a.e. close to $t_2 + \pi$, i.e., $\dot{\bar{x}}(t) + \dot{z}(t) \geq 0$, and therefore $\dot{\bar{x}}(t) \geq -\dot{z}(t) > 0$ for $t \in]t_2 + \pi, t_2 + \pi + \delta]$, with a suitable $\delta > 0$. The same argument as before then shows that in fact $\dot{\bar{x}}(t) > 0$ for $t \in]t_2 + \pi, t_2 + 2\pi[$, since there $\dot{z}(t) < 0$. Again this as well enforces $\dot{x}_0(t) \geq 0$ and $\dot{x}(t) \geq 0$ for $t \in]t_2 + \pi, t_2 + 2\pi[$. The argument may be infinitely continued to imply that all three functions x, x_0, and $\bar{x}$ are non-decreasing in $t \in]t_0, \infty[$, but this cannot hold besides if x, x_0, and $\bar{x}$ are constant in $t \in]t_0, \infty[$; cf. Lemma 3.4.5, and recall that the functions are almost periodic. However, then $A \sin(t + \theta_0) = \bar{A} \sin(t + \bar{\theta}) = 0$ for $t \in]t_0, \infty[$ leads to $A = \bar{A} = 0$, a contradiction. Hence (3.78) indeed is verified.

By Cor. 3.3.1, $\mathcal{S}_{\mathrm{ap}}$ moreover is convex. In addition, $\mathcal{S}_{\mathrm{ap}}$ may be seen to be closed w.r. to the C_b^1-norm. It thus follows that

$$\mathcal{S}_{\mathrm{ap}} = \left\{ x_0 + A\cos(\cdot + \theta_0) : A \in [A_2, A_3] \right\},$$

for some $A_2 \leq 0 \leq A_3$, the latter since $x_0 \in \mathcal{S}_{\mathrm{ap}}$. Hence (3.76) is obtained with $x_* = x_0 + A_2 \cos(\cdot + \theta_0) \in \mathcal{S}_{\mathrm{ap}}$ and $A_* = A_3 - A_2 > 0$.

To finally verify the asymptotic stability of $\mathcal{S}_{\mathrm{ap}}$, let x be a solution of (3.17) on $[0, \infty[$ (or on any other interval $[t_0, \infty[$). From Thm. 3.1.3 we obtain $x \in C_b^1([0, \infty[)$. Denoting $\Delta_{\mathcal{E}, A}(t) = (x(t) - x_A(t))^2 + (\dot{x}(t) - \dot{x}_A(t))^2$ for $t \in [0, \infty[$, we see that $\Delta_{\mathcal{E}, A}$ is decreasing, whence $\Delta_{\mathcal{E}, A}(t) \to e_{\infty, A}$ as $t \to \infty$ for some $e_{\infty, A} \geq 0$. We need to show that $e_{\infty, A} = 0$ for some $A \in [0, A_*]$. To see this, we note that the map $t \mapsto (p(t), x_*(t), \dot{x}_*(t), \cos(t + \theta_0), \sin(t + \theta_0)) \in \mathbb{R}^5$ is almost periodic. Hence Lemma 3.4.2 applies to yield some sequence $\tau_j \to \infty$ such that as $j \to \infty$

$$p_{\tau_j} \to p, \quad x_{*, \tau_j} \to x_*, \quad \dot{x}_{*, \tau_j} \to \dot{x}_*,$$
$$\cos(\cdot + \tau_j + \theta_0) \to \cos(\cdot + \theta_0), \quad \text{and} \quad \sin(\cdot + \tau_j + \theta_0) \to \sin(\cdot + \theta_0),$$

all uniformly on $\mathbb{R}$. Define $x_j(t) = x(t + \tau_j)$ for $t \in [-\tau_j, \infty[$. Then $\ddot{x}_j(t) + x_j(t) + \operatorname{Sgn} \dot{x}_j(t) \ni p_{\tau_j}(t)$ a.e. in $[-\tau_j, \infty[$, thus the argument of the proof of Thm. 3.1.4 can be used to see that there is a solution $\bar{x}$ of (3.17) such that $x_j \to \bar{x}$ and $\dot{x}_j \to \dot{\bar{x}}$ as $j \to \infty$, at least on a subsequence, and uniformly on any compact subinterval of $\mathbb{R}$; note that neither the subsequence nor the limit function does depend on $A \in [0, A_*]$. Next we remark that for $t \in \mathbb{R}$

$$
\begin{aligned}
&|x_A(t + \tau_j) - x_A(t)| \\
&\quad \leq |x_*(t + \tau_j) - x_*(t)| + A |\cos(t + \tau_j + \theta_0) - \cos(t + \theta_0)| \\
&\quad \leq |x_{*,\tau_j} - x_*|_\infty + A_* |\cos(\cdot + \tau_j + \theta_0) - \cos(\cdot + \theta_0)|_\infty \to 0, \quad j \to \infty,
\end{aligned}
$$

and similarly for the derivatives. For $t \in \mathbb{R}$ fixed, $x_j(t)$ is defined for $j \in \mathbb{N}$ large enough, and $x(t + \tau_j) = x_j(t) \to \bar{x}(t)$ as $j \to \infty$, again similarly for the derivatives. Hence we can set $t + \tau_j$ in the definition of $\Delta_{\varepsilon,A}$ and pass to the limit $j \to \infty$ to find that

$$
\left(\bar{x}(t) - x_A(t)\right)^2 + \left(\dot{\bar{x}}(t) - \dot{x}_A(t)\right)^2 \equiv e_{\infty,A}, \quad t \in \mathbb{R}, \qquad (3.79)
$$

for every $A \in [0, A_*]$. In particular, we obtain $\bar{x} \in C_b^1(\mathbb{R})$. Since x_A is nonconstant, we are thus in a situation where Cor. 3.1.1 applies to $\bar{x}$ and x_A, since (3.79) says that (3.32) is satisfied. Thus $\ddot{z} + z = 0$ a.e. in $\mathbb{R}$ for $z = \bar{x} - x_A$, and therefore $\bar{x}(t) = x_A(t) + B \cos(t + \varphi)$, where $B, \varphi \in \mathbb{R}$ will depend on A. As the almost periodic functions comprise a vector space, we infer that $\bar{x}$ is an almost periodic solution of (3.17), i.e., $\bar{x} \in S_{\mathrm{ap}}$. From (3.76) it then follows that $\bar{x} = x_{\bar{A}}$ for some $\bar{A} \in [0, A_*]$, hence we find $e_{\infty,\bar{A}} = 0$ according to (3.79). This completes the proof of the theorem. $\qquad \square$

3.4 Appendix: Almost periodic functions

Some background material concerning almost periodic functions is presented here; see BESICOVITCH [24] or CORDUNEANU [57] for more details.

Definition 3.4.1. *A function $p \in C_b(\mathbb{R})$ is called almost periodic, if the closure of $\{p_\tau : \tau \in \mathbb{R}\}$ w.r. to $|\cdot|_\infty$ is compact.*

Here $C_b(\mathbb{R})$ denotes the space of continuous and uniformly bounded functions $p : \mathbb{R} \to \mathbb{R}$ with norm $|p|_\infty = \max_{t \in \mathbb{R}} |p(t)|$, and $p_\tau(t) = p(t + \tau)$ for $t \in \mathbb{R}$. Almost periodic functions with values in $\mathbb{R}^n$ (in particular, in $\mathbb{C}$) are defined analogously. There are alternative useful characterizations of almost periodic functions.

Lemma 3.4.1. *The following are equivalent for a function $p \in C_b(\mathbb{R}; \mathbb{C})$.*

(a) p is almost periodic.
(b) For every $\varepsilon > 0$ there exists $\delta > 0$ such that for any interval $I \subset \mathbb{R}$ of length δ there is $t_* \in I$ with $|p(t + t_*) - p(t)| \le \varepsilon$ for all $t \in \mathbb{R}$.
(c) For every $\varepsilon > 0$ there exists a trigonometric polynomial

$$T_\varepsilon(t) = \sum_{j=1}^{J} \alpha_j e^{i\lambda_j t},$$

with suitable $\alpha_j \in \mathbb{C}$ and $\lambda_j \in \mathbb{R}$, such that $|p(t) - T_\varepsilon(t)| \le \varepsilon$ for all $t \in \mathbb{R}$.

Proof: Cf. CORDUNEANU [57, p. 14 ff]. □

Example 3.4.1. (a) Every periodic function is almost periodic.
(b) Let $(\alpha_j) \subset]0, \infty[$ be such that $\sum_{j=1}^{\infty} \alpha_j < \infty$, and suppose $(\lambda_j) \subset \mathbb{R}$ is bounded. Then the function $p(t) = \sum_{j=1}^{\infty} \alpha_j e^{i\lambda_j t}$ is almost periodic, by Lemma 3.4.1(c). The same holds for the real-valued

$$p(t) = \sum_{j=1}^{\infty} \alpha_j \sin(\lambda_j t), \quad t \in \mathbb{R},$$

and note that in addition $p \in C^\infty(\mathbb{R})$. ◇

The function $p(t) = \sin(t) + \sin(\sqrt{2}t)$ provides an example of an almost periodic function that is not periodic. The next lemma contains a useful property of almost periodic functions which is obvious for periodic functions.

Lemma 3.4.2. *Let $p \in C_b(\mathbb{R})$ be almost periodic. Then there exists $\tau_j^{\pm} \to \pm\infty$ such that $p_{\tau_j^{\pm}} \to p$ as $j \to \infty$, uniformly on $\mathbb{R}$.*

Proof: This is a consequence of the characterization (b) from Lemma 3.4.1, since we can fix $\varepsilon_j \searrow 0$, choose the corresponding $\delta_j > 0$, e.g. define $I_j = [j, j + \delta_j]$, and then set $\tau_j^+ = t_{*,j}$. □

Lemma 3.4.3. *Let $p, q : \mathbb{R} \to \mathbb{C}$ be almost periodic functions. Then the mean*

$$\langle p, q \rangle_{ap} = \lim_{T \to \infty} \frac{1}{T} \int_0^T p(t) \overline{q(t)} \, dt$$

does exist, and even

$$\langle p, q \rangle_{ap} = \lim_{T \to \infty} \frac{1}{T} \int_\tau^{\tau+T} p(t) \overline{q(t)} \, dt$$

uniformly w.r. to $\tau \in \mathbb{R}$.

Proof: Since p and $\bar{q}$ are almost periodic, so is $\psi = p\bar{q}$. Hence the result follows from the fact that for almost periodic functions φ the mean value $\lim_{T\to\infty} \frac{1}{T} \int_\tau^{\tau+T} \varphi(t)\,dt$ does exist, uniformly in τ; see CORDUNEANU [57, Thm. 1.12]. $\qquad\square$

Example 3.4.2. We have

$$\langle \sin(\eta t), e^{it} \rangle_{\text{ap}} = \begin{cases} -\frac{i}{2} & : \quad \eta = 1 \\ 0 & : \quad \eta \neq 1 \end{cases},$$

as well as

$$\langle |\sin t|, 1 \rangle_{\text{ap}} = \frac{2}{\pi}. \tag{3.80}$$

Indeed, for $\eta = 1$ we calculate

$$\frac{1}{T} \int_0^T \sin t\, e^{-it}\, dt = \frac{1}{2T} \left(\sin^2 T - i[T - \sin T \cos T] \right) \to -\frac{i}{2}, \quad T \to \infty,$$

whereas for $\eta \neq 1$

$$\frac{1}{T} \int_0^T \sin(\eta t) e^{-it}\, dt = \frac{e^{-it}}{(1 - \eta^2)T} \left(\eta \cos(\eta t) + i \sin(\eta t) \right) \Big|_{t=0}^{t=T} \to 0$$

as $T \to \infty$. Finally,

$$\frac{1}{2k\pi} \int_0^{2k\pi} |\sin t|\, dt = \frac{2}{\pi},$$

whence (3.80) follows, as we know that $\frac{1}{T} \int_0^T |\sin t|\, dt \to \langle |\sin t|, 1 \rangle_{\text{ap}}$ as $T \to \infty$. $\qquad\diamond$

Lemma 3.4.4. *Assume that $p \in C_b(\mathbb{R})$ is almost periodic. Then every $\psi \in \overline{\{p_\tau : \tau \in \mathbb{R}\}}^{|\cdot|_\infty}$ as well is almost periodic, and moreover*

$$\inf_{t\in\mathbb{R}} \psi(t) = \inf_{t\in\mathbb{R}} p(t), \quad \sup_{t\in\mathbb{R}} \psi(t) = \sup_{t\in\mathbb{R}} p(t), \quad \text{and} \quad \langle \psi, e^{it} \rangle_{\text{ap}} = \langle p, e^{it} \rangle_{\text{ap}}.$$

Proof: Let $(\tau_j) \subset \mathbb{R}$ be such that $|p_{\tau_j} - \psi|_\infty \to 0$ as $j \to \infty$. Observing the estimate

$$|\psi(t + t_*) - \psi(t)| \leq |\psi(t + t_*) - p_{\tau_j}(t + t_*)| + |p(t + t_* + \tau_j) - p(t + \tau_j)|$$
$$+ |p_{\tau_j}(t) - \psi(t)|$$
$$\leq 2 |p_{\tau_j} - \psi|_\infty + |p(\cdot + t_*) - p|_\infty,$$

it follows from Lemma 3.4.1(b) and p being almost periodic that also ψ is almost periodic. Next, as $p_{\tau_j} \to \psi$ uniformly, we moreover have e.g. $\inf \psi = \lim_{j \to \infty} (\inf p_{T_j}) = \lim_{j \to \infty} (\inf p) = \inf p$. The last assertion $\langle \psi, e^{it} \rangle_{ap} = \langle p, e^{it} \rangle_{ap}$ is a consequence of the fact that for almost periodic functions φ the mean value $\varphi \mapsto \lim_{T \to \infty} \frac{1}{T} \int_0^T \varphi(t)\, dt$ is continuous w.r. to uniform convergence; cf. CORDUNEANU [57, Thm. 1.13]. $\qquad \square$

Lemma 3.4.5. *If $p \in C_b(\mathbb{R})$ is almost periodic and decreasing, then p is constant.*

Proof: By considering $\bar{p}(t) = p(t) - \inf_{\mathbb{R}} p$ if necessary, we may suppose that $p(t) \geq 0$ in $\mathbb{R}$. Assume $p(t_1) \geq p(t_2) + 2\varepsilon$ for some $t_1 < t_2$ and some $\varepsilon > 0$. We choose the associated $\delta > 0$ from Lemma 3.4.1(b), and then for $I = [-(t_2 - t_1) - \delta, -(t_2 - t_1)]$ an $t_* \in I$ such that $|p(t + t_*) - p(t)| \leq \varepsilon$ for $t \in \mathbb{R}$. Setting $t = t_1 - t_*$ this implies $p(t_1) \leq \varepsilon + p(t_1 - t_*)$. On the other hand, $t_1 - t_* \geq t_2$, and hence $p(t_1 - t_*) \leq p(t_2)$ since p is decreasing. This yields the contradiction $p(t_1) \leq \varepsilon + p(t_2)$. $\qquad \square$

4. Lyapunov exponents for non-smooth dynamical systems

For smooth dynamical systems, Lyapunov exponents are a well-established tool for making qualitative predictions about the long-term behaviour of the system, since they describe the exponential convergence or divergence of nearby trajectories. To explain this, we consider $\dot{y} = f(y)$ with $f \in C^1(\mathbb{R}^d, \mathbb{R}^d)$, and we assume the equation generates a (global) semiflow $\varphi : [0, \infty[\times \mathbb{R}^d \to \mathbb{R}^d$, i.e., $y(\cdot) = \varphi(\cdot, y_0)$ solves the differential equation and has $y(0) = y_0$. Let y and $\bar{y}$ be two fixed solutions with $y(0) = y_0$ and $\bar{y}(0) = \bar{y}_0$. To study how small changes $|\bar{y}_0 - y_0|$ of the initial values will affect the difference of the solutions at a later time $t > 0$, we write

$$\bar{y}(t) - y(t) = \varphi(t, \bar{y}_0) - \varphi(t, y_0) \approx \partial_y \varphi(t, y_0)(\bar{y}_0 - y_0) = \partial_y \varphi(t, y_0) z_0 \, .$$

Hence the quantity

$$\lambda^+(y_0, z_0) = \limsup_{t \to \infty} \left(\frac{1}{t} \ln \frac{|\partial_y \varphi(t, y_0) z_0|}{|z_0|} \right) = \limsup_{t \to \infty} \left(\frac{1}{t} \ln |\partial_y \varphi(t, y_0) z_0| \right)$$

somehow describes the evolution of small perturbations of the initial value, and it is called an upper Lyapunov exponent; "upper" refers to the lim sup. Letting $T(t, y) = T_f(t, y) = \partial_y \varphi(t, y)$, we alternatively have

$$\lambda^+(y_0, z_0) = \limsup_{t \to \infty} \left(\frac{1}{t} \ln |T(t, y_0) z_0| \right) . \tag{4.1}$$

Observe that this linearization $T : [0, \infty[\times \mathbb{R}^d \to \mathcal{L}(\mathbb{R}^d)$ defines what is called a cocycle for the semiflow $(\varphi^t)_{t \geq 0}$ [with $\varphi^t(y) = \varphi(t, y)$], i.e.,

$$T(t + s, y) = T(t, \varphi^s(y)) T(s, y) \quad \text{for} \quad t, s \in [0, \infty[\quad \text{and} \quad y \in \mathbb{R}^3, \tag{4.2}$$

as follows by differentiating the semiflow identity $\varphi(t + s, y) = \varphi(t, \varphi(s, y))$ w.r. to y.

In order that $\lambda^+(y_0, z_0)$ be really related to the long-term behaviour (or the stability) of the system under consideration, the $\limsup_{t \to \infty}$ in (4.1) has to exist as a $\lim_{t \to \infty}$. Therefore one tries to obtain information in which cases this $\limsup_{t \to \infty}$ in fact may be replaced with $\lim_{t \to \infty}$.

The easiest example for such situation is provided by the linear systems $\dot{y} = Ay$ with $A \in \mathcal{L}(\mathbb{R}^d)$. Then $\varphi(t, y_0) = e^{At} y_0$ and $T(t, y) = e^{At}$. Through transformation of A to Jordan's canonical form it may be seen that $\lambda^+(y_0, z_0)$ in fact exists as a limit for all $y_0, z_0 \in \mathbb{R}^d$, and, moreover, that

$$\{\lambda^+(y_0, z_0) : y_0, z_0 \in \mathbb{R}^d\} = \{\operatorname{Re} \lambda^{(i)} : 1 \leq i \leq k\},$$

with $\lambda^{(i)}, 1 \leq i \leq k$, being the $k \leq d$ different eigenvalues of A; see REITMANN [194, Bsp. 23.1]. This also enlightens a connection of the $\lambda^+(y_0, z_0)$ to the stability of the system, since e.g. $\operatorname{Re} \lambda^{(i)} < 0$ for $1 \leq i \leq k$ implies that $y = 0$ is attracting all trajectories at an exponential rate.

For general cocycles and corresponding flows, a powerful tool to prove the existence of the $\lim_{t \to \infty}$ in (4.1) is given by Oseledets' multiplicative ergodic theorem from OSELEDETS [167] (henceforth abbreviated MET); see also L. ARNOLD [10], L. ARNOLD/WIHSTUTZ [11], or KRENGEL [111]. For the purpose of motivation, here we state the MET only in a quite reduced form, while in Thm. 4.3.1 of the appendix, see p. 136, the full theorem and also the definitions of the relevant terms from ergodic theory may be found.

Theorem. *Let φ be a measurable semiflow on $\mathbb{R}^d$, T be a measurable cocycle for $(\varphi^t)_{t \geq 0}$ as above, and let μ be an ergodic probability measure which is invariant w.r. to the semiflow. Then, if two integrability conditions are satisfied, there exists an invariant set $\Gamma \subset \mathbb{R}^d$ with $\mu(\Gamma) = 1$, and we find $k \in \{1, \ldots, d\}$ and distinct numbers $\lambda^{(1)}, \ldots, \lambda^{(k)} \in \mathbb{R}$ (where $\lambda^{(1)}$ may be $-\infty$), such that for all $y_0 \in \Gamma$ and $y \in \mathbb{R}^d$*

$$\lim_{t \to \infty} \left(\frac{1}{t} \ln |T(t, y_0)y| \right) = \lambda^{(i)}$$

for some $1 \leq i \leq k$. Thus in particular the $\limsup_{t \to \infty}$ in (4.1) exists as a $\lim_{t \to \infty}$ for μ-a.e. $y_0 \in \mathbb{R}^d$ and all $z_0 = y \in \mathbb{R}^d$.

Note that the assertion of the theorem is intimately connected to μ; if $\mu = \delta_{y_0}$ is a Dirac measure, then from the theorem we can conclude the existence of $\lim_{t \to \infty}$ in (4.1) only for this particular y_0.

Since in non-smooth dynamical systems (like the pendulum with dry friction (1.2), see the introduction) the semiflow can no longer be expected to be smooth w.r. to y, the first basic question to obtain Lyapunov exponents for such systems by application of the MET is:

How to define a canonical cocycle for the non-smooth system ?

The construction of this cocycle will be carried out in Sect. 4.1 and described there in greater detail, so that we restrict ourselves here to a sketch. In principle, we rely on partitioning $\mathbb{R}^d$ in suitable manifolds, the validity regions of different smooth systems, which yield the complete non-smooth system when being pieced together. For example, denoting by $y = (x, v, \tau)$ a typical point in $\mathbb{R}^3$, (1.4) means by definition of Sgn that $\dot{y} = f_+(y) = (v, -x + b^-(\tau), 1)$ in $\{y \in \mathbb{R}^3 : v > 0\}$ and $\dot{y} = f_-(y) = (v, -x + b^+(\tau), 1)$ in $\{y \in \mathbb{R}^3 :$

$v < 0\}$ with $b^{\pm}(\tau) = \gamma \sin(\eta\tau) \pm 1$; the situation is more complicated along the surface $\{y \in \mathbb{R}^3 : v = 0\}$.

For the construction of the cocycle assume for the moment that every initial value $y_0 \in \mathbb{R}^d$ generates a sequence of switching times $(t_i(y_0))_{i \in \mathbb{N}}$ from one manifold to the next, and so forth, along the trajectory starting in y_0. Then, heuristically, $\varphi^t(\cdot)$ may be expected to be C^1 for $t \in]t_i(y_0), t_{i+1}(y_0)[$ resp. for $t \in]t_{i+1}(y_0), t_{i+2}(y_0)[$. Hence $T(t, y_0) = \partial_y \varphi^t(y_0)$ is well-defined for $t \in]t_i(y_0), t_{i+1}(y_0)[$ and for $t \in]t_{i+1}(y_0), t_{i+2}(y_0)[$, so that we have to investigate the transition of the linearization at the switching time $t_{i+1}(y_0)$. A formula of the type

$$\lim_{t \to t_{i+1}(y_0)^+} \partial_y \varphi^t(y_0) = A_{i+1}(y_0) \left(\lim_{t \to t_{i+1}(y_0)^-} \partial_y \varphi^t(y_0) \right)$$

for this transition will be derived in Sect. 4.1.2, and under suitable additional assumptions the transition operator $A_{i+1}(y_0)$ can be given in very explicit form. Thus we will obtain in Sect. 4.1.3 a canonical cocycle for the underlying non-smooth system by defining $T(t, y_0) = \partial_y \varphi^t(y_0)$ for $t \in]t_{i+1}(y_0), t_{i+2}(y_0)[$, and at the transition time $t = t_{i+1}(y_0)$, $T(t_{i+1}(y_0), y_0) = A_{i+1}(y_0) \left(\lim_{t \to t_{i+1}(y_0)^-} \partial_y \varphi^t(y_0) \right)$.

In fact matters are somewhat more technical, and in particular it is not the complete truth that $\varphi^t(\cdot)$ is C^1 for fixed $t \in]t_i(y_0), t_{i+1}(y_0)[$, as may already be seen through the argument that for y near to y_0 then also $t \in]t_i(y), t_{i+1}(y)[$ is required in order to be in the same manifold of the partition, and hence to have the same governing right-hand side. This amounts in imposing additional assumptions which will be made precise in Sect. 4.1.1. These assumptions will lead to a restriction of admissible initial values to some forward invariant "good" set $G \subset \mathbb{R}^d$. To have $y_0 \in G$ essentially means that the generated trajectory should never meet some exceptional sets in the different manifolds, and those exceptional sets will be part of the respective lower-dimensional boundaries. Therefore it will be possible in Sect. 4.1.4 to prove that in fact $\lambda^d(\mathbb{R}^d \setminus G) = 0$, with λ^d being Lebesgue measure in $\mathbb{R}^d$.

Hence we obtain a well-defined cocycle for every non-smooth system which satisfies several natural conditions described in Sect. 4.1. Assuming the existence of a (physically relevant ergodic) invariant measure, we thus can apply the MET to obtain Lyapunov exponents for the corresponding non-smooth system. Since for complicated systems with practical relevance it is in general almost impossible to calculate or estimate the Lyapunov exponents directly, they often have to be calculated numerically; see Sect. 9.1 for remarks in this direction.

Since a priori it is not clear that the Lyapunov exponents are really related to the behaviour of the system, the next basic issue is:

Do these exponents correctly reflect the dynamical behaviour of the non-smooth system ?

It is the purpose of Sect. 4.2 to give a positive answer to this question in case of the pendulum with dry friction (1.2), so that one can also be confident for more complicated systems. We start in Sect. 4.2.1 by explaining where the semiflow for (1.2) comes from, whereas in Sect. 4.2.2 we verify the general assumptions imposed in Sect. 4.1 in case of the pendulum with dry friction. In Sect. 4.2.3 we first recall results from Sect. 3.2.1 concerning the existence and stability of periodic solutions to (1.2), which we also want to summarize here briefly. In the most interesting resonant case $\eta = 1$ the situation is as follows: for $\gamma \in [0,1]$ exactly the equilibria $x_0 \in [\gamma - 1, -\gamma + 1]$ of (1.2) are the periodic solutions. (Note that it is not possible for a forced smooth equation $\ddot{x} + g(x, \dot{x}) = p(t)$ to have a fixed point, whereas here we have a whole interval.) If $\gamma \in]1, 4/\pi[$, then a unique periodic solution x_γ appears, it is non-trivial and stable, and of a very special form: there are $\tau_1^\gamma \in]0, \pi/2[$ and $\tau_2^\gamma \in]\pi/2, \pi[$ such that x_γ is given through

$$\begin{cases} x_\gamma(\tau) \equiv -1 + \gamma \sin \tau_1^\gamma & \text{for} \quad \tau \in [0, \tau_1^\gamma], \\ \dot{x}_\gamma(\tau) > 0 & \text{for} \quad \tau \in]\tau_1^\gamma, \tau_2^\gamma[, \\ x_\gamma(\tau) \equiv 1 - \gamma \sin \tau_1^\gamma & \text{for} \quad \tau \in [\tau_2^\gamma, \pi], \\ x_\gamma(\tau) = -x_\gamma(\tau + \pi) & \text{for} \quad \tau \in \mathbb{R} \end{cases}$$

Therefore the existence of x_γ is closely related to the non-smoothness of the system, since x_γ has $\dot{x}_\gamma \equiv 0$ in $[0, \tau_1^\gamma]$, $[\tau_2^\gamma, \pi + \tau_1^\gamma]$, $[\pi + \tau_2^\gamma, 2\pi]$, and so on. Moreover, we have seen that no periodic solutions of (1.2) can exist for $\gamma \in]4/\pi, \infty[$. These results are extended in Sect. 4.2.3 by proving that in fact the interval of fixed points $[\gamma - 1, -\gamma + 1]$ for $\gamma \in [0, 1]$ and the periodic solution x_γ for $\gamma \in]1, 4/\pi[$ are asymptotically stable *uniformly* w.r. to initial values in bounded sets, and also uniformly w.r. to phase shifts of the forcing. Moreover, for $\gamma \in]4/\pi, \infty[$ all solutions are unbounded, again uniformly w.r. to initial values in bounded sets and phase shifts. So the dynamical behaviour of the resonant system is completely understood, and to answer the above question we have to see how this is reflected in the corresponding Lyapunov exponents.

As will be explained in Sects. 4.2.4(ii) and 4.2.4(iii), it is a consequence of the stability results just mentioned that for $\dot{y} \in F(y)$ from (1.4), considered on $\mathbb{R}^2 \times S^1$, the ergodic measures are as follows: if $\gamma \in [0, 1]$, then the only ergodic measures are the Dirac measures δ_{x_0} with $x_0 \in [\gamma - 1, -\gamma + 1]$, whereas it is uniform distribution on the graph of x_γ in case that $\gamma \in]1, 4/\pi[$. Moreover, there are no invariant measures at all for $\gamma \in]4/\pi, \infty[$. Therefore we apply the MET with the cocycle constructed and these ergodic measures to obtain as Lyapunov exponents $-\infty, 0, 0$ for $\gamma \in [0, 1]$ resp. $-\infty, \pi^{-1} \ln |\cos(\tau_2^\gamma - \tau_1^\gamma)|, 0$ for $\gamma \in]1, 4/\pi[$, whereas for $\gamma \in]4/\pi, \infty[$ the MET is not applicable. It is clear that for $\gamma \in [0, 4/\pi[$ these exponents correctly reflect the dynamical behaviour of the underlying resonant system, and they also agree with numerically found values; see MICHAELI [140]. Moreover, it should be remarked that the exponents are continuous at $\gamma = 1$; see Rem. 4.2.1.

It is of course true that in fact this gives nothing new, since the respective long-term behaviour of the system is already know. But at least it serves as a hint that for predictions of the dynamics one can rely on Lyapunov exponents also in non-smooth dynamical systems like (1.2).

In the context of (1.2), another related basic question is:

$$\text{Does system (1.2) show "chaotic behaviour" ?}$$

In smooth systems it is argued that positive Lyapunov exponents indicate "chaotic behaviour" (we do not make this term more precise), so that we are interested whether such positive exponents will also appear for (1.2), not especially at resonance. By analysis of the cocycle we will show in Sect. 4.2.4 that in fact all upper Lyapunov exponents $\lambda^+(y_0, y) = \limsup_{t \to \infty} \left(\frac{1}{t} \ln |T(t, y_0)y| \right)$ are *non-positive*, i.e., whenever the MET is applicable, none of the exponents will be larger than zero. Since we argued above that application of the MET to the constructed cocycle will yield Lyapunov exponents which correctly reflect the dynamical behaviour, we thus conclude that the pendulum with dry friction (1.2) *does not* show chaotic behaviour. This result is not restricted to the special forcing $\gamma \sin(\eta \tau)$, but carries over to all forcings which lead to a well-defined cocycle. The last Sect. 4.3 of this chapter is an appendix, where we first collect some results from ergodic theory in Sect. 4.3.1, whereas Sect. 4.3.2 contains technical lemmas.

Of course there are several questions left open concerning Lyapunov exponents for non-smooth dynamical systems, and also concerning (1.2). It is a main point to clarify the existence of invariant measures under some general assumptions, whereas in concrete examples this problem mostly will be reduced to finding forward invariant sets, since the semiflow induced by differential inclusion often is continuous in dissipative cases; whence, as soon as we have detected a compact forward invariant set, the existence of an invariant measure follows from the theorem of KRYLOV and BOGOLUBOV; see KATOK/HASSELBLATT [105, Thm. 4.1.1]. A much more difficult task is then to single out a "physically relevant" ergodic measure. Moreover, in the general context, some relation should be made between Lyapunov exponents and dynamical properties of the system, like those mentioned in ECKMANN/RUELLE [74, p. 632] for smooth systems; see MICHAELI [140], where some results in this direction have been obtained.

This Chap. 4 follows KUNZE [115]. In KUNZE/MICHAELI [120] we already introduced a "formal system" like here in Sect. 4.1.1 to describe non-smooth dynamical systems as (1.2) and to define the corresponding Lyapunov exponents. Although correct, this approach had the drawback that some assumptions imposed unfortunately cannot be verified for concrete systems; see the related comments before Cor. 4.1.6 in Sect. 4.1.1.

4.1 General theory

In this section we will define a canonical cocycle for non-smooth systems which satisfy several assumptions. By application of the MET Thm. 4.3.1 we thus will obtain Lyapunov exponents for such systems.

To motivate the assumptions which will be needed, we first have a look back on (1.4) from the introduction. We already noted there that (1.4) implies $\dot{y} = f_+(y)$ in $\{y \in \mathbb{R}^3 : v > 0\}$ and $\dot{y} = f_-(y)$ in $\{y \in \mathbb{R}^3 : v < 0\}$, and we will see later that also the surface $\{y \in \mathbb{R}^3 : v = 0\}$ may be partitioned into finitely many parts, according to which ODE $\dot{y} = f(y)$ is governing the flow of the inclusion $\dot{y} \in F(y)$, at least for some time, when starting in this surface. So we may summarize what has to be expected for a non-smooth system like (1.4):

There are manifolds $M_k \subset \mathbb{R}^d$ for $k \in I$, I being some finite index set, and a manifold $M_\infty \subset \mathbb{R}^d$ together with C^1-functions f_k and f_∞ acting as right-hand sides, such that the semiflow is $\varphi^t(y_0) = \Phi_{f_k}(t, y_0)$ in $[0, t_1(y_0)[$, where for suitable $g \in C^1(\mathbb{R}^d, \mathbb{R}^d)$, $\Phi_g(\cdot, y_0)$ denotes the global flow generated by $\dot{y} = g(y)$. In addition, $t_1(y_0) > 0$ is the first switching time to a different manifold after starting in y_0. The manifold M_∞ will play a special role: it cannot be left if once reached, and thus $t_1(y_0) = \infty$ for $y_0 \in M_\infty$. In the example of the pendulum with dry friction we have

$$M_\infty = \{y \in \mathbb{R}^3 : v = 0, \gamma - 1 \le x \le 1 - \gamma\} \quad \text{for} \quad \gamma \in [0, 1],$$

corresponding to the interval $[\gamma - 1, 1 - \gamma]$ of stationary solutions, and we let $M_\infty = \emptyset$ for $\gamma \in]1, \infty[$. There are several natural assumptions to be made: if $y_0 \in M_k$, then the trajectory should not leave M_k until the first switching time, i.e., $\varphi^t(y_0) \in M_k$ for $t \in [0, t_1(y_0)[$, and also the consistency condition $t_1(\varphi^t(y_0)) = t_1(y_0) - t$ for $t \in [0, t_1(y_0)[$ should hold. Moreover, $\varphi(\cdot, y_0)$ can be expected to be continuous.

Since we want define a cocycle for non-smooth dynamical systems by imitating $T(t, y_0) = \partial_y \varphi^t(y_0)$ from the smooth case, we have to remark at this point that the requirements mentioned so far are not enough to advance further in this direction. To explain this, note that if e.g. $y_0 \in M_k$ and $\xi_1(y_0) = \varphi^{t_1(y_0)}(y_0) \in M_l$, then $\varphi^t(y_0) = \Phi_{f_l}(t - t_1(y_0), \xi_1(y_0))$ for $t \in [t_1(y_0), t_1(y_0) + t_1(y_1)[$ by the above assumptions, because we have to solve $\dot{y} = f_l(y)$ and start in $\xi_1(y_0)$ at time $t_1(y_0)$. As we do not know anything so far about differentiability of $t_1(\cdot)$ and $\xi_1(\cdot)$, we have to include further restrictions.

In concrete non-smooth examples it cannot be expected to have $t_1 \in C^1(\mathbb{R}^d)$, as may be seen from Fig. 4.1 below, where typical example trajectories of the autonomous system (1.4) of the pendulum with dry friction for $\gamma \in]0, 1[$ are shown. Here τ resp. x are displayed in vertical resp. horizontal direction, whereas the v-axis is perpendicular, i.e., the plane $\{v = 0\}$ coincides with the plane of the paper with $\{v > 0\}$ lying behind. Since the right-hand side of (1.4) has a "1" in the third line, the semiflow is of the form $\varphi^t(y_0) = (*, *, \tau_0 + t)$ for $y_0 = (\tau_0, x_0, v_0) \in \mathbb{R}^3$. Note that we do not take time mod $(2\pi/\eta)$ for the moment being, because this simplifies notation and figures. Although we did not specify precisely the different manifolds for the system (1.4) up to now, comparing both shown trajectories it is clear that $t_1(\cdot)$ is discontinuous at y_0, since some y close to y_0 will have a considerably smaller first switching time than y_0. Figure 4.1 also makes clear that we can do better if the reference trajectory of y_0 does not meet some "exceptional sets" E_k, which for the present purpose of motivation may be thought of to be both boundary curves $\{y \in \mathbb{R}^3 : v = 0, x = b^\pm(\tau)\}$; in fact these exceptional sets are smaller. The corresponding requirement for the first arrival point $\xi_1(y_0)$ at a different manifold M_l after start in $y_0 \in M_k$ is that there should be a M_k-neighborhood of y_0 which is sent to M_l by $\xi_1(\cdot)$. Therefore, on every

$$D_k = \Big\{ y \in M_k : \varphi^t(y) \notin E_k \; \forall t \in [0, t_1(y)], \text{ and}$$
$$l \in I \cup \{\infty\}, \xi_1(y) \in M_l \Rightarrow \forall\, t \in [0, t_1(y)[\; \exists\, \varepsilon > 0 :$$
$$\xi_1 \big(B_\varepsilon(\varphi^t(y)) \cap M_k \big) \subset M_l \Big\}, \tag{4.3}$$

it makes sense to assume $t_1(\cdot)$ to be differentiable.

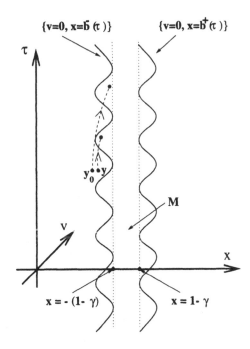

Fig. 4.1. Example trajectories of (1.4) for $\gamma \in]0, 1[$

Here are some additional remarks related to the definition of D_k: we want to find a "good" set $G \subset \mathbb{R}^d$ of initial values y_0 such that for trajectories starting in y_0 the cocycle $T(t, y_0)$ can be constructed meaningfully. So an additional requirement is to have $t_i(y_0) \to \infty$ as $i \to \infty$ (where the $t_i(y_0)$ are the consecutive switching times) in case that the trajectory does not arrive in M_∞, since the positive real axis has to be partitioned by the switching times to define $T(t, y_0)$ for all $t \in [0, \infty[$. In case that a trajectory arrives at M_∞, no further switching will take place. Note that the assumption $t_i(y_0) \to \infty$ excludes the possibility of infinitely many switchings in a finite time, which nevertheless is possible for some models, like for a rotor touching a rigid boundary; see the extensive study in SZCZYGIELSKI [213], and also A. MÜLLER [153].

The set G will simply be obtained through the requirement that if a trajectory arrives at the manifold M_k at some time $t_i(y_0)$, then the arrival point should be already in D_k. Since we need to have G forward invariant in order to verify the cocycle property (4.2), it is thus not enough to write, in the definition of D_k, "$\varphi^{t_1(y_0)}(y_0) \notin E_k$" instead of "$\varphi^t(y) \notin E_k$ for all $t \in [0, t_1(y)]$", and for the same reason we have the second condition instead of just "$\exists\, \varepsilon > 0 : \xi_1(B_\varepsilon(y) \cap M_k) \subset M_l$". In concrete examples, both the condition on $\xi_1(\cdot)$ not to separate neighborhoods and the condition on the semiflow not to meet E_k in (4.3) will be somehow related, but in general they are not the same, as we will see in the example of the pendulum with dry friction in Sect. 4.2.

4.1.1 Preliminaries

After the preceding motivation, we will impose the following hypotheses: We are given a partition $\mathbb{R}^d = \left(\bigcup_{k=1}^{K} M_k \right) \cup M_\infty$ of $\mathbb{R}^d$ in mutually disjoint C^1-manifolds. Moreover, we suppose that we have an exceptional set $E_k \subset \overline{M_k}^{\mathbb{R}^d}$ and a given function $f_k \in C^1(\mathbb{R}^d, \mathbb{R}^d)$ for every $k \in I \cup \{\infty\} = \{1, \ldots, K\} \cup \{\infty\}$. Let $\varphi : [0, \infty[\times \mathbb{R}^d \to \mathbb{R}^d$ be a measurable semiflow, i.e., $(t, y) \mapsto \varphi^t(y) = \varphi(t, y)$ is (Borel-) measurable, $\varphi^0 = \mathrm{id}_{\mathbb{R}^d}$ and $\varphi^{t+s} = \varphi^t \circ \varphi^s$ for all $t, s \in [0, \infty[$. Let the function $t_1(\cdot) : \mathbb{R}^d \to\,]0, \infty]$ be given. In the case that $M_\infty \neq \emptyset$ we fix some $y_\infty \in M_\infty$ and define

$$\xi_0(y) = y,\ y \in \mathbb{R}^d, \qquad \xi_1(y) = \begin{cases} \varphi^{t_1(y)}(y) & : \ y \in \mathbb{R}^d \setminus M_\infty \\ y_\infty & : \ y \in M_\infty \end{cases} \quad \text{and}$$

$$\xi_{i+1}(y) = \xi_1(\xi_i(y)) \quad \text{for} \quad i \in \mathbb{N}, \ y \in \mathbb{R}^d \tag{4.4}$$

as well as

$$t_0(y) = 0, \quad t_{i+1}(y) = t_1(\xi_i(y)) + t_i(y) \geq t_i(y) \quad \text{for} \quad i \in \mathbb{N}, \ y \in \mathbb{R}^d. \tag{4.5}$$

We assume that

(A1) $M_\infty = \{y \in \mathbb{R}^d : t_1(y) = \infty\}$,

hence $t_{i+1}(y) = \infty$ iff $\xi_i(y) \in M_\infty$ or $t_i(y) = \infty$. We also assume that

(A2) $y \in \mathbb{R}^d \setminus M_\infty$, $t \in [0, t_1(y)[\Rightarrow t_1(\varphi^t(y)) + t = t_1(y)$,
(A3) $k \in I \cup \{\infty\}$, $y \in M_k \Rightarrow \varphi^t(y) \in M_k$ for $t \in [0, t_1(y)[$,
(A4) $\varphi(\cdot, y)$ is continuous for $y \in \mathbb{R}^d$,
(A5) $\varphi^t(y) = \Phi_{f(\cdot, y)}(t, y)$ for $y \in \mathbb{R}^d$ and $t \in [0, t_1(y)[$,
(A6) $y \in \mathbb{R}^d$, $t_i(y) < \infty$ for all $i \in \mathbb{N} \Rightarrow 0 = t_0(y) < t_1(y) < t_2(y) \ldots \to \infty$,
(A7) $D_\infty = \{y \in M_\infty : \varphi^t(y) \notin E_\infty \,\forall t \in [0, \infty[\} = M_\infty \setminus E_\infty \neq \emptyset$, if $M_\infty \neq \emptyset$, and
(A8) $t_1 \in C^1(D_k)$ for $k \in I$, with D_k from (4.3).

We start with some elementary relationships between these objects.

Lemma 4.1.1. *Let* $(A1) - (A7)$ *be satisfied. Then the following assertions hold.*

(a) *For* $i, j \in \mathbb{N}_0$ *and* $y \in \mathbb{R}^d$ *we have* $\xi_{i+j}(y) = \xi_i(\xi_j(y))$.
(b) *For* $i \in \mathbb{N}_0$, $j \in \mathbb{N}$ *and* $y \in \mathbb{R}^d$ *we have* $t_j(\xi_i(y)) + t_i(y) = t_{i+j}(y)$.
(c) *For* $i \in \mathbb{N}_0$ *and* $y \in \mathbb{R}^d$ *we have*

$$\xi_i(y) = \begin{cases} \varphi^{t_i(y)}(y) & : \quad t_i(y) < \infty \\ y_\infty & : \quad t_i(y) = \infty \end{cases}$$

(d) *For* $i \in \mathbb{N}$, $j \in \mathbb{N}_0$, $y \in \mathbb{R}^d$ *with* $t_j(y) < \infty$ *and* $t \in [t_j(y), t_{j+1}(y)[$ *we have* $\xi_i(\varphi^t(y)) = \xi_{i+j}(y)$.
(e) *For* $i \in \mathbb{N}_0$, $y \in \mathbb{R}^d$ *with* $t_i(y) < \infty$ *and* $t \in [t_i(y), t_{i+1}(y)[$ *we have* $t_j(\varphi^t(y)) + t = t_{i+j}(y)$ *for* $j \in \mathbb{N}$.
(f) *For* $i \in \mathbb{N}_0$ *and* $y \in \mathbb{R}^d$ *with* $t_i(y) < \infty$ *there exists* $k \in I \cup \{\infty\}$ *such that* $\varphi([t_i(y), t_{i+1}(y)[, y) \subset M_k$.
(g) *For* $i \in \mathbb{N}_0$, $y \in \mathbb{R}^d$ *with* $t_i(y) < \infty$ *and* $t \in [t_i(y), t_{i+1}(y)[$

$$\varphi^t(y) = \Phi_{f(\cdot, \varphi^{t_i(y)}(y))}\left(t - t_i(y), \varphi^{t_i(y)}(y)\right). \tag{4.6}$$

If also $t_{i+1}(y) < \infty$, *then (4.6) holds for* $t = t_{i+1}(y)$, *too.*
(h) *For* $i \in \mathbb{N}$ *and* $y \in \mathbb{R}^d$: $t_i(y) < \infty \iff t_j(y) < \infty$ *for all* $j = 1, \ldots, i \iff \xi_j(y) \notin M_\infty$ *for all* $j = 0, \ldots, i - 1$.

Proof: Ad (a): This is obvious by (4.4). Ad (b): For $j = 1$ this is just (4.5). If (b) holds for j, then by (a) $t_{j+1}(\xi_i(y)) + t_i(y) = t_1(\xi_j(\xi_i(y))) + t_j(\xi_i(y)) + t_i(y) = t_1(\xi_{i+j}(y)) + t_{i+j}(y) = t_{i+j+1}(y)$. Ad (c): For $i = 0, 1$ the claim holds by (4.4) and (4.5). For $i \Rightarrow i+1$ we note that in the case $t_{i+1}(y) < \infty$ we obtain due to (4.4), (c) for i and (4.5), $\xi_{i+1}(y) = \xi_1(\xi_i(y)) = \varphi^{t_1(\xi_i(y))}(\varphi^{t_i(y)}(y)) = \varphi^{t_{i+1}(y) - t_i(y)}(\varphi^{t_i(y)}(y)) = \varphi^{t_{i+1}(y)}(y)$. On the other hand, if $t_{i+1}(y) = \infty$, then $\xi_i(y) \in M_\infty$ or $t_i(y) = \infty$, and the latter also implies $\xi_i(y) = y_\infty \in M_\infty$ by induction hypotheses. Thus in every case $\xi_{i+1}(y) = \xi_1(\xi_i(y)) = y_\infty$ by definition. Ad (d): We first consider the case $j = 0$, i.e., $t \in [0, t_1(y)[$. Suppose first that $t_1(y) < \infty$. Then by $(A2)$ we have $\xi_1(\varphi^t(y)) = \varphi^{t_1(\varphi^t(y))}(\varphi^t(y)) = \varphi^{t_1(y)-t}(\varphi^t(y)) = \varphi^{t_1(y)}(y) = \xi_1(y)$, and hence $\xi_i(\varphi^t(y)) = \xi_i(y)$ for $i \in \mathbb{N}$ by (a). In case that $t_1(y) = \infty$ we have $y \in M_\infty$, hence $\xi_1(y) = y_\infty$, and also $\varphi^t(y) \in M_\infty$ by $(A3)$, and thus $\xi_1(\varphi^t(y)) = y_\infty = \xi_1(y)$, and again the claim for arbitrary $i \in \mathbb{N}$ follows from (a). So (d) holds for $j = 0$ and all $i \in \mathbb{N}$. If $j \in \mathbb{N}$ is arbitrary and $t \in [t_j(y), t_{j+1}(y)[$ with $t_j(y) < \infty$, we first consider the case $t_{j+1}(y) < \infty$. Then $\bar{y} = \xi_j(y)$ has $t_1(\bar{y}) = t_{j+1}(y) - t_j(y)$ by (b), hence $s = t - t_j(y) \in [0, t_1(\bar{y})[$. So we obtain by the part already proved and by (a), $\xi_i(\varphi^t(y)) = \xi_i(\varphi^s(\bar{y})) = \xi_i(\bar{y}) = \xi_{i+j}(y)$. On the other hand, if $t_{j+1}(y) = \infty$ and $\bar{y}$ is defined as before, then $t_j(y) < \infty$ implies $\bar{y} \in M_\infty$. Since by (c) and $(A3)$ also $\varphi^t(y) = \varphi^{t-t_j(y)}(\bar{y}) \in M_\infty$, we again arrive at $\xi_1(\varphi^t(y)) = y_\infty = \xi_1(\bar{y}) = \xi_{j+1}(y)$, and the claim follows from (a). Ad (e): We first prove the claim for $i = 0$ and all $j \in \mathbb{N}$. For that, the case $j = 1$ is just $(A2)$. To obtain $j \Rightarrow j + 1$ we first assume $t_1(y) < \infty$. Then (d) and the assumption for j imply $t_{j+1}(\varphi^t(y)) + t = t_1(\xi_j(\varphi^t(y))) + t_j(\varphi^t(y)) + t = t_1(\xi_j(y)) + t_j(y) = t_{j+1}(y)$. If $t_1(y) = \infty$, then $t_j(y) = \infty$ for all $j \geq 1$, and also $y \in M_\infty$, hence $\varphi^t(y) \in M_\infty$ for $t \in [0, \infty[$, and thus $t_j(\varphi^t(y)) = \infty$ for $j \in \mathbb{N}$. Hence (e) holds for $i = 0$ and all $j \in \mathbb{N}$. If $i \in \mathbb{N}_0$, $j \in \mathbb{N}$, $y \in \mathbb{R}^d$ with $t_i(y) < \infty$ and $t \in [t_i(y), t_{i+1}(y)[$ are arbitrary, we first consider the case $t_{i+1}(y) < \infty$. Then $\bar{y} = \xi_i(y)$ has $t_1(\bar{y}) = t_{i+1}(y) - t_i(y)$ by (4.5), hence $\bar{t} = t - t_i(y) \in [0, t_1(\bar{y})[$ and $\varphi^{\bar{t}}(\bar{y}) = \varphi^t(y)$ by (c). So by the part already proved and (b), $t_j(\varphi^t(y)) + t = t_j(\varphi^{\bar{t}}(\bar{y})) + t = t_j(\bar{y}) - \bar{t} + t = t_j(\xi_i(y)) + t_i(y) = t_{i+j}(y)$. On the other hand, if $t_{i+1}(y) = \infty$, then $t_{i+j}(y) = \infty$ for all $j \in \mathbb{N}$, and as in (d), $\varphi^t(y) \in M_\infty$, hence $t_j(\varphi^t(y)) = \infty$ for $j \in \mathbb{N}$. Ad (f) & (g): Again we first assume that $t_{i+1}(y) < \infty$. With $\bar{y} = \xi_i(y)$ and $\bar{t} = t - t_i(y)$ we have $[t_i(y), t_{i+1}(y)[= [0, t_1(\bar{y})[+ t_i(y)$. Hence $\varphi([t_i(y), t_{i+1}(y)[, y) = \varphi([0, t_1(\bar{y})[, \bar{y})$, and thus (f) is a consequence of $(A3)$. This also yields, because of $(A5)$,

$$\varphi^t(y) = \varphi^{\bar{t}}(\bar{y}) = \Phi_{f(\cdot, \bar{y})}(\bar{t}, \bar{y}) = \Phi_{f(\cdot, \varphi^{t_i(y)}(y))}\left(t - t_i(y), \varphi^{t_i(y)}(y)\right).$$

Moreover, $(A4)$ and the continuity of every $\Phi_g(\cdot, z)$ show that (4.6) also holds for $t = t_{i+1}(y)$. In the case that $t_{i+1}(y) = \infty$, again $\bar{y} = \xi_i(y) = \varphi^{t_i(y)}(y) \in M_\infty$, hence $\varphi^t(y) = \varphi^{t-t_i(y)}(\bar{y}) \in M_\infty$ for $t \in [t_i(y), \infty[$ by $(A3)$, so that we can choose $k = \infty$ in (f). Concerning (g), we have $\varphi^s(\bar{y}) = \Phi_{f_\infty}(s, \bar{y})$ for $s \in [0, \infty[$ by $(A5)$. Consequently we take $s = t - t_i(y)$ to obtain the claim of (g) for $t_{i+1}(y) = \infty$. Ad (h): The first equivalence is because of $t_{p+1}(y) \geq t_p(y)$ for all $p \in \mathbb{N}_0$. Concerning the second, "$\Rightarrow$" follows from (4.5). "$\Leftarrow$:" If $t_{j+1}(y) = \infty$ for $j \in \{0, \ldots, i-1\}$, then also $t_j(y) = \infty$ by assumption and (4.5). Repeating this argument, we arrive at $t_1(y) = \infty$, contradicting $y = \xi_0(y) \notin M_\infty$. Therefore the proof of Lemma 4.1.1 is complete. $\square$

For later reference we also state explicitly the obvious

Lemma 4.1.2. *Let (A8) hold. Then for every $k \in I$ and $R > 0$ there exists $T = T(k, R) > 0$ such that $t_1(y) \le T$ for $y \in \bar{B}_R(0) \cap D_k$.*

Now we make the following

General Hypothesis. Throughout the rest of Sect. 4.1 we always will assume that $(A1) - (A8)$ from above are satisfied. Only additional assumptions will be mentioned in the statement of results.

Our next step is to introduce the "good" set G of initial values y_0 for which the cocycle $T(t, y_0)$ can be defined. As already mentioned in the introductory paragraph before Sect. 4.1.1, G will be obtained by taking those y_0 for which $\xi_i(y_0) \in M_k$ already implies $\xi_i(y_0) \in D_k$. Hence we have to distinguish between initial values y_0 which lead to trajectories never reaching $D_\infty \subset M_\infty$, cf. Lemma 4.1.3, and those whose trajectory arrive at D_∞, cf. Lemma 4.1.4. In the latter case we also have to pay attention at which step i we first have $\xi_i(y_0) \in D_\infty$.

Lemma 4.1.3. *Let $G^0_{<\infty} = \mathbb{R}^d$, and for $i \in \mathbb{N}_0$*

$$G^{i+1}_{<\infty} = G^i_{<\infty} \cap \Big\{y \in \mathbb{R}^d : t_{i+1}(y) < \infty, \text{ and}$$

$$\xi_i(y) \in M_k \text{ for } k \in I \Rightarrow \xi_i(y) \in D_k\Big\}.$$

Then $G^{i+1}_{<\infty} \subset G^i_{<\infty}$ and

$$G^i_{<\infty} = \Big\{y \in \mathbb{R}^d : t_i(y) < \infty, \text{ and}$$

$$0 \le j \le i - 1, \ \xi_j(y) \in M_k \text{ for } k \in I$$
$$\Rightarrow \varphi^t(y) \notin E_k \text{ for } t \in [t_j(y), t_{j+1}(y)], \text{ and}$$
$$\xi_{j+1}(y) \in M_l \text{ for } l \in I \cup \{\infty\} \Rightarrow \forall \, t \in [t_j(y), t_{j+1}(y)[\, \exists \, \varepsilon > 0 :$$
$$\xi_1\big(B_\varepsilon(\varphi^t(y)) \cap M_k\big) \subset M_l\Big\} .$$

Moreover,

$$G_{<\infty} = \bigcap_{i=0}^{\infty} G^i_{<\infty}$$

$$= \Big\{y \in \mathbb{R}^d : t_i(y) < \infty \text{ for all } i \in \mathbb{N}, \text{ and}$$

$$i \in \mathbb{N}_0, \ k \in I, \ \xi_i(y) \in M_k \Rightarrow \varphi^t(y) \notin E_k \text{ for } t \in [t_i(y), t_{i+1}(y)],$$
$$\text{and } l \in I, \ \xi_{i+1}(y) \in M_l \Rightarrow \forall \, t \in [t_i(y), t_{i+1}(y)[\, \exists \, \varepsilon > 0 :$$
$$\xi_1\big(B_\varepsilon(\varphi^t(y)) \cap M_k\big) \subset M_l\Big\}$$

is forward invariant, i.e., $\varphi^t(G_{<\infty}) \subset G_{<\infty}$ for all $t \in [0, \infty[$.

Proof: The characterizations of $G^i_{<\infty}$ and $G_{<\infty}$ are immediate. To show the invariance of the latter, let $y \in G_{<\infty}$ and $\bar{y} = \varphi^t(y)$. Then $t \in [t_j(y), t_{j+1}(y)[$ for some $j \in \mathbb{N}_0$ by (A6). Hence by Lemma 4.1.1(e), $t_p(\bar{y}) = t_{j+p}(y) - t < \infty$ for $p \in \mathbb{N}$. If $i \in \mathbb{N}_0$, then choose $k \in I$ with $\xi_i(\bar{y}) \in M_k$ and fix $s \in [t_i(\bar{y}), t_{i+1}(\bar{y})]$. In case that $i = 0$ we have $s + t \in [t, t_{j+1}(y)] \subset [t_j(y), t_{j+1}(y)]$. Since $\varphi^t(y) = \xi_0(\bar{y}) \in M_k$, Lemma 4.1.1(f) and the mutual disjointness of the manifolds implies that also $\xi_j(y) \in M_k$, and hence $\varphi^s(\bar{y}) = \varphi^{s+t}(y) \notin E_k$, because of $y \in G_{<\infty}$. On the other hand, if $i \in \mathbb{N}$ we obtain $s \in [t_{i+j}(y), t_{i+j+1}(y)] - t$ and by means of Lemma 4.1.1(d), $\xi_{i+j}(y) = \xi_i(\bar{y}) \in M_k$. Since $y \in G_{<\infty}$ we find again $\varphi^s(\bar{y}) = \varphi^{s+t}(y) \notin E_k$. The second condition is checked in the same way, and thus we have shown $\bar{y} \in G_{<\infty}$. $\quad\square$

Lemma 4.1.4. *For $i \in \mathbb{N}_0$ define*

$$G^i_\infty = G^i_{<\infty} \cap \{y \in \mathbb{R}^d : \xi_i(y) \in D_\infty\}$$

$$= \Big\{ y \in \mathbb{R}^d : t_i(y) < \infty, \text{ and}$$

$0 \le j \le i - 1$, $\xi_j(y) \in M_k$ for $k \in I \Rightarrow \varphi^t(y) \notin E_k$ for $t \in [t_j(y), t_{j+1}(y)]$, and $\xi_{j+1}(y) \in M_l$ for $l \in I \cup \{\infty\} \Rightarrow \forall\, t \in [t_j(y), t_{j+1}(y)[\;\exists\, \varepsilon > 0 :$

$$\xi_1\big(B_\varepsilon(\varphi^t(y)) \cap M_k\big) \subset M_l, \text{ and } \varphi^t(y) \in D_\infty \text{ for } t \in [t_i(y), \infty[\Big\}.$$

Then $G^i_\infty \cap G^p_\infty = \emptyset$ for $i \ne p$, and $G_\infty = \bigcup_{i \in \mathbb{N}_0} G^i_\infty$ is forward invariant.

Proof: The characterization of G^i_∞ is clear. If $y \in G^i_\infty \cap G^p_\infty$ for some $p > i$, then in particular $\xi_i(y) = \varphi^{t_i(y)}(y) \in M_\infty$, hence $t_{i+1}(y) = \infty$ by (4.5), and thus $t_p(y) \ge t_{i+1}(y) = \infty$, a contradiction. To prove the invariance, let $\bar{y} = \varphi^t(y)$ for fixed $y \in G^i_\infty$ and $t \ge 0$. If $t \ge t_i(y)$, then $\bar{y} \in D_\infty = G^0_\infty \subset G_\infty$, since for $s \ge 0$ we have $t + s \ge t_i(y)$, hence $\varphi^s(\bar{y}) = \varphi^{s+t}(y) \in D_\infty$. If $t \in [t_p(y), t_{p+1}(y)[$ with $p \in \{0, \ldots, i-1\}$, then $\bar{y} \in G^{i-p}_\infty \subset G_\infty$: by Lemma 4.1.1 (e) we have $t_{i-p}(\bar{y}) = t_i(y) - t < \infty$. Fix $0 \le j \le i - p - 1$ with $\xi_j(\bar{y}) = \xi_{j+p}(y) \in M_k$, $k \in I$, and $\xi_{i+1}(\bar{y}) = \xi_{j+p+1}(y) \in M_l$, $l \in I \cup \{\infty\}$. Then for $s \in [t_j(\bar{y}), t_{j+1}(\bar{y})] = [t_{j+p}(y), t_{j+p+1}(y)] - t$ it is a consequence of $y \in G^i_\infty$ and $0 \le j + p \le i - 1$ that $\varphi^s(\bar{y}) = \varphi^{s+t}(y) \notin E_k$. Moreover, for $s \in [t_j(\bar{y}), t_{j+1}(\bar{y})[$ in the same way $s + t \in [t_{j+p}(y), t_{j+p+1}(y)[$. Hence $\xi_{j+p+1}(y) \in M_l$ and the definition of G^i_∞ implies that for some $\varepsilon > 0$,

$$\xi_1\big(B_\varepsilon(\varphi^s(\bar{y})) \cap M_k\big) = \xi_1\big(B_\varepsilon(\varphi^{s+t}(y)) \cap M_k\big) \subset M_l.$$

Finally, for fixed $s \in [t_{i-p}(\bar{y}), \infty[= [t_i(y), \infty[-t$ it follows again from the definition of G^i_∞ that $\varphi^s(\bar{y}) = \varphi^{s+t}(y) \in D_\infty$. Hence we have shown $\bar{y} \in G_\infty$ in every case. $\quad\square$

We also emphasize two facts which were obtained in the above proof.

Corollary 4.1.1. *For $i \in \mathbb{N}$*

$$\xi_{i-1}\left(G^i_{<\infty}\right) \subset G^1_{<\infty}.$$

Proof: This follows from an inspection of the proof to Lemma 4.1.4: For $y \in G^i_{<\infty}$ and $t = t_{i-1}(y)$ in this proof, we have $\bar{y} = \xi_{i-1}(y)$ and $p = i - 1$, hence $\bar{y} \in G^{i-p}_{<\infty} = G^1_{<\infty}$.   □

Corollary 4.1.2. *For $i \in \mathbb{N}_0$, $y \in G^i_\infty$, $p \in \{0,\ldots,i\}$, and $t \in [t_p(y), t_{p+1}(y)[$*

$$\varphi^t(y) \in G^{i-p}_\infty.$$

Proof: Again this follows from the proof of Lemma 4.1.4: If $p = i$, then $t \geq t_i(y)$, and it was shown that in this case $\varphi^t(y) \in G^0_\infty$. On the other hand, if $p \in \{0,\ldots,i-1\}$, then we obtained above $\varphi^t(y) = \bar{y} \in G^{i-p}_\infty$.   □

The next corollary serves to introduce the set G of "good" initial values, for which later in Thm. 4.1.1 the cocycle can be defined.

Corollary 4.1.3. *The set*

$$G = G_{<\infty} \cup G_\infty$$

is forward invariant, with $G_{<\infty}$ from Lemma 4.1.3 and G_∞ from Lemma 4.1.4.

Now we turn to the investigation of the differentiability of $t_i(\cdot)$ and $\xi_i(\cdot)$ on $G^i_{<\infty}$.

Lemma 4.1.5. *For $i \in \mathbb{N}$ and $k \in I$*

$$t_i \in C^1(G^i_{<\infty} \cap M_k, \mathbb{R}) \quad \text{and} \quad \xi_i \in C^1(G^i_{<\infty} \cap M_k, \mathbb{R}^d). \qquad (4.7)$$

In particular, $t_i \in C^1(G^i_\infty \cap M_k, \mathbb{R})$ and $\xi_i \in C^1(G^i_\infty \cap M_k, \mathbb{R}^d)$.

Proof: The claim will be proved by induction. For that, we will also show that

$$i \in \mathbb{N}, \quad y_0 \in G^i_{<\infty} \cap M_k \quad \text{for} \quad k \in I, \quad \xi_i(y_0) \in M_l \quad \text{for} \quad l \in I \cup \{\infty\},$$
$$\text{and } \eta > 0 \quad \Longrightarrow \quad \exists \rho > 0: \ \xi_i\big(B_\rho(y_0) \cap G^i_{<\infty} \cap M_k\big) \subset B_\eta(\xi_i(y_0)) \cap M_l \qquad (4.8)$$

is valid, and we will use the abbreviation $\Sigma_{i,k} = G^i_{<\infty} \cap M_k$. If $i = 1$, then we have $\Sigma_{1,k} = D_k$ for $k \in I$ by definition, and hence $t_1 \in C^1(\Sigma_{1,k}, \mathbb{R})$ by $(A7)$. Thus for $y_0 \in \Sigma_{1,k}$ we find $\delta_1 > 0$ and $\tau_1 \in C^1(B_{\delta_1}(y_0), \mathbb{R})$ such that $\tau_1(y) = t_1(y)$ for $y \in B_{\delta_1}(y_0) \cap \Sigma_{1,k}$. Hence $\phi_1 \in C^1(B_{\delta_1}(y_0), \mathbb{R}^d)$, where $\phi_1(y) = \Phi_{f_k}(\tau_1(y), y)$. Moreover, by definition of $\Sigma_{1,k}$ and by (4.6) we have $\phi_1(y) = \xi_1(y)$ for $y \in B_{\delta_1}(y_0) \cap \Sigma_{1,k}$, and thus $\xi_1 \in C^1(\Sigma_{1,k}, \mathbb{R}^d)$. In particular, if $\eta > 0$ is given, then $\phi_1(B_{\rho_1}(y_0)) \subset B_\eta(\phi_1(y_0))$ for some $\rho_1 \in (0, \delta_1]$. If $l \in I \cup \{\infty\}$ is such that $\xi_1(y_0) \in M_l$, then by definition of $G^1_{<\infty}$ there exists $\rho_2 > 0$ with $\xi_1(B_{\rho_2}(y_0) \cap M_k) \subset M_l$. Hence with $\rho = \rho_1 \wedge \rho_2$ we obtain $\xi_1(B_\rho(y_0) \cap \Sigma_{1,k}) \subset B_\eta(\xi_1(y_0)) \cap M_l$, and therefore (4.7) and (4.8) for $i = 1$.

If (4.7) and (4.8) already hold for i and $y_0 \in \Sigma_{i+1,k} \subset \Sigma_{i,k}$ as well as $\eta > 0$ are fixed, then by Cor. 4.1.1, $\bar{y}_0 = \xi_i(y_0) \in G^1_{<\infty} \cap M_l = \Sigma_{1,l}$ for some $l \in I$ and $\xi_1(\bar{y}_0) = \xi_{i+1}(y_0) \in M_p$ for some $p \in I \cup \{\infty\}$. Hence, by assumption, there are $\rho_1 > 0$, $\bar{\tau}_1 \in C^1(B_{\rho_1}(\bar{y}_0), \mathbb{R})$ and $\bar{\phi}_1 \in C^1(B_{\rho_1}(\bar{y}_0), \mathbb{R}^d)$ such that $\bar{\tau}_1(\bar{y}) = t_1(\bar{y})$ and $\bar{\phi}_1(\bar{y}) = \xi_1(\bar{y})$ for $\bar{y} \in B_{\rho_1}(\bar{y}_0) \cap \Sigma_{1,l}$. Moreover, by (4.8) applied to $i = 1$ and $\bar{y}_0$, we may suppose that ρ_1 is chosen small enough to ensure also $\xi_1(B_{\rho_1}(\bar{y}_0) \cap \Sigma_{1,l}) \subset B_\eta(\xi_1(\bar{y}_0)) \cap M_p$. Next, since $y_0 \in \Sigma_{i,k}$, by assumption we find $\rho_2 > 0$, $\tau_i \in C^1(B_{\rho_2}(y_0), \mathbb{R})$ and $\phi_i \in C^1(B_{\rho_2}(y_0), \mathbb{R}^d)$ satisfying $\tau_i(y) = t_i(y)$ and $\phi_i(y) = \xi_i(y)$ for $y \in B_{\rho_2}(y_0) \cap \Sigma_{i,k}$, and we can assume that $\phi_i(y) \in B_{\rho_1}(\phi_i(y_0)) = B_{\rho_1}(\bar{y}_0)$ for $y \in B_{\rho_2}(y_0)$. By (4.8) for i we find a $\rho_3 > 0$ with $\xi_i(B_{\rho_3}(y_0) \cap \Sigma_{i,k}) \subset B_{\rho_1}(\bar{y}_0) \cap M_l$. Let $\delta_{i+1} = \rho_2 \wedge \rho_3 > 0$, $\tau_{i+1}(y) = \bar{\tau}_1(\phi_i(y)) + \tau_i(y)$, and $\phi_{i+1}(y) = \bar{\phi}_1(\phi_i(y))$ for $y \in B_{\delta_{i+1}}(y_0)$. Then $\tau_{i+1}(y) = t_{i+1}(y)$ and $\phi_{i+1}(y) = \xi_{i+1}(y)$ in $B_{\delta_{i+1}}(y_0) \cap \Sigma_{i+1,k}$. Furthermore, $\tau_{i+1} \in C^1(B_{\delta_{i+1}}(y_0), \mathbb{R})$ and $\phi_{i+1} \in C^1(B_{\delta_{i+1}}(y_0), \mathbb{R}^d)$. Finally, let $\rho = \rho_3 > 0$. By Cor. 4.1.1 we have

$$\xi_i(B_\rho(y_0) \cap \Sigma_{i+1,k}) \subset \xi_i(G^{i+1}_{<\infty}) \subset G^1_{<\infty},$$

and we also obtain

$$\xi_i(B_\rho(y_0) \cap \Sigma_{i+1,k}) \subset \xi_i(B_{\rho_3}(y_0) \cap \Sigma_{i,k}) \subset B_{\rho_1}(\bar{y}_0) \cap M_l,$$

and thus $\xi_i(B_\rho(y_0) \cap \Sigma_{i+1,k}) \subset B_{\rho_1}(\bar{y}_0) \cap \Sigma_{1,l}$ what in turn yields

$$\xi_{i+1}(B_\rho(y_0) \cap \Sigma_{i+1,k}) = \xi_1(\xi_i(\ldots)) \subset \xi_1(B_{\rho_1}(\bar{y}_0) \cap \Sigma_{1,l}) \subset B_\eta(\xi_{i+1}(y_0)) \cap M_p,$$

i.e., (4.8) for $i + 1$. $\square$

Corollary 4.1.4. *For $i \in \mathbb{N}_0$, $k \in I$, $y_0 \in G^{i+1}_{<\infty} \cap M_k$, and $t \in]t_i(y_0), t_{i+1}(y_0)[$ there exists $r > 0$ such that the following holds:*

(a) If $y \in B_r(y_0) \cap G^{i+1}_{<\infty} \cap M_k$, then $t \in]t_i(y), t_{i+1}(y)[$.
(b) $\varphi^t(\cdot) \in C^1(B_r(y_0) \cap G^{i+1}_{<\infty} \cap M_k, \mathbb{R}^d)$.

Proof: Since $y_0 \in \Sigma_{i+1,k} \subset \Sigma_{i,k}$, by Lemma 4.1.5 we find $\delta > 0$, $\tau_i, \tau_{i+1} \in C^1(B_\delta(y_0), \mathbb{R})$ and $\phi_i \in C^1(B_\delta(y_0), \mathbb{R}^d)$ such that $\tau_i(y) = t_i(y)$ and $\phi_i(y) = \xi_i(y)$ for $y \in B_\delta(y_0) \cap \Sigma_{i,k}$ as well as $\tau_{i+1}(y) = t_{i+1}(y)$ for $y \in B_\delta(y_0) \cap \Sigma_{i+1,k}$.

Moreover, if $\bar{y}_0 = \xi_i(y_0) \in M_l$ for some $l \in I$, we may also assume that $\xi_i(B_\delta(y_0) \cap \Sigma_{i,k}) \subset M_l$ by (4.8). Then by continuity of τ_i and τ_{i+1} there exists $r \in]0, \delta]$ with $|\tau_i(y) - t_i(y_0)| \leq (t - t_i(y_0))/2$ and $|\tau_{i+1}(y) - t_{i+1}(y_0)| \leq (t_{i+1}(y_0) - t)/2$ for $y \in B_r(y_0)$. Therefore (a) holds. If we define $\theta(y) = \Phi_{f_l}(t - \tau_i(y), \phi_i(y))$, then we obtain $\theta \in C^1(B_r(y_0), \mathbb{R}^d)$, and by Lemma 4.1.1 (g) & (c),

$$\varphi^t(y) = \Phi_{f(\cdot, \xi_i(y))}(t - t_i(y), \xi_i(y)) = \Phi_{f_l}(t - \tau_i(y), \phi_i(y)) = \theta(y)$$

for $y \in B_r(y_0) \cap \Sigma_{i+1,k} \subset B_r(y_0) \cap \Sigma_{i,k}$. □

Next we state a similar corollary for the case of $y_0 \in G_\infty^i$, i.e., for the transition from $\mathbb{R}^3 \setminus M_\infty$ to $D_\infty \subset M_\infty$ at time $t_i(y_0)$. Note in particular that only $y_0 \in G_\infty^i$ is required here, compared to the assumption $y_0 \in G_{<\infty}^i$ from Cor. 4.1.4, since we know that $t_{i+1}(y) = \infty$ for $y \in G_\infty^i$, and we thus only have to ensure that $t_i(\cdot)$ varies continuously.

Corollary 4.1.5. *For $i \in \mathbb{N}_0$, $k \in I \cup \{\infty\}$, $y_0 \in G_\infty^i \cap M_k$, and $t \in]t_p(y_0), t_{p+1}(y_0)[$ for a suitable $p \in \{0, \ldots, i\}$, there exists $r > 0$ such that the following holds:*

(a) If $y \in B_r(y_0) \cap G_\infty^i \cap M_k$, then $t \in]t_p(y), t_{p+1}(y)[$.
(b) $\varphi^t(\cdot) \in C^1(B_r(y_0) \cap G_\infty^i \cap M_k, \mathbb{R}^d)$.

Proof: First we consider the case $p \in \{0, \ldots, i-1\}$. Then $y_0 \in G_\infty^i \subset G_{<\infty}^i \subset G_{<\infty}^{p+1}$ implies that by Cor. 4.1.4(a), (b) we find $r > 0$ such that $y \in B_r(y_0) \cap G_{<\infty}^{p+1} \cap M_k$ yields $t \in]t_p(y), t_{p+1}(y)[$ and such that $\varphi^t(\cdot) \in C^1(B_r(y_0) \cap G_{<\infty}^{p+1} \cap M_k, \mathbb{R}^d)$. Since $G_\infty^i \subset G_{<\infty}^{p+1}$, this gives (a) and (b). On the other hand, if $p = i$, then $t_{p+1}(y) = \infty$ for $y \in G_\infty^i$. Thus we may argue analogously as in the proof of Cor. 4.1.4, by Lemma 4.1.5. □

4.1.2 Transition of the linearization at switching times

In this section we derive the form of the transition operators $A_{i+1}(y_0) : \mathcal{L}(\mathbb{R}^d) \to \mathcal{L}(\mathbb{R}^d)$ of the linearized flow at switching times $t_{i+1}(y_0)$ for suitable initial values y_0. An analogous formula, see in particular (4.18) below, was already obtained in the early papers AIZERMAN/GANTMAKHER [3, 4, 5], and later in P.C. MÜLLER [154], by formal investigation of nearby trajectories; it was then used to calculate Lyapunov exponents for non-smooth systems by BOCKMAN [27]. We remark that if the underlying system is smooth, then all f_k's are the same, and formula (4.11) yields $A_{i+1}(y_0) = \text{id}_{\mathcal{L}(\mathbb{R}^d)}$, i.e., the transition is smooth as well.

We start with transitions from $\mathbb{R}^3 \setminus M_\infty$ to $\mathbb{R}^3 \setminus M_\infty$ at some time $t_{i+1}(y_0)$.

Lemma 4.1.6. *For* $i \in \mathbb{N}_0$ *and* $y_0 \in G^{i+2}_{<\infty}$ *choose* $l, p \in I$ *with* $\xi_i(y_0) \in M_l$ *and* $\xi_{i+1}(y_0) \in M_p$. *Let*

$$\Gamma^i_l(t, y) = \Phi_{f_l}(t - t_i(y), \xi_i(y)) \quad \text{for} \quad (t, y) \in \mathbb{R} \times \mathbb{R}^d. \tag{4.9}$$

Then for $t \in [t_{i+1}(y_0), t_{i+2}(y_0)[$

$$\begin{aligned}
\partial_y \varphi^t(y_0) = \Big[&\partial_y \Phi_{f_p}(t - t_{i+1}(y_0), \xi_{i+1}(y_0)) \, f_l(\xi_{i+1}(y_0)) \\
&- f_p\big(\Phi_{f_p}(t - t_{i+1}(y_0), \xi_{i+1}(y_0))\big) \Big] \langle \nabla t_{i+1}(y_0), \cdot \rangle \\
&+ \partial_y \Phi_{f_p}(t - t_{i+1}(y_0), \xi_{i+1}(y_0)) \, \partial_y \Gamma^i_l(t_{i+1}(y_0), y_0) \,. \tag{4.10}
\end{aligned}$$

In addition, define $A_{i+1}(y_0) : \mathcal{L}(\mathbb{R}^d) \to \mathcal{L}(\mathbb{R}^d)$ *through*

$$A_{i+1}(y_0)(B) = \big[f_l(\xi_{i+1}(y_0)) - f_p(\xi_{i+1}(y_0)) \big] \langle \nabla t_{i+1}(y_0), \cdot \rangle + B, \ B \in \mathcal{L}(\mathbb{R}^d). \tag{4.11}$$

Then

$$\lim_{t \to t_{i+1}(y_0)^+} \partial_y \varphi^t(y_0) = A_{i+1}(y_0) \left(\lim_{t \to t_{i+1}(y_0)^-} \partial_y \varphi^t(y_0) \right).$$

Proof: The following formal calculations are justified by Lemma 4.1.5, (4.8), and Cor. 4.1.4(a), (b). By Lemma 4.1.1(g) we obtain for $t \in]t_{i+1}(y_0), t_{i+2}(y_0)[$

$$\begin{aligned}
\partial_y \varphi^t(y_0) &= \partial_t \Phi_{f_p}(t - t_{i+1}(y_0), \xi_{i+1}(y_0)) \big(-\nabla t_{i+1}(y_0) \big) \\
&\quad + \partial_y \Phi_{f_p}(t - t_{i+1}(y_0), \xi_{i+1}(y_0)) \, \xi'_{i+1}(y_0) \\
&= -f_p\big(\Phi_{f_p}(t - t_{i+1}(y_0), \xi_{i+1}(y_0))\big) \nabla t_{i+1}(y_0) \\
&\quad + \partial_y \Phi_{f_p}(t - t_{i+1}(y_0), \xi_{i+1}(y_0)) \, \xi'_{i+1}(y_0). \tag{4.12}
\end{aligned}$$

Thus, since $\Phi_{f_p}(0, y) = y$ and $\partial_y \Phi_{f_p}(0, y) = \mathrm{id}_{\mathbb{R}^d}$ for all $y \in \mathbb{R}^d$,

$$\lim_{t \to t_{i+1}(y_0)^+} \partial_y \varphi^t(y_0) = -f_p(\xi_{i+1}(y_0)) \langle \nabla t_{i+1}(y_0), \cdot \rangle + \xi'_{i+1}(y_0). \tag{4.13}$$

We have $\partial_t \Gamma^i_l(t, y) = f_l(\Gamma^i_l(t, y))$. Also, for $t \in]t_i(y_0), t_{i+1}(y_0)[$ and y in a suitable $G^1_{<\infty} \cap M_k$ – neighborhood of y_0, it follows from (4.8) and Lemma 4.1.1(g) that $\varphi^t(y) = \Gamma^i_l(t, y)$ as well as

$$\begin{aligned}
\xi_{i+1}(y) = \varphi^{t_{i+1}(y)}(y) &= \Phi_{f(\cdot, \varphi^{t_i(y)}(y))}\big(t_{i+1}(y) - t_i(y), \varphi^{t_i(y)}(y)\big) \\
&= \Phi_{f_l}(t_{i+1}(y) - t_i(y), \xi_i(y)) = \Gamma^i_l(t_{i+1}(y), y).
\end{aligned}$$

Therefore

$$\begin{aligned}
\xi'_{i+1}(y_0) &= \partial_t \Gamma^i_l(t_{i+1}(y_0), y_0) \nabla t_{i+1}(y_0) + \partial_y \Gamma^i_l(t_{i+1}(y_0), y_0) \\
&= f_l(\xi_{i+1}(y_0)) \langle \nabla t_{i+1}(y_0), \cdot \rangle + \partial_y \Gamma^i_l(t_{i+1}(y_0), y_0). \tag{4.14}
\end{aligned}$$

Inserting this into (4.12) we obtain (4.10). In addition, taking into account that

$$\partial_y \varphi^t(y_0) = \partial_y \Gamma_l^i(t, y_0) \to \partial_y \Gamma_l^i(t_{i+1}(y_0), y_0) \quad \text{as} \quad t \to t_{i+1}(y_0)^-, \quad (4.15)$$

the claimed form of the transition operator $A_{i+1}(y_0)$ follows from (4.13) and (4.14).

Next we consider transitions from $\mathbb{R}^3 \setminus M_\infty$ to $D_\infty \subset M_\infty$ at time $t_{i+1}(y_0)$

Lemma 4.1.7. *For $i \in \mathbb{N}_0$ and $y_0 \in G_\infty^{i+1}$ choose $l \in I$ such that $\xi_i(y_0) \in M_l$. Then all assertions of Lemma 4.1.6 hold with $p = \infty$.*

Proof: Because of Cor. 4.1.5, the proof of Lemma 4.1.6 carries over. Note that here $\xi_{i+1}(y_0) \in M_\infty$, i.e., $p = \infty$ in the above notation, and $t_{i+2}(y_0) = \infty$. □

Sometimes it is possible to find a form of $\partial_y \varphi^t(y_0)$ and of the transition operators in Lemma 4.1.6 resp. Lemma 4.1.7 which is easier to handle. It will turn out that for this the following additional assumption is sufficient.

(A9) For every $k \in I$ there exist $h_k \in C^1(\mathbb{R}^d, \mathbb{R})$ such that

$$y \in M_k, \quad \xi_1(y) \notin E_k \quad \Longrightarrow \quad \begin{cases} h_k(\xi_1(y)) = 0, \\ \langle \nabla h_k(\xi_1(y)), f_k(\xi_1(y)) \rangle \neq 0 \end{cases} \quad (4.16)$$

So condition (4.16) means that for a trajectory starting in $y \in M_k$ the function h_k being zero indicates the next switching to a different manifold. In addition, the trajectory should not arrive tangentially at $\{h_k = 0\}$.

This is the right place to comment on the assumptions imposed in KUNZE/MICHAELI [120] to obtain a "formal system" like in Sect. 4.1.1 to describe non-smooth dynamical systems appropriately. In condition $(A3)$ of this paper it was required that for $y \in M_k$, $h_k(\Phi_{f_k}(t, y)) = 0$ if and only if $t = t_1(y)$. This cannot be verified in concrete systems, as may be seen from the example of the pendulum with dry friction (1.4). In this case, the function $h_k(x, v, \tau) = v$ being zero means that a trajectory starting e.g. in $y_0 \in M_k = \{y \in \mathbb{R}^3 : v < 0\}$ reaches the surface $\{y \in \mathbb{R}^3 : v = 0\}$. Of course this will happen at time $t = t_1(y_0)$, but since here the flow $\Phi_{f_k}(\cdot, y_0)$ induced by the right-hand side f_k on M_k is oscillatory, we will find (infinitely many) $t_* > t_1(y_0)$ such that also $\Phi_{f_k}(t_*, y_0) \in \{y \in \mathbb{R}^3 : v = 0\}$. Thus the "only if" in the former $(A3)$ does not hold, and it was exactly this condition which was needed to show the differentiability of $t_1(\cdot)$ in this paper. To summarize, although the results in KUNZE/MICHAELI [120] are correct, the developed "formal system" was not appropriate for applications.

After this short interlude, we turn back to calculate a simpler form of the transition operators $A_{i+1}(y_0)$ from Lemma 4.1.6 under assumption $(A9)$. In the following we identify $\mathcal{L}(\mathbb{R}^d)$ with $\mathbb{R}^{d \times d}$. We will find that now $A_{i+1}(y_0) : \mathbb{R}^{d \times d} \to \mathbb{R}^{d \times d}$ is linear, i.e., given through a matrix $A_{i+1}(y_0) \in \mathbb{R}^{d \times d}$.

Corollary 4.1.6. *Let, in addition to the general assumptions* $(A1) - (A8)$, *also* $(A9)$ *be satisfied. Fix* $i \in \mathbb{N}_0$ *and* $y_0 \in G^{i+2}_{<\infty} \cup G^{i+1}_{\infty}$, *and choose* $l \in I$ *and* $p \in I \cup \{\infty\}$ *such that* $\xi_i(y_0) \in M_l$ *and* $\xi_{i+1}(y_0) \in M_p$. *Then for* $t \in [t_{i+1}(y_0), t_{i+2}(y_0)[$

$$
\partial_y \varphi^t(y_0)
$$

$$
= \left(\frac{1}{\langle \nabla h_l, \, f_l \rangle (\xi_{i+1}(y_0))} \left(\left[f_p(\Phi_{f_p}(t - t_{i+1}(y_0), \xi_{i+1}(y_0))) \right. \right. \right.
$$

$$
\left. \left. - \partial_y \Phi_{f_p}(t - t_{i+1}(y_0), \xi_{i+1}(y_0)) \, f_l(\xi_{i+1}(y_0)) \right]_\kappa \cdot \left[\nabla h_l(\xi_{i+1}(y_0)) \right]_\nu \right)_{1 \le \kappa, \nu \le d}
$$

$$
+ \partial_y \Phi_{f_p}(t - t_{i+1}(y_0), \xi_{i+1}(y_0)) \bigg) \circ \left(\lim_{s \to t_{i+1}(y_0)^-} \partial_y \varphi^s(y_0) \right). \qquad (4.17)
$$

Here $\partial_y \varphi^t(y_0)$ *at* $t = t_{i+1}(y_0)$ *is understood to be the limit from the right, and* $[\ldots]_\kappa$ *and* $[\ldots]_\nu$ *are the* κ^{th} *and* ν^{th} *component of the respective d-vectors.*

In particular, we also find that $A_{i+1}(y_0) \in \mathbb{R}^{d \times d}$ *is the matrix*

$$
A_{i+1}(y_0) = \frac{1}{\langle \nabla h_l, \, f_l \rangle (\xi_{i+1}(y_0))} \left(\left[f_p(\xi_{i+1}(y_0)) - f_l(\xi_{i+1}(y_0)) \right]_\kappa \right.
$$

$$
\left. \left[\nabla h_l(\xi_{i+1}(y_0)) \right]_\nu \right)_{1 \le \kappa, \nu \le d} + \mathrm{id}_{\mathbb{R}^d}.
$$

$$
\qquad (4.18)
$$

Proof: Since $G^{i+2}_{<\infty} \cup G^{i+1}_{\infty} \subset G^{i+1}_{<\infty}$, in fact we find $l \in I$ with $\bar{y}_0 = \xi_i(y_0) \in M_l$. In addition, $\xi_1(\bar{y}_0) = \xi_{i+1}(y_0) \notin E_l$ by definition of $G^{i+1}_{<\infty}$, cf. Lemma 4.1.3. In particular the denominators in (4.17) and (4.18) are not zero, by the second condition in (4.16) of $(A9)$. We continue now to argue formally, which is again justified by Lemma 4.1.5, (4.8), Cor. 4.1.4, and Cor. 4.1.5, in both of the respective cases $y_0 \in G^{i+2}_{<\infty}$ and $y_0 \in G^{i+1}_{\infty}$. For y in a suitable $G^{i+2}_{<\infty}$-neighborhood resp. G^{i+1}_{∞}-neighborhood U of y_0 we also have $\xi_i(y) \in M_l$, $t_{i+1}(y) < \infty$, and $\xi_{i+1}(y) \notin E_l$. Then $\Phi_{f_l}(t_{i+1}(y) - t_i(y), \xi_i(y)) = \varphi(t_{i+1}(y), y) = \xi_{i+1}(y) = \xi_1(\xi_i(y))$ by Lemma 4.1.1(g) and (c). Let $g(t, y) = h_l(\Phi_{f_l}(t - t_i(y), \xi_i(y))) = h_l(\Gamma^i_l(t, y))$ for $t \in \mathbb{R}$ and $y \in \mathbb{R}^d$, cf. (4.9). We conclude that

$$
\partial_t g(t, y) = \langle \nabla h_l, \, f_l \rangle (\Phi_{f_l}(t - t_i(y), \xi_i(y))), \quad \text{and}
$$

$$
\partial_y g(t, y) = \nabla h_l(\Gamma^i_l(t, y)) \, \partial_y \Gamma^i_l(t, y).
$$

From the first condition of (4.16) in assumption $(A9)$ applied to $\xi_i(y)$ we obtain

$$
g(t_{i+1}(y), y) = h_l(\xi_1(\xi_i(y)) = 0 \quad \text{for} \quad y \in U,
$$

hence $\nabla t_{i+1}(y) = - \left(\partial_t g(t_{i+1}(y), y) \right)^{-1} \partial_y g(t_{i+1}(y), y)$. Consequently,

$$\nabla t_{i+1}(y_0) = -\frac{\nabla h_l\big(\Gamma_l^i(t_{i+1}(y_0), y_0)\big)\,\partial_y\,\Gamma_l^i(t_{i+1}(y_0), y_0)}{\langle h_l,\, f_l\rangle(\xi_{i+1}(y_0))}$$
$$= -\frac{\nabla h_l(\xi_{i+1}(y_0))\,\partial_y\,\Gamma_l^i(t_{i+1}(y_0), y_0)}{\langle h_l,\, f_l\rangle(\xi_{i+1}(y_0))}.$$

This can be inserted into (4.10) resp. (4.11) to yield (4.17) resp. (4.18), since $\lim_{s\to t_{i+1}(y_0)-}\partial_y\,\varphi^s(y_0) = \partial_y\,\Gamma_l^i(t_{i+1}(y_0), y_0)$ by (4.15). $\qquad\square$

4.1.3 Construction of a cocycle

By means of Lemma 4.1.6 and Lemma 4.1.7 we can now construct a canonical cocycle. Instead of defining $T : [0, \infty[\times\mathbb{R}^d \to \mathcal{L}(\mathbb{R}^d)$ we could also restrict T to $[0, \infty[\times G$, since G is forward invariant.

Theorem 4.1.1. *Let the forward invariant G be as in Cor. 4.1.3. Define $T : [0, \infty[\times\mathbb{R}^d \to \mathcal{L}(\mathbb{R}^d)$ as follows*

$$T(t,y) = \begin{cases} \partial_y\,\varphi^t(y) & : y \in G_{<\infty}\,,\ \forall\,i \in \mathbb{N}\,:\ t \neq t_i(y) \\[2mm] A_{i+1}(y)\Big(\partial_y\Gamma_l^i(t_{i+1}(y), y)\Big) & : y \in G_{<\infty}\,, \\ & \quad \exists i \in \mathbb{N}_0\,:\ t = t_{i+1}(y) \\ & \quad [\text{with } l \in I \text{ such that } \xi_i(y) \in M_l] \\[2mm] \partial_y\,\varphi^t(y) & : \exists i_0 \in \mathbb{N}_0\,:\ y \in G_\infty^{i_0}\,, \\ & \quad t \neq t_1(y), \ldots, t_{i_0}(y) \\[2mm] A_{i+1}(y)\Big(\partial_y\Gamma_l^i(t_{i+1}(y), y)\Big) & : \exists i_0 \in \mathbb{N}_0\,:\ y \in G_\infty^{i_0}, \\ & \quad \exists i \in \{0, \ldots, i_0 - 1\}\,:\ t = t_{i+1}(y) \\ & \quad [\text{with } l \in I \text{ such that } \xi_i(y) \in M_l] \\[2mm] 0 & : y \notin G \end{cases}$$

with Γ_l^i from (4.9). Then T is a cocycle for $(\varphi^t)_{t\geq 0}$.

Proof: We first note that T is well-defined, because $G_\infty^i \cap G_\infty^j = \emptyset$ for $i \neq j$, by Lemma 4.1.4, and also $G_\infty^{i_0} \subset G_{<\infty}^{i_0}$. The latter implies that $A_1(y), \ldots, A_{i_0-1}(y)$ are defined for $y \in G_\infty^{i_0}$ through Lemma 4.1.6, and $A_{i_0}(y)$ is defined through Lemma 4.1.7.

We have to show that

$$T(t + s, y) = T(t, \varphi^s(y))\,T(s, y) \quad \text{for} \quad t, s \in [0, \infty[\quad \text{and} \quad y \in \mathbb{R}^d, \quad (4.19)$$

and again we argue formally. First we remark that (4.19) is true for $y \notin G$. Note also that (4.19) holds if $t = 0$ or if $s = 0$, since G is forward invariant and $T(0, y) = \mathrm{id}_{\mathbb{R}^d}$ for all $y \in G$, by the 1st and 3rd line of the definition.

So fix $t, s > 0$ and $y \in G$. According to the definition of G we distinguish cases. Case 1: $y \in G_{<\infty}$. Choose $i, j \in \mathbb{N}$ such that $s \in [t_i(y), t_{i+1}(y)[$ and $t+s \in [t_j(y), t_{j+1}(y)[$. Assume first that in addition $s \neq t_i(y)$ and $t+s \neq t_j(y)$, i.e., $s \neq t_k(y)$ and $t + s \neq t_k(y)$ for all $k \in \mathbb{N}$. Then also $t \neq t_k(\varphi^s(y))$ for all $k \in \mathbb{N}$ by Lemma 4.1.1(e), and hence the invariance of $G_{<\infty}$ and Cor. 4.1.4 imply that we can differentiate $\varphi^{t+s}(y) = \varphi^t(\varphi^s(y))$ to obtain (4.19) by definition of T for such s and $t + s$. Next, if $s = t_i(y)$ or $t + s = t_j(y)$, then in all cases $s + \tau \in]t_i(y), t_{i+1}(y)[$ and $t + (s + \tau) \in]t_j(y), t_{j+1}(y)[$ for $\tau > 0$ sufficiently small. Consequently, $T(t + s + \tau, y) = T(t, \varphi^{s+\tau}(y)) \, T(s + \tau, y)$ by the part already shown. Since $T(\cdot, y)$ is right-continuous by definition of T and by Lemma 4.1.6, cf. (4.15), it is hence sufficient to prove that for fixed $t > 0$ and $y_0 \in G_{<\infty}$ the function $T(t, \cdot)$ is continuous at y_0 on the set $D_{y_0} = \{z \in G_{<\infty} : t_k(z) \leq t_k(y_0) \text{ for all } k \in \mathbb{N}\}$. Indeed, if this is shown, then we can take, due to the invariance of $G_{<\infty}$, $y_0 = \varphi^s(y) \in G_{<\infty}$ and $z_\tau = \varphi^{s+\tau}(y) \in G_{<\infty}$, because

$$t_k(z_\tau) = t_k(\varphi^{s+\tau}(y)) = t_{i+k}(y) - (s + \tau) \leq t_{i+k}(y) - s$$
$$= t_k(\varphi^s(y)) = t_k(y_0),$$

for $k \in \mathbb{N}$ by Lemma 4.1.1(e). The continuity of $\varphi(\cdot, y)$, cf. $(A4)$, yields $z_\tau \to \varphi^s(y)$ as $\tau \to 0^+$, and therefore $T(t, \varphi^{s+\tau}(y)) \to T(t, \varphi^s(y))$ as $\tau \to 0^+$. This gives (4.19) in the limit.

To prove the continuity of $T(t, \cdot)$ at $y_0 \in G_{<\infty}$ on D_{y_0}, we may assume that $t = t_{i+1}(y_0)$ for some $i \in \mathbb{N}_0$, since otherwise Cor. 4.1.4 implies that $T(t, \cdot) = \partial_y \varphi^t(\cdot)$ is continuous even in a whole $G_{<\infty}$-neighborhood (not restricted to D_{y_0}) of y_0. So let $t = t_{i+1}(y_0)$ and choose $l, p \in I$ such that $\xi_i(y_0) \in M_l$ and $\xi_{i+1}(y_0) \in M_p$. Hence by definition and by (4.13)

$$T(t, y_0) = A_{i+1}(y_0)\left(\partial_y \Gamma_l^i(t_{i+1}(y_0), y_0)\right)$$
$$= - f_p(\xi_{i+1}(y_0)) \langle \nabla t_{i+1}(y_0), \cdot \rangle + \xi'_{i+1}(y_0). \qquad (4.20)$$

Because of $t_i(y_0) < t = t_{i+1}(y_0) < t_{i+2}(y_0)$, we have $t \in [t_{i+1}(z), t_{i+2}(z)[$ for $z \in D_{y_0}$ sufficiently near to y_0 by definition of D_{y_0} and by Lemma 4.1.5. For $z \in G_{<\infty}$ near to y_0 we also have $\xi_i(z) \in M_l$ and $\xi_{i+1}(z) \in M_p$ by (4.8). Since thus either

$$T(t, z) = A_{i+1}(z)\left(\partial_y \Gamma_l^i(t_{i+1}(z), z)\right)$$

if $t = t_{i+1}(z)$ or, cf. (4.12),

$$T(t, z) = \partial_y \varphi^t(z) = - f_p\big(\Phi_{f_p}(t_{i+1}(y_0) - t_{i+1}(z), \xi_{i+1}(z))\big) \langle \nabla t_{i+1}(z), \cdot \rangle$$
$$+ \partial_y \Phi_{f_p}(t_{i+1}(y_0) - t_{i+1}(z), \xi_{i+1}(z)) \xi'_{i+1}(z)$$

if $t \in]t_{i+1}(z), t_{i+2}(z)[$, the claim hence follows from (4.20), the regularity of the functions involved, from $\Phi_{f_p}(0, \xi_{i+1}(y_0)) = \xi_{i+1}(y_0)$ and according to $\partial_y \Phi_{f_p}(0, \xi_{i+1}(y_0)) = \mathrm{id}_{\mathbb{R}^d}$.

<u>Case 2:</u> $y \in G_\infty$. Then $y \in G_\infty^{i_0}$ for a unique $i_0 \in \mathbb{N}_0$ by definition of G_∞, and since $t_{i_0+1}(y) = \infty$ we can choose $i, j \in \{0, \ldots, i_0\}$ with $s \in [t_i(y), t_{i+1}(y)[$ and $t + s \in [t_j(y), t_{j+1}(y)[$. In case that $s \neq t_i(y)$ and $t + s \neq t_j(y)$ we find again by means of Lemma 4.1.1(e) that also $t \neq t_k(y)$ for $k \in \{0, \ldots, i_0\}$. Hence Cor. 4.1.5 implies that once more $\varphi^{t+s}(y) = \varphi^t(\varphi^s(y))$ may be differentiated to yield (4.19). Now we can continue to argue as in Case 1, and we thus only have to show that

$$T(t, \varphi^{s+\tau}(y)) \to T(t, \varphi^s(y)) \quad \text{as} \quad \tau \to 0^+, \tag{4.21}$$

since again $T(\cdot, y)$ is right-continuous, by Lemma 4.1.6 and Lemma 4.1.7. To prove (4.21) we may assume in addition that $i \neq i_0$. Indeed, if $i = i_0$, then $\varphi^s(y) \in G_\infty^0 = D_\infty$ by Cor. 4.1.2, and therefore also $\varphi^{s+\tau}(y) \in D_\infty$ for $\tau \geq 0$ by definition of D_∞ and (A3). Since $T(t, z) = \partial_y \varphi(t, z) = \partial_y \Phi_{f_\infty}(t, z)$ for $t \in [0, \infty[$ and $z \in G_\infty^0 = D_\infty$, (4.21) follows from the continuity of $\varphi(\cdot, y)$ in case that $i = i_0$. So we suppose $i \leq i_0 - 1$ and let $y_0 = \varphi^s(y)$, as well as $D_{y_0} = \{z \in G_\infty^{i_0-i} : t_k(z) \leq t_k(y_0) \text{ for } k \in \{0, \ldots, i_0 - i\}\}$. Note that $y_0 \in D_{y_0}$ and $\varphi^{s+\tau}(y) \in G_\infty^{i_0-i}$ for $\tau > 0$ small, by Cor. 4.1.2. Also $t_k(\varphi^{s+\tau}(y)) = t_{i+k}(y) - (s + \tau) \leq t_{i+k}(y) - s = t_k(\varphi^s(y)) = t_k(y_0)$ for $\tau > 0$ small and $k \in \{0, \ldots, i_0 - i\}$ by Lemma 4.1.1(e), i.e., $\varphi^{s+\tau}(y) \in D_{y_0}$. Hence, to show (4.21), it is again enough to prove the continuity of $T(t, \cdot)$ at y_0 on D_{y_0}. Because $t_{i_0-i+1}(y_0) = t_{i_0+1}(y) = \infty$, and as a consequence of Cor. 4.1.5, we may assume that $t = t_{k_0+1}(y_0)$ for some $k_0 \in \{0, \ldots, i_0 - i - 1\}$. Since we can repeat the arguments from Case 1 above, by means of Lemma 4.1.6 and Lemma 4.1.7, we only have to show that

$$z \in D_{y_0}, \quad z \text{ sufficiently close to } y_0 \implies t \in [t_{k_0+1}(z), t_{k_0+2}(z)[. \tag{4.22}$$

For this, we first remark that $k_0 + 1 \leq i_0 - i$, and thus $t < t_{k_0+1}(z)$ is impossible, because then the definition of D_{y_0} would imply $t < t_{k_0+1}(z) \leq t_{k_0+1}(y_0) = t$. To prove $t < t_{k_0+2}(z)$ for $z \in D_{y_0}$ sufficiently close to y_0, we start with the case $k_0 \leq i_0 - i - 2$. Then (4.22) follows from Lemma 4.1.5: Since $t_{k_0+2}(\cdot)$ in particular is continuous on a $G_{<\infty}^{k_0+2}$-neighborhood of y_0, and since $k_0 + 2 \leq i_0 - i$, we have $G_{<\infty}^{k_0+2} \supset G_{<\infty}^{i_0-i} \supset G_\infty^{i_0-i} \supset D_{y_0}$. Therefore $t_{k_0+2}(\cdot)$ is also continuous on a D_{y_0}-neighborhood of y_0. Thus we have to consider finally the case that $k_0 = i_0 - i - 1$. Then $D_{y_0} \subset G_\infty^{i_0-i}$ implies that for $z \in D_{y_0}$ we must have $\xi_{i_0-i}(z) \in D_\infty \subset M_\infty$, and therefore $t_{k_0+2}(z) = t_{i_0-i+1}(z) = t_1(\xi_{i_0-i}(z)) + t_{i_0-i}(z) = \infty$, cf. (4.5). This completes the proof of (4.22), and hence also the proof of the theorem. □

Remark 4.1.1. By (4.15) and Thm. 4.1.1 in fact we have

$$
T(t,y) = \begin{cases}
\partial_y \varphi^t(y) & : \ y \in G_{<\infty}, \\
& \quad \forall\, i \in \mathbb{N} : \ t \neq t_i(y) \\[2mm]
A_{i+1}(y)\Big(\lim_{s \to t_{i+1}(y)^-} T(s,y)\Big) & : \ y \in G_{<\infty}, \\
& \quad \exists\, i \in \mathbb{N}_0 : \ t = t_{i+1}(y) \\
& \quad [\text{with } l \in I \ \text{such that } \xi_i(y) \in M_l] \\[2mm]
\partial_y \varphi^t(y) & : \ \exists\, i_0 \in \mathbb{N}_0 : \ y \in G_\infty^{i_0}, \\
& \quad t \neq t_1(y), \dots, t_{i_0}(y) \\[2mm]
A_{i+1}(y)\Big(\lim_{s \to t_{i+1}(y)^-} T(s,y)\Big) & : \ \exists\, i_0 \in \mathbb{N}_0 : \ y \in G_\infty^{i_0}, \\
& \quad \exists\, i \in \{0, \dots, i_0 - 1\} : \ t = t_{i+1}(y) \\
& \quad [\text{with } l \in I \ \text{such that } \xi_i(y) \in M_l] \\[2mm]
0 & : \ y \notin G
\end{cases}
$$

and both the $\partial_y \varphi^t(y)$-terms and the matrices $A_{i+1}(y)$ can be calculated by means of Cor. 4.1.6 in case that additionally $(A9)$ holds. So we have obtained a complete description of the cocycle. $\diamondsuit$

Next we show that the cocycle T satisfies a certain integral equation. This integral representation may be used in certain cases to verify the integrability conditions (4.108) from the MET Thm. 4.3.1; see Cor. 4.1.8 below.

Corollary 4.1.7. *For $y \in G$ and $t \in [t_i(y), t_{i+1}(y)[$,*

$$
T(t,y) = Q_i(y) + \int_0^t P_i(s,y)\, F(s,y)\, T(s,y)\, ds. \tag{4.23}
$$

Here $i \in \mathbb{N}_0$, if $y \in G_{<\infty}$, and $i \in \{0, \dots, i_0\}$ for $y \in G_\infty^{i_0} \subset G_\infty$. Moreover, we define $Q_i(y) = (A_i(y) \cdot \ldots \cdot A_1(y))\,(\mathrm{id}_{\mathbb{R}^d}) \in \mathcal{L}(\mathbb{R}^d)$, $P_i(s,y) = A_i(y) \cdot \ldots \cdot A_{j+2}(y) \cdot A_{j+1}(y) : \mathcal{L}(\mathbb{R}^d) \to \mathcal{L}(\mathbb{R}^d)$ for $s \in [t_j(y), t_{j+1}(y)[$, and $P_i(s,y) = \mathrm{id}_{\mathcal{L}(\mathbb{R}^d)}$ for $j \geq i$. In addition, we let $F(s,y) = f'(\varphi^s(y), \xi_j(y))$ if $s \in [t_j(y), t_{j+1}(y)[$.

Proof: We first consider the case $i = 0$, i.e., $t \in [0, t_1(y)[$ for some fixed $y \in G$. Then $Q_0(y) = \mathrm{id}_{\mathbb{R}^d}$ and $F(s,y) = f'(\varphi^s(y), y)$ for $s \in [0, t_1(y)[$. If we let $f_k = f(\cdot, y)$, $k \in I \cup \{\infty\}$, we obtain by $(A5)$ and (4.9), $\varphi^s(y) = \Phi_{f_k}(s,y) = \Gamma_k^0(s,y)$ for $s \in [0, t_1(y)[$. Thus,

$$
\partial_t T(s,y) = \partial_t \big(\partial_y \Gamma_k^0(s,y)\big) = \partial_y \big(\partial_t \Gamma_k^0(s,y)\big) = \partial_y \big(f_k(\Gamma_k^0(s,y))\big)
$$
$$
= f_k'(\Gamma_k^0(s,y))\, \partial_y \Gamma_k^0(s,y) = F(s,y)\, T(s,y) \in \mathcal{L}(\mathbb{R}^d). \tag{4.24}
$$

Since $T(0,y) = \mathrm{id}_{\mathbb{R}^d}$ this may be integrated to give (4.23) for $i = 0$.

If (4.23) already holds for some $i \in \mathbb{N}$, then in particular by definition of $T(t_i(y), y)$ and by assumption

$$T(t_i(y), y) = A_i(y) \left(\lim_{t \to t_i(y)^-} T(t, y) \right)$$

$$= A_i(y) Q_{i-1}(y) + \int_0^{t_i(y)} A_i(y) P_{i-1}(s, y) F(s, y) T(s, y) \, ds.$$

(4.25)

Moreover, if $f_l = f(\cdot, \xi_i(y))$, $l \in I \cup \{\infty\}$, then $\varphi^s(y) = \Gamma_l^i(s, y)$ for $s \in [t_i(y), t_{i+1}(y)[$, and hence the differential equation for T from (4.24) is also valid for $s \in [t_i(y), t_{i+1}(y)[$. Integrating this from $t_i(y)$ to t and using (4.25) we thus obtain for $t \in [t_i(y), t_{i+1}(y)[$

$$T(t, y) = A_i(y) Q_{i-1}(y) + \int_0^{t_i(y)} A_i(y) P_{i-1}(s, y) F(s, y) T(s, y) \, ds$$

$$+ \int_{t_i(y)}^t F(s, y) T(s, y) \, ds$$

$$= Q_i(y) + \int_0^t P_i(s, y) F(s, y) T(s, y) \, ds,$$

by definition of Q_i, P_i and F.   □

We remark that (4.23) is similar to formulas for solutions of certain impulsive equations, cf. e.g. BAINOV/ZABREIKO/KOSTADINOV [20]. The following corollary is analogous to KUNZE/MICHAELI [120, Cor. 3].

Corollary 4.1.8. *Suppose that $(A1) - (A9)$ hold. If*

$$\exists c > 0 : |f_k'|_\infty \leq c, \quad k \in I \cup \{\infty\},$$

and if $\|A_{i+1}(y)\|_d \leq 1$ for all $y \in G \setminus G_\infty^0$ and i, where $i \in \mathbb{N}_0$ for $y \in G_{<\infty}$, and $i \in \{0, \ldots, i_0\}$ for $y \in G_\infty^{i_0+1}$ with $i_0 \in \mathbb{N}_0$, then for every $t_* > 0$ there exists $C(t_*) > 0$ such that

$$\sup \left\{ \|T(t, y)\|_3 : t \in [0, t_*], y \in \mathbb{R}^d \right\} \leq C(t_*).$$

In particular, the integrability conditions (4.108) in the MET Thm. 4.3.1 are satisfied for every probability measure $\mu \in \mathcal{P}_{\text{inv}}(\varphi)$.

Proof: This follows from Cor. 4.1.7, since

$$\|T(t, y)\|_3 \leq 1 + c \int_0^t \|T(s, y)\|_3 \, ds$$

for $t \in [0, t_*]$. Thus $C(t_*) = \exp(ct_*)$ is suitable, by Gronwall's lemma.   □

4.1.4 Sets of measure zero

Of course one is interested to have a set G of "good" initial values as large as possible for the definition of the cocycle T in Thm. 4.1.1. In this section we will give a criterion to ensure that $\lambda^d(\mathbb{R}^d \setminus G) = 0$, and it will turn out that this criterion is easy to satisfy in applications, cf. the example of the pendulum with dry friction in Sect. 4.2.

In addition to $(A1) - (A8)$ from Sect. 4.1.1 we consider the following assumption.

$(A10)$ The dimension d is $d \geq 2$, and there exists a set $C_{I\cup\{\infty\}} \subset \mathbb{R}^d$ such

that $C_{I\cup\{\infty\}} \subset \bigcup\limits_{k\in I\cup\{\infty\}} C_k$. Here for $k \in I \cup \{\infty\}$, either $C_k = \emptyset$, or C_k is

a manifold of dimension $\leq d - 2$ and class C^1, which can be described by a single chart $\alpha_k : \mathbb{R}^{d-2} \to C_k$. In addition it is supposed that

$$D_k \supset \{y \in M_k : \varphi^t(y) \notin C_{I\cup\{\infty\}} \ \forall t \in [0, t_1(y)]\}, \quad k \in I, \qquad (4.26)$$

and

$$D_\infty \supset \{y \in M_\infty : \varphi^t(y) \notin C_{I\cup\{\infty\}} \ \forall t \in [0, \infty[\}. \qquad (4.27)$$

The set $C_{I\cup\{\infty\}}$ may be thought of to be a union of boundary curves of the different manifolds, and the condition means that if the flow does not hit these boundary curves, then it should neither hit the exceptional set E_k, nor there should be neighborhoods separated by ξ_1, cf. the definition of D_k at (4.3) and in $(A7)$. Of course, in a concrete example the C_k will be related to the E_k, but there is no need to impose assumptions on this a priori. We also remark that we allow $\{y \in M_k : \varphi^t(y) \notin C_{I\cup\{\infty\}} \ \forall t \in [0, t_1(y)]\}$ to be empty, since in that case the flow transports M_k to the lower-dimensional $C_{I\cup\{\infty\}}$.

We start with some auxiliary results.

Lemma 4.1.8. *Let $(A1) - (A8)$ be satisfied. If $k \in I$ and $N \subset \mathbb{R}^d$ is such that $\lambda^d(N) = 0$, then*

$$\lambda^d\left(\{y \in D_k : \xi_1(y) \in N\}\right) = 0.$$

Proof: Fix $R > 0$ and choose $T = T(k, R)$ with $y \in \overline{B}_R(0) \cap D_k \Rightarrow t_1(y) \leq T$, cf. Lemma 4.1.2. Since $t_1(y) < \infty$ and $f(\cdot, y) = f_k$ for $y \in D_k$, we have $\xi_1(y) = \varphi(t_1(y), y) = \Phi_{f_k}(t_1(y), y)$ by Lemma 4.1.1(g). Since every $\Phi_{f_k}(t, \cdot)$ is a diffeomorphism with inverse $\Phi_{f_k}(-t, \cdot)$, we obtain

$$\{y \in \overline{B}_R(0) \cap D_k : \xi_1(y) \in N\} \subset \Phi_{f_k}([-T, 0] \times N).$$

Since it follows from Lemma 4.3.3(b) that $\lambda^d(\Phi_{f_k}([-T, 0] \times N)) = 0$, the claim is obtained by letting $R \to \infty$. $\qquad\square$

Lemma 4.1.9. *Let* $(A1) - (A8)$ *and* $(A10)$ *be satisfied. If* $k \in I$, *then*

$$\lambda^d(M_k \setminus D_k) = 0.$$

Proof: By (4.26) of $(A10)$

$$M_k \setminus D_k \subset \{y \in M_k : \exists t \in [0, t_1(y)] : \varphi^t(y) \in C_{I \cup \{\infty\}}\}$$
$$\subset \bigcup_{l \in I \cup \{\infty\}} \{y \in \mathbb{R}^d : \exists t \in [0, \infty[: \Phi_{f_k}(t, y) \in C_l\},$$

where the last inclusion follows from Lemma 4.1.1(g) and $(A10)$. Thus $\lambda^d(M_k \setminus D_k) = 0$ by Lemma 4.3.3(a). □

Lemma 4.1.10. *Let* $(A1) - (A8)$ *and* $(A10)$ *be satisfied. Then*

$$\mathbb{R}^d \setminus G \subset \bigcup_{i \in \mathbb{N}_0} \Big\{ y \in \mathbb{R}^d : t_i(y) < \infty, \text{ and}$$
$$\exists t \in [t_i(y), t_{i+1}(y)[: \varphi^t(y) \in C_{I \cup \{\infty\}} \Big\}.$$

Proof: If $y \in \mathbb{R}^d$ is not an element of the right-hand side, then for every $i \in \mathbb{N}_0$ the following alternative holds: either $t_i(y) = \infty$, or $t_i(y) < \infty$ and $\varphi^t(y) \notin C_{I \cup \{\infty\}}$ for all $t \in [t_i(y), t_{i+1}(y)[$. We will show that $y \in G$, and for this we first claim that

$$i_0 \in \mathbb{N}_0, \quad t_{i_0}(y) < \infty \implies y \in G^{i_0}_{<\infty}. \tag{4.28}$$

This is clear for $i_0 = 0$, since $G^0_{<\infty} = \mathbb{R}^d$, cf. Lemma 4.1.3. If (4.28) already holds for some $i_0 \in \mathbb{N}_0$ and $t_{i_0+1}(y) < \infty$, then also $t_{i_0}(y) < t_{i_0+1}(y) < \infty$, and thus by assumption $y \in G^{i_0}_{<\infty}$. This also implies that $\xi_{i_0}(y) \in M_\infty$ is impossible, since $t_1(\xi_{i_0}(y)) = t_{i_0+1}(y) - t_{i_0}(y)$ by (4.5). So choose $k \in I$ such that, cf. Lemma 4.1.1(c), $\xi_{i_0}(y) = \varphi(t_{i_0}(y), y) \in M_k$. By the above alternative we have $\varphi^s(y) \notin C_{I \cup \{\infty\}}$ for $s \in [0, t_{i_0+2}(y)[$, and hence in particular for $s \in [t_{i_0}(y), t_{i_0+1}(y)]$. Therefore, if $t \in [0, t_1(\xi_{i_0}(y))] = [0, t_{i_0+1}(y) - t_{i_0}(y)]$, then we have $\varphi^t(\xi_{i_0}(y)) = \varphi(t + t_{i_0}(y), y) \notin C_{I \cup \{\infty\}}$, i.e., $\xi_{i_0}(y) \in D_k$ by (4.26) of $(A10)$. This gives $y \in G^{i_0+1}_{<\infty}$ by the definition in Lemma 4.1.3, and consequently we have shown (4.28).

To prove that $y \in G$, we first consider <u>Case 1:</u> $t_i(y) < \infty$ for all $i \in \mathbb{N}_0$. Then $y \in G^i_{<\infty}$ for all $i \in \mathbb{N}_0$ by (4.28), and thus $y \in G_{<\infty} \subset G$. <u>Case 2:</u> $t_{i_0+1}(y) = \infty$ for some first $i_0 \in \mathbb{N}_0$. Then $t_{i_0}(y) < \infty$, and therefore $y \in G^{i_0}_{<\infty}$ by (4.28). Additionally, $t_0(y) < t_1(y) < \ldots t_{i_0}(y) < \infty$, and thus $\varphi^s(y) \notin C_{I \cup \{\infty\}}$ for $s \in [t_k(y), t_{k+1}(y)[$, with $k \in \{0, \ldots, i_0\}$, by the above alternative. Since $t_{i_0+1}(y) = \infty$, this means $\varphi^s(y) \notin C_{I \cup \{\infty\}}$ for $s \in [0, \infty[$. By $(A1)$ we also must have $\xi_{i_0}(y) \in M_\infty$, because $t_1(\xi_{i_0}(y)) = t_{i_0+1}(y) - t_{i_0}(y) = \infty$, cf. (4.5). Since $\varphi^t(\xi_{i_0}(y)) = \varphi(t + t_{i_0}(y), y) \notin C_{I \cup \{\infty\}}$ for $t \in [0, \infty[$, (4.27) of assumption $(A10)$ gives $\xi_{i_0}(y) \in D_\infty$, and therefore, cf. Lemma 4.1.4, $y \in G^{i_0}_\infty \subset G$ as was to be shown. □

By means of the previous lemma we can prove

Theorem 4.1.2. *Let* $(A1) - (A8)$ *and* $(A10)$ *hold. Then* $\lambda^d(\mathbb{R}^d \setminus G) = 0$.

Proof: By Lemma 4.1.10 it is sufficient to show $\lambda^d(A_i) = 0$ for every $i \in \mathbb{N}_0$, with

$$A_i = \{y \in \mathbb{R}^d : t_i(y) < \infty \text{ and } \exists t \in [t_i(y), t_{i+1}(y)[: \varphi^t(y) \in C_{I \cup \{\infty\}}\}.$$

This will be achieved by induction, and for this we first claim that

$$A_{i+1} \subset \xi_1^{-1}(A_i) \quad \text{for} \quad i \in \mathbb{N}_0. \tag{4.29}$$

Indeed, if $y \in A_{i+1}$, then $t_i(\xi_1(y)) = t_{i+1}(y) - t_1(y)$ by Lemma 4.1.1(b), and thus $t_i(\xi_1(y)) < \infty$. By assumption we find $\bar{t} \in [t_{i+1}(y), t_{i+2}(y)[$ with $\varphi^{\bar{t}}(y) \in C_{I \cup \{\infty\}}$. Therefore $t = \bar{t} - t_1(y)$ has $t \in [t_{i+1}(y) - t_1(y), t_{i+2}(y) - t_1(y)[= [t_i(\xi_1(y)), t_{i+1}(\xi_1(y))[$ and $\varphi^t(\xi_1(y)) = \varphi^{t+t_1(y)}(y) \in C_{I \cup \{\infty\}}$. Hence (4.29) is satisfied. Next we remark that

$$A_0 \subset \bigcup_{l \in I \cup \{\infty\}} \bigcup_{k \in I \cup \{\infty\}} \{y \in \mathbb{R}^d : \exists t \in [0, \infty[: \Phi_{f_l}(t, y) \in C_k\}$$

by $(A5)$ and $(A10)$. As a consequence of $f_k \in C^1(\mathbb{R}^d, \mathbb{R}^d)$ this yields $\lambda^d(A_0) = 0$, by Lemma 4.3.3(a) and assumption $(A10)$ on the C_k.

Suppose that we have already shown $\lambda^d(A_i) = 0$ for some $i \in \mathbb{N}_0$. Because in particular $t_1(y) < \infty$ for $y \in A_{i+1}$, we have $A_{i+1} \cap M_\infty = \emptyset$, and thus by (4.29)

$$A_{i+1} \subset \{y \in A_{i+1} : \xi_1(y) \in A_i\} = \bigcup_{k \in I} \{y \in A_{i+1} \cap M_k : \xi_1(y) \in A_i\}$$

$$\subset \bigcup_{k \in I} \{y \in D_k : \xi_1(y) \in A_i\} \cup \left(\bigcup_{k \in I} (M_k \setminus D_k) \right)$$

Hence also $\lambda^d(A_{i+1}) = 0$ by Lemma 4.1.8 and Lemma 4.1.9, and this completes the proof of the theorem. □

4.1.5 Remarks on the periodic case

Up to now we always considered the semiflow φ being defined on $\mathbb{R}^d$, i.e., $\varphi : [0, \infty[\times \mathbb{R}^d \to \mathbb{R}^d$. In case that this semiflow is induced by a ω-periodically forced $(d - 1)$-dimensional non-smooth system (as in the example of the pendulum with dry friction (1.2)), the state space $\mathbb{R}^d$ equivalently may be replaced by the d-dimensional metric manifold $\mathbb{R}^d_\omega = \mathbb{R}^{d-1} \oplus S^1_\omega \subset \mathbb{R}^{d+1}$, where S^1_ω is S^1 parameterized through $[0, \omega] \ni \tau \mapsto e^{2\pi i \tau / \omega} \in S^1$. It is more or less obvious that all results from Sects. 4.1.1-4.1.4 have an analogue when the system is considered from this point of view, and we want to give some more comments (without proofs) concerning this subject.

Since the system was made autonomous here artificially by introducing time as a new variable additionally to the $(d-1)$ "space"-variables, every right-hand side $f_k \in C^1(\mathbb{R}^d, \mathbb{R}^d)$, $k \in I \cup \{\infty\}$, is of the form $f_k(y) = (*, \ldots, *, 1)$, resulting from the additional "$\dot{t} = 1$". By $(A5)$ this implies that $\varphi : [0, \infty[\times \mathbb{R}^d \to \mathbb{R}^d$ has to have the form $\varphi^t(y_0) = (\varphi_1(t, y_0), \ldots, \varphi_{d-1}(t, y_0), \tau_0 + t)$ for $t \in [0, \infty[$ and $y_0 = (\xi_0, \tau_0) \in \mathbb{R}^{d-1} \times \mathbb{R}$. By definition of the cocycle $T : [0, \infty[\times \mathbb{R}^d \to \mathcal{L}(\mathbb{R}^d)$ in Thm. 4.1.1, cf. also Lemma 4.1.6 and Lemma 4.1.7, this in turn implies for $y_0 \in G$

$$T(t, y_0) = \begin{pmatrix} T_1(t, y_0) \\ \vdots \\ T_{d-1}(t, y_0) \\ 0 \ \cdots \ 0 \ 1 \end{pmatrix} \in \mathbb{R}^{d \times d}, \qquad (4.30)$$

with $T_j(t, y_0) \in \mathbb{R}^d$ being the jth line of $T(t, y_0)$. By the assumed periodicity in the time component, in addition $\varphi^t(\xi_0, \tau_0 + k\omega) = \varphi^t(\xi_0, \tau_0) + (0, \ldots, 0, k\omega)$, hence also $T(t, \xi_0, \tau_0 + k\omega) = T(t, \xi_0, \tau_0)$, for $k \in \mathbb{Z}$. Therefore the semiflow and the cocycle carry over well-defined to $\mathbb{R}^d_\omega$. To be more precise, let $\varphi_{\text{per}} : [0, \infty[\times \mathbb{R}^d_\omega \to \mathbb{R}^d_\omega$ be defined as

$$\varphi^t_{\text{per}}(\xi_0, e^{2\pi i \tau_0/\omega}) = \left(\varphi_1(t, \xi_0, \tau_0), \ldots, \varphi_{d-1}(t, \xi_0, \tau_0), e^{2\pi i [\tau_0 + t]/\omega}\right) \qquad (4.31)$$

for $t \in [0, \infty[$ and $(\xi_0, e^{2\pi i \tau_0/\omega}) \in \mathbb{R}^{d-1} \oplus S^1_\omega = \mathbb{R}^d_\omega$. Then φ_{per} is a well-defined semiflow on $\mathbb{R}^d_\omega$ which inherits all properties of φ. To transfer T to $\mathbb{R}^d_\omega$, let $\mathcal{T}(\mathbb{R}^d_\omega) = \mathbb{R}^{d-1} \oplus S^1$ denote the tangent bundle of $\mathbb{R}^d_\omega$ with fibers $\mathcal{T}_{y_0}(\mathbb{R}^d_\omega) = \mathbb{R}^{d-1} \oplus (\mathbb{R} i e^{2\pi i \tau_0/\omega})$ for $y_0 = (\xi_0, e^{2\pi i \tau_0/\omega}) \in \mathbb{R}^d_\omega$. These fibers have a natural norm given through $\|(x_1, \ldots, x_{d-1}, x_d i e^{2\pi i \tau_0/\omega})\|_{\mathcal{T}_{y_0}(\mathbb{R}^d_\omega)} = |(x_1, \ldots, x_d)|_{\mathbb{R}^d}$, and $T_{\text{per}}(t, y_0) \in \mathcal{L}\left(\mathcal{T}_{y_0}(\mathbb{R}^d_\omega), \mathcal{T}_{\varphi^t_{\text{per}}(y_0)}(\mathbb{R}^d_\omega)\right)$ is defined as

$$T_{\text{per}}(t, y_0)(x_1, \ldots, x_{d-1}, x_d i e^{2\pi i \tau_0/\omega}) = \begin{pmatrix} \langle T_1(t, \xi_0, \tau_0), (x_1, \ldots, x_d) \rangle_{\mathbb{R}^d} \\ \vdots \\ \langle T_{d-1}(t, \xi_0, \tau_0), (x_1, \ldots, x_d) \rangle_{\mathbb{R}^d} \\ x_d i e^{2\pi i [\tau_0 + t]/\omega} \end{pmatrix}$$

$$\in \mathcal{T}_{\varphi^t_{\text{per}}(y_0)}(\mathbb{R}^d_\omega).$$

This T_{per} is also well-defined, a cocycle for $(\varphi^t_{\text{per}})_{t \geq 0}$, and inherits the properties of T. In particular,

$$\left\| T_{\text{per}}(t, y_0)(x_1, \ldots, x_{d-1}, x_d i e^{2\pi i \tau_0/\omega}) \right\|_{\mathcal{T}_{\varphi^t_{\text{per}}(y_0)}(\mathbb{R}^d_\omega)}$$
$$= \left| T(t, \xi_0, \tau_0)(x_1, \ldots, x_d) \right|_{\mathbb{R}^d} \qquad (4.32)$$

by (4.30), meaning that it is equivalent to calculate Lyapunov exponents for T or for T_{per}, cf. the MET Thm. 4.3.1.

To summarize, although we were not too precise in this section, it should be clear that in periodically forced problems we can switch from φ and T to φ_{per} and T_{per} whenever the need arises. Since the formulation in $\mathbb{R}^d$ usually is more convenient, this will only be the case when we are looking for invariant measures w.r. to the semiflow, since by the unboundedness of $\varphi^t(y_0) = (\varphi_1(t, y_0), \ldots, \varphi_{d-1}(t, y_0), \tau_0 + t)$ in the last component, $\mathcal{P}_{inv}(\varphi)$ will always be empty, while $\mathcal{P}_{inv}(\varphi_{per})$ is the natural set to consider here.

4.2 An application: A pendulum with dry friction

We investigate in detail the example

$$\ddot{x}(\tau) + x(\tau) + \operatorname{sgn} \dot{x}(\tau) = \gamma \sin(\eta \tau) \quad \text{for} \quad \tau \in \mathbb{R}, \tag{1.1}$$

see the introduction. This equation will be considered dynamically correct as the differential inclusion

$$\ddot{x}(\tau) + x(\tau) \in \gamma \sin(\eta \tau) - \operatorname{Sgn} \dot{x}(\tau) \quad \text{a.e.}, \tag{1.2}$$

with the corresponding autonomous version $\dot{y} \in F(y)$ a.e., that is

$$\left. \begin{array}{c} \dfrac{d}{dt} \begin{pmatrix} x \\ v \\ \tau \end{pmatrix} \in F(x, v, \tau) \text{ a.e.}, \quad y = (x, v, \tau) \in \mathbb{R}^3, \quad \text{where} \\[2em] F(x, v, \tau) = \{v\} \times \{\gamma \sin(\eta \tau) - x - w : w \in \operatorname{Sgn} v\} \times \{1\} \end{array} \right\} \tag{1.4}$$

Note that time is denoted by t in (1.4), i.e., solutions are functions $y = y(t) = (x(t), v(t), \tau(t))$ of "proper time", whereas the former time variable τ from (1.2) now also depends on t, but in a rather simple way: $\tau(t) = \tau_0 + t$, with $\tau_0 = \tau(0)$, due to the equation $\frac{d\tau}{dt} = 1$. Therefore we will indicate both differentiation w.r. to t and w.r. to τ by a dot.

4.2.1 Definition of the semiflow

We start by recalling from Ex. 2.2.3 how (1.4) does generate a semiflow $\varphi : [0, \infty[\times \mathbb{R}^3 \to \mathbb{R}^3$.

Theorem 4.2.1. *For all $y_0 = (x_0, v_0, \tau_0) \in \mathbb{R}^3$ there exists a unique solution $\varphi(\cdot, y_0) : [0, \infty[\to \mathbb{R}^3$ of (1.4) with data $\varphi(0, y_0) = y_0$, or equivalently, a unique solution $x(\cdot; y_0) : [\tau_0, \infty[\to \mathbb{R}$ of (1.2) with $x(\tau_0; y_0) = x_0$ and $\dot{x}(\tau_0; y_0) = v_0$. Here $x(\cdot; y_0) \in C^1([\tau_0, \infty[)$ as well as $\ddot{x}(\cdot; y_0) \in L^1_{loc}([\tau_0, \infty[)$. Moreover, (1.4) induces a semiflow $\varphi : [0, \infty[\times \mathbb{R}^3 \to \mathbb{R}^3$, where*

$$\varphi^t(y_0) = \big(x(\tau_0 + t; y_0), \dot{x}(\tau_0 + t; y_0), \tau_0 + t\big) \quad \text{for} \quad t \in [0, \infty[. \qquad (4.33)$$

In addition, $\varphi^t(x_0, v_0, \tau_0 + k(2\pi/\eta)) = \varphi^t(y_0) + (0, 0, k(2\pi/\eta))$ for $k \in \mathbb{N}_0$, and the semiflow is locally Lipschitz continuous on $[0, \infty[\times \mathbb{R}^3$.

Since we intend to apply the results from Sect. 4.1, we already note that in particular we have obtained in Thm. 4.2.1 that

$$\varphi : [0, \infty[\times \mathbb{R}^3 \to \mathbb{R}^3 \text{ is a measurable semiflow which satisfies}$$
$$(A4) \text{ in Sect. 4.1.1.} \qquad (4.34)$$

4.2.2 Verification of $(A1) - (A10)$

In this section we will show that the pendulum example (1.4) fits into the framework built up in Sect. 4.1.1. To have a non-empty and non-trivial manifold M_∞, we only consider the case

$$\gamma \in]0, 1[\qquad (4.35)$$

in (1.4) resp. (1.2). Then

$$\mathcal{S}_\gamma = [\gamma - 1, -\gamma + 1] \subset \mathbb{R} \qquad (4.36)$$

is the set of stationary solutions of (1.2). We remark that $\gamma = 1$ is excluded in order not to have M_∞ only a line; see Fig. 4.1 in Sect. 4.1. Although in what follows some arguments and estimates will rely on $\gamma \leq 1$, for $\gamma \in]1, \infty[$ the approach to describe the system as in Sect. 4.1.1 and the verification of assumptions $(A1) - (A10)$ are similar. In case of $\gamma \in]1, \infty[$ there are no equilibria.

Here and in the following we always let $y = (x, v, \tau)$ denote a typical point in $\mathbb{R}^3$. We let

$$b^\pm(\tau) = \gamma \sin(\eta\tau) \pm 1, \qquad (4.37)$$

and define $\tau_{\text{mod}} = \tau \mod (2\pi/\eta)$ for $\tau \in \mathbb{R}$. Then $b^\pm(\tau) = b^\pm(\tau_{\text{mod}})$. Let

$$M_+ = M_1 = \{ y \in \mathbb{R}^3 : v > 0 \} \cup \{ y \in \mathbb{R}^3 : v = 0, x < b^-(\tau) \}$$
$$\cup \{ y \in \mathbb{R}^3 : v = 0, x = b^-(\tau),$$
$$\tau_{\text{mod}} \in [0, \pi/2\eta[\cup [3\pi/2\eta, 2\pi/\eta[\},$$

$$M_- = M_2 = \{ y \in \mathbb{R}^3 : v < 0 \} \cup \{ y \in \mathbb{R}^3 : v = 0, x > b^+(\tau) \}$$
$$\cup \{ y \in \mathbb{R}^3 : v = 0, x = b^+(\tau), \tau_{\text{mod}} \in [\pi/2\eta, 3\pi/2\eta[\},$$

$$M_{\text{left}} = M_3 = \{ y \in \mathbb{R}^3 : v = 0, x = b^-(\tau), \tau_{\text{mod}} \in]\pi/2\eta, 3\pi/2\eta[\}$$
$$\cup \{ y \in \mathbb{R}^3 : v = 0, b^-(\tau) < x < \gamma - 1 \},$$

$$M_{\text{right}} = M_4 = \{ y \in \mathbb{R}^3 : v = 0, x = b^+(\tau),$$
$$\tau_{\text{mod}} \in]3\pi/2\eta, 2\pi/\eta[\cup [0, \pi/2\eta[\}$$
$$\cup \{ y \in \mathbb{R}^3 : v = 0, -\gamma + 1 < x < b^+(\tau) \},$$

$$M_\infty = \{ y \in \mathbb{R}^3 : v = 0, \gamma - 1 \leq x \leq -\gamma + 1 \}, \qquad (4.38)$$

In particular the M_k are mutually disjoint, and also $\mathbb{R}^3 = \bigcup_{k=1}^4 M_k \cup M_\infty$. The subscripts "left" resp. "right" indicate that the respective manifold is lying to the left-hand resp. right-hand side of M_∞. In fact, M_{left} and M_{right} need not be distinguished from the dynamical point of view, but this turns out to make the description easier to follow.

Next, we define the right-hand sides f_k with $b^\pm$ from (4.37) as

$$f_+(x,v,\tau) = (v, -x + b^-(\tau), 1), \quad f_-(x,v,\tau) = (v, -x + b^+(\tau), 1), \quad \text{and}$$
$$f_{\text{left}}(x,v,\tau) = f_{\text{right}}(x,v,\tau) = f_\infty(x,v,\tau) = (0,0,1) \quad \text{for} \quad (x,v,\tau) \in \mathbb{R}^3,$$
$$(4.39)$$

and we set $f_1 = f_+$, $f_2 = f_-$, $f_3 = f_{\text{left}}$, and $f_4 = f_{\text{right}}$.

Finally, for $k \in \{1,2,3,4\} \cup \{\infty\}$ the exceptional sets $E_k \subset \overline{M_k}^{\mathbb{R}^3}$ are given through

$$E_+ = E_1 = \{y \in \mathbb{R}^3 : v = 0, x = b^-(\tau),$$
$$\tau_{\text{mod}} \in [0, \pi/2\eta] \cup [3\pi/2\eta, 2\pi/\eta[\,\},$$

$$E_- = E_2 = \{y \in \mathbb{R}^3 : v = 0, x = b^+(\tau), \tau_{\text{mod}} \in [\pi/2\eta, 3\pi/2\eta]\,\},$$

$$E_{\text{left}} = E_3 = \{y \in \mathbb{R}^3 : v = 0, x = b^-(\tau),\ \tau_{\text{mod}} \in]\pi/2\eta, 3\pi/2\eta[\,\},$$

$$E_{\text{right}} = E_4 = \{y \in \mathbb{R}^3 : v = 0, x = b^+(\tau),$$
$$\tau_{\text{mod}} \in [0, \pi/2\eta[\,\cup\,]3\pi/2\eta, 2\pi/\eta[\,\},$$

$$E_\infty = \{\gamma - 1, -\gamma + 1\} \times \{0\} \times \mathbb{R}. \qquad (4.40)$$

We start with the investigation of M_∞.

Lemma 4.2.1. *Let* $t_1(y) = \infty$ *for* $y \in M_\infty$. *Then for every* $y = (x,0,\tau) \in M_\infty$, $\varphi^t(y) = (x, 0, \tau + t)$ *for* $t \in [0, t_1(y)[$. *Thus in particular* $\varphi^t(y) \in M_\infty$ *and* $\varphi^t(y) = \Phi_{f_\infty}(t,y)$ *for* $t \in [0, t_1(y)[$.

Proof: This follows from the uniqueness of solutions of (1.4) to the right and from the fact that $x - \gamma \sin(\eta[\tau + t]) \in [x - \gamma, x + \gamma] \subset [-1,1]$ for $t \in [0, \infty[$, because we have $\gamma - 1 \le x \le -\gamma + 1$ by definition of M_∞. □

In the following we will always consider $\arcsin : [-1, 1] \to [-\pi/2, \pi/2]$.

Lemma 4.2.2. *For* $y = (x, 0, \tau) \in M_{\text{left}}$ *let*

$$\sigma_{\text{left}}(x) = \frac{1}{\eta}\left(\frac{\pi}{2} + \arcsin\left(\frac{x+1}{\gamma}\right)\right),$$

and define

$$t_1(y) = \begin{cases} \sigma_{\text{left}}(x) - \tau_{\text{mod}} - \pi/2\eta & : \quad \tau_{\text{mod}} \in [0, \pi/2\eta[\\ \sigma_{\text{left}}(x) - \tau_{\text{mod}} + 3\pi/2\eta & : \quad \tau_{\text{mod}} \in]\pi/2\eta, 2\pi/\eta[\end{cases} . \qquad (4.41)$$

Analogously, for $y = (x, 0, \tau) \in M_{\text{right}}$ we let

$$\sigma_{\text{right}}(x) = \frac{1}{\eta}\left(\frac{\pi}{2} - \arcsin\left(\frac{x-1}{\gamma}\right)\right),$$

and define

$$t_1(y) = \begin{cases} \sigma_{\text{right}}(x) - \tau_{\text{mod}} + \pi/2\eta & : \ \tau_{\text{mod}} \in [0, 3\pi/2\eta[\\ \sigma_{\text{right}}(x) - \tau_{\text{mod}} + 5\pi/2\eta & : \ \tau_{\text{mod}} \in]3\pi/2\eta, 2\pi/\eta[\end{cases} \qquad (4.42)$$

Then for $y \in M_{\text{left}}$

(a) $t_1(y) \in]0, 2\pi/\eta[$,
(b) $\varphi^t(y) = (x, 0, \tau + t)$ for $t \in [0, t_1(y)[$, and hence in particular for $t \in [0, t_1(y)[$ we have $\varphi^t(y) \in M_{\text{left}}$, $\varphi^t(y) = \Phi_{f_{\text{left}}}(t, y)$, and $t_1(\varphi^t(y)) + t = t_1(y)$. In addition,

$$\varphi(t_1(y), y) \in \{y \in \mathbb{R}^3 : v = 0, \ x = b^-(\tau),$$
$$\tau_{\text{mod}} \in [0, \pi/2\eta[\cup]3\pi/2\eta, 2\pi/\eta[\} \subset M_+ . \qquad (4.43)$$

An analogous statement holds for $y \in M_{\text{right}}$. Also $(x, 0, \tau) \in M_{\text{left}} \Leftrightarrow (-x, 0, \tau + \pi/\eta) \in M_{\text{right}}$, and in this case $t_1(x, 0, \tau) = t_1(-x, 0, \tau + \pi/\eta)$.

Proof: We only consider $y \in M_{\text{left}}$ and remark that the function $\sigma_{\text{left}} :]-\gamma-1, \gamma-1[\to]0, \pi/\eta[$ measures the distance in (vertical) τ-direction of the boundary curve given through $x = b^-(\tau)$, $\tau_{\text{mod}} \in]3\pi/2\eta, 2\pi/\eta[\cup[0, \pi/2\eta[$, to the line $\tau = 3\pi/2\eta$. If we treat as an example the case $\tau_{\text{mod}} \in [0, \pi/2\eta[$, then we obtain (a), since $b^-(\tau) < x < \gamma-1 \Rightarrow \sin(\eta\tau) < (x+1)/\gamma < 1 \Rightarrow \tau_{\text{mod}} < \eta^{-1}\arcsin((x+1)/\gamma) < \pi/2\eta \Rightarrow 0 < t_1(y) < \pi/2\eta - \tau_{\text{mod}} < 2\pi/\eta$. Concerning (b), it follows from the uniqueness of solutions of (1.4) that the flow moves just on a straight line upwards until it hits the curve $x = b^-(\tau)$, because, again for $\tau_{\text{mod}} \in [0, \pi/2\eta[$ and for $t \in [0, t_1(y)[$, we obtain $\tau + t \in [\tau, \sigma_{\text{left}}(x) - \pi/2\eta[= [\tau, \eta^{-1}\arcsin((x+1)/\gamma)[\Rightarrow \gamma\sin(\eta[\tau+t]) - x \in [\gamma\sin(\eta\tau) - x, 1] \subset [-1, 1]$, the latter being a consequence of $x < \gamma-1$ and $\gamma \le 1$. Thus $\frac{d}{dt}(x, 0, \tau+t) = (0, 0, 1) \in F(\varphi^t(y))$ for $t \in [0, t_1(y)[$, hence $\varphi^t(y) = (x, 0, \tau+t)$. From this we clearly get $\varphi^t(y) \in M_{\text{left}}$ and $\varphi^t(y) = \Phi_{f_{\text{left}}}(t, y)$ in $[0, t_1(y)[$. Moreover, if $\tau_{\text{mod}} \in [0, \pi/2\eta[$ and $t \in [0, t_1(y)[$, then $(\tau+t)_{\text{mod}} = \tau_{\text{mod}} + t \in [0, \pi/2\eta[$, and therefore $t_1(x, 0, \tau+t) + t = \sigma_{\text{left}}(x) - (\tau+t)_{\text{mod}} - \pi/2\eta + t = t_1(y)$. Since also (4.43) is clear from the definition of $t_1(y)$, we thus have proved (b).

To show the relation between M_{left} and M_{right}, we again consider as an example only $y = (x, 0, \tau) \in M_{\text{left}}$ with $\tau_{\text{mod}} \in [0, \pi/2\eta[$. Then it follows from $b^-(\tau) < x < \gamma - 1$ that $-\gamma + 1 < -x < b^+(\tau + \pi/\eta)$, since $b^+(\tau + \pi/\eta) = -b^-(\tau)$, and hence $(-x, 0, \tau + \pi/\eta) \in M_{\text{right}}$. Also, $(\tau + \pi/\eta)_{\text{mod}} = \tau_{\text{mod}} + \pi/\eta \in [\pi/\eta, 3\pi/2\eta[$, and therefore $\sigma_{\text{left}}(x) = \sigma_{\text{right}}(-x)$ yields $t_1(x, 0, \tau) = \sigma_{\text{left}}(x) - \tau_{\text{mod}} - \pi/2\eta = \sigma_{\text{right}}(-x) - (\tau + \pi/\eta)_{\text{mod}} + \pi/2\eta = t_1(-x, 0, \tau + \pi/\eta)$ as was to be verified. $\qquad \square$

The definition of $t_1(\cdot)$ on M_+ or M_- is more complicated, since trajectories starting in e.g. $y_0 \in M_+$ may return to the surface $\{v = 0\}$ at some time $s_1(y_0)$ without passing to a different manifold afterwards, cf. the trajectory of y_0 in Fig. 4.1 in Sect. 4.1. We start with a preparatory lemma.

Lemma 4.2.3. *For $y_0 \in M_+$ there exists $\varepsilon > 0$ such that*

$$\varphi(]0, \varepsilon] \times \{y_0\}) \subset \{y \in \mathbb{R}^3 : v > 0\} \subset M_+ . \tag{4.44}$$

Define

$$s_1(y_0) = \sup \left\{ \varepsilon > 0 : \varphi(]0, \varepsilon] \times \{y_0\}) \subset \{y \in \mathbb{R}^3 : v > 0\} \right\}. \tag{4.45}$$

Then the following holds:

(a) $s_1(y_0) \in]0, \infty[$.
(b) $\varphi(t, y_0) \in \{y \in \mathbb{R}^3 : v > 0\}$ for $t \in]0, s_1(y_0)[$, and $\varphi(s_1(y_0), y_0) \in \{y \in \mathbb{R}^3 : v = 0, x \geq b^-(\tau)\}$.
(c) Let $y_1 = (x_1, 0, \tau_1) = \varphi(s_1(y_0), y_0)$. If $y_1 \in M_+$, then

$$y_1 \in \{y \in \mathbb{R}^3 : v = 0, \ x = b^-(\tau), \ \tau_{\text{mod}} \in [0, \pi/2\eta [\cup]3\pi/2\eta, 2\pi/\eta [\} . \tag{4.46}$$

In addition, if also $y_2 = (x_2, 0, \tau_2) = \varphi(s_1(y_1), y_1) \in M_+$, then $s_1(y_1) \geq \pi \wedge (\pi/\eta)$.
(d) If $y_0 = (x_0, 0, \tau_0) \in \{y \in \mathbb{R}^3 : v = 0, x = b^-(\tau), \tau_{\text{mod}} \in [0, \pi/2\eta [\cup]3\pi/2\eta, 2\pi/\eta [\} \subset M_+$, then

$$s_1(y_0) > \begin{cases} \pi \wedge \left(\pi/2\eta - (\tau_0)_{\text{mod}} \right) & : \ (\tau_0)_{\text{mod}} \in [0, \pi/2\eta [\\ \pi \wedge \left(5\pi/2\eta - (\tau_0)_{\text{mod}} \right) & : \ (\tau_0)_{\text{mod}} \in]3\pi/2\eta, 2\pi/\eta [\end{cases} .$$

(e) If $\gamma \in]0, 1]$ and $\varepsilon_0 \in]0, 2\gamma[$, then there exist $\varepsilon_1 > 0$ and $\delta_1 > 0$ with the following property:

$$y_0 = (x_0, 0, \tau_0) \in \{ y \in \mathbb{R}^3 : v = 0, \ x = b^-(\tau) \leq \gamma - 1 - \varepsilon_0,$$
$$\tau_{\text{mod}} \in [0, \pi/2\eta [\cup]3\pi/2\eta, 2\pi/\eta [\}$$
$$\implies \quad \varepsilon_1 \in]0, s_1(y_0)[, \quad \varphi(\varepsilon_1, y_0) \in \{y \in \mathbb{R}^3 : v \geq \delta_1\}.$$

(f) $\varphi(s_1(y_0), y_0) \in \mathbb{R}^3 \setminus \{y \in \mathbb{R}^3 : v = 0, x = b^-(\tau), \tau_{\text{mod}} \in]\pi/2\eta, 3\pi/2\eta [\}$.
(g) For $y_0 \in M_+ \setminus \{y \in \mathbb{R}^3 : v = 0, x = b^-(\tau)\}$ there exists $\delta > 0$ such that $s_1(y) > \delta$ for every $y \in B_\delta(y_0) \cap M_+$.

An analogous statement holds for $y_0 \in M_-$.

Proof: We consider only $y_0 = (x_0, v_0, \tau_0) \in M_+$ and prove (4.44) first. If $v_0 > 0$, then the claim follows from the continuity of $\varphi(\cdot, y_0)$. The second case to handle is $y_0 = (x_0, 0, \tau_0) \in M_+$ with $x_0 < b^-(\tau_0)$. We denote by

$x(\tau)$, $\tau \in [\tau_0, \infty[$, the corresponding solution of (1.2) with $x(\tau_0) = x_0$ and $\dot{x}(\tau_0) = 0$. Define $\Delta(\tau) = b^-(\tau) - x(\tau)$. Then $\Delta \in C^1([\tau_0, \infty[)$ and $\Delta(\tau_0) > 0$. Since (1.2) is equivalent to

$$
\left.
\begin{array}{lll}
\ddot{x}(\tau) + x(\tau) = b^-(\tau) & \text{a.e. in } \{\tau \in [\tau_0, \infty[: \dot{x}(\tau) > 0\}, \\
\ddot{x}(\tau) + x(\tau) = b^+(\tau) & \text{a.e. in } \{\tau \in [\tau_0, \infty[: \dot{x}(\tau) < 0\}, \\
\ddot{x}(\tau) + x(\tau) \in [b^-(\tau), b^+(\tau)] & \text{a.e. in } \{\tau \in [\tau_0, \infty[: \dot{x}(\tau) = 0\},
\end{array}
\right\}
$$
$$(4.47)$$

we have $\ddot{x}(\tau) \geq \Delta(\tau)$ a.e. in $[\tau_0, \infty[$, and thus $\dot{x}(\tau) = \dot{x}(\tau_0) + \int_{\tau_0}^{\tau} \ddot{x}(\sigma)\,d\sigma \geq \int_{\tau_0}^{\tau} \Delta(\sigma)\,d\sigma > 0$ locally to the right of τ_0. The last case to analyze is $y_0 = (x_0, 0, \tau_0) \in M_+$ with $x_0 = b^-(\tau_0)$ and $(\tau_0)_{\text{mod}} \in [0, \pi/2\eta[\,\cup\,[3\pi/2\eta, 2\pi/\eta[$. Consider the solution $z(\cdot)$ on $[\tau_0, \infty[$ of the initial value problem $\ddot{z} + z = b^-(\tau)$, $z(\tau_0) = x_0$, $\dot{z}(\tau_0) = 0$, i.e.,

$$
z(\tau) = x_0 \cos(\tau - \tau_0) + \int_{\tau_0}^{\tau} \sin(\tau - \sigma)\, b^-(\sigma)\, d\sigma.
$$

Then it follows from $x_0 = b^-(\tau_0)$, by deriving a second order ODE for $\dot{z}$, that

$$
\dot{z}(\tau) = \gamma\eta \int_{\tau_0}^{\tau} \sin(\tau - \sigma)\, \cos(\eta\sigma)\, d\sigma \quad \text{for} \quad \tau \in [\tau_0, \infty[. \tag{4.48}
$$

Hence $(\tau_0)_{\text{mod}} \in [0, \pi/2\eta[\,\cup\,[3\pi/2\eta, 2\pi/\eta[$ implies $\dot{z}(\tau) > 0$ in some $]\tau_0, \tau_0+\varepsilon]$. By uniqueness to the right of solutions to (1.2) we obtain $x(\tau) = z(\tau)$ until the first time $\tau_1 > \tau_0$ with $z(\tau_1) = 0$, so at least in $]\tau_0, \tau_0+\varepsilon]$. Transferred back to (1.4) this gives (4.44), since in $]0, \varepsilon]$ we have $\varphi(t, y_0) = (z(\tau_0 + t), \dot{z}(\tau_0 + t), \tau_0 + t)$.

Therefore $s_1(y_0) > 0$ is well-defined, and by definition and from the continuity of $\varphi(\cdot, y_0)$ it follows that $\varphi(t, y_0) \in \{y \in \mathbb{R}^3 : v > 0\}$ for $t \in]0, s_1(y_0)[$ and $\varphi(s_1(y_0), y_0) \in \{y \in \mathbb{R}^3 : v = 0\}$. Let $y_1 = (x_1, v_1, \tau_1) = \varphi(s_1(y_0), y_0)$ denote the point of arrival at the surface $\{v = 0\}$ of the trajectory at time $s_1(y_0)$. Then $v_1 = 0$ and $\tau_1 = \tau_0 + s_1(y_0)$, and also $x_1 \geq b^-(\tau_1)$. To see the latter, we assume on the contrary that $x_1 < b^-(\tau_1)$. The corresponding solution $x(\cdot)$ of (1.2) satisfies $\ddot{x} + x = b^-(\tau)$ in $]\tau_0, \tau_1[$, $\dot{x}(\tau) > 0$ in $]\tau_0, \tau_1[$, $x(\tau_0) = x_0$, $\dot{x}(\tau_0) = v_0$, $x(\tau_1) = x_1$, and $\dot{x}(\tau_1) = v_1 = 0$. Then $\ddot{x}(\tau) = b^-(\tau) - x(\tau) \geq \delta_0 > 0$ in some interval $]\tau_1 - \delta_1, \tau_1[\subset]\tau_0, \tau_1[$, and therefore $0 = \dot{x}(\tau_1) = \dot{x}(\tau_1 - \delta_1) + \int_{\tau_1-\delta_1}^{\tau_1} \ddot{x}(\sigma)\, d\sigma \geq \delta_0\delta_1$, a contradiction. Hence $x_1 \geq b^-(\tau_1)$, and thus (b) is proved. Moreover, $s_1(y_0)$ is finite, since otherwise, by (4.44), there would exist $\tau_1 \geq \tau_0$ and a solution $x(\cdot)$ on $[\tau_1, \infty[$ of $\ddot{x} + x = b^-(\tau)$, $x(\tau_1) = x_1$ and $\dot{x}(\tau_1) = v_1 > 0$ with $\dot{x}(\tau) > 0$ in $[\tau_1, \infty[$, a contradiction to Lemma 4.3.2. This proves (a).

Concerning (c), let $y_1 = (x_1, 0, \tau_1) = \varphi(s_1(y_0), y_0)$ be as above. Then $x_1 \geq b^-(\tau_1)$ by (b). Suppose now in addition that $y_1 \in M_+$ and assume $(\tau_1)_{\text{mod}} = 3\pi/2\eta$. Then $x(\tau) < x(\tau_1) = b^-(\tau_1) \leq b^-(\tau)$ in $]\tau_0, \tau_1[$, and consequently we obtain the contradiction $0 = \dot{x}(\tau_1) = \dot{x}(\tau_0) + \int_{\tau_0}^{\tau_1} \ddot{x}(\sigma)\, d\sigma > 0$. Because of $y_1 \in M_+$ and $v_1 = 0$, (4.46) follows.

$$D_k = \left\{ y \in M_k : \varphi^t(y) \notin E_k \ \forall t \in [0, t_1(y)] , \text{ and} \right.$$
$$l \in I \cup \{\infty\}, \xi_1(y) \in M_l \Rightarrow \forall \, t \in [0, t_1(y)[\ \exists \, \varepsilon > 0 :$$
$$\left. \xi_1 \big(B_\varepsilon(\varphi^t(y)) \cap M_k \big) \subset M_l \right\},$$

and we let $D_+ = D_1$, $D_- = D_2$, $D_{\text{left}} = D_3$, and $D_{\text{right}} = D_4$, cf. the definition of the manifolds M_k in (4.38) and the definition of the exceptional sets $E_k \subset M_k$ in (4.40).

We also need to introduce

$$C_+ = \{y \in \mathbb{R}^3 : v = 0, x = b^-(\tau)\}, \quad C_- = \{y \in \mathbb{R}^3 : v = 0, x = b^+(\tau)\},$$
$$C_{\text{left}} = \{y \in \mathbb{R}^3 : v = 0, x = \gamma - 1\}, \quad C_{\text{right}} = \{y \in \mathbb{R}^3 : v = 0, x = -\gamma + 1\},$$
$$C_\infty = \emptyset, \quad \text{and} \quad C_{I \cup \{\infty\}} = C_+ \cup C_- \cup C_{\text{left}} \cup C_{\text{right}} . \tag{4.60}$$

Since here the dimension of the system is three, and of course C_+, C_-, C_{left}, and C_{right} are C^1-manifolds of dimension one, each of which can be described with a single global chart, in particular the first part of assumption $(A10)$ from Sect. 4.1.4 holds.

Next we recall from $(A7)$ in Sect. 4.1.1 that $D_\infty = \{y \in M_\infty : \varphi^t(y) \notin E_\infty \ \forall t \in [0, \infty[\}$, with E_∞ from (4.40). Because the flow in M_∞ is just a straight line in τ-direction by Lemma 4.2.1, we obtain

Lemma 4.2.8. *Assumption $(A7)$ from Sect. 4.1.1 is satisfied, and*

$$D_\infty \supset \{y \in M_\infty : \varphi^t(y) \notin C_{I \cup \{\infty\}} \ \forall t \in [0, \infty[\} .$$

Next we consider D_{left} and D_{right}.

Lemma 4.2.9. *We have*

$$D_{\text{left}} = M_{\text{left}} \setminus E_{\text{left}} \supset \{y \in M_{\text{left}} : \varphi^t(y) \notin C_{I \cup \{\infty\}} \ \forall t \in [0, t_1(y)]\} , \tag{4.61}$$

and $t_1(\cdot) \in C^1(D_{\text{left}})$. An analogous statement holds for D_{right}.

Proof: We only consider the case of D_{left}. By Lemma 4.2.2(b),

$$D_{\text{left}} = \left\{ y \in M_{\text{left}} : \varphi^t(y) \notin E_{\text{left}} \ \forall t \in [0, t_1(y)] , \text{ and} \right.$$
$$\left. \forall \, t \in [0, t_1(y)[\ \exists \, \varepsilon > 0 : \xi_1 \big(B_\varepsilon(\varphi^t(y)) \cap M_{\text{left}} \big) \subset M_+ \right\} ,$$

because $\xi_1(y) = \varphi(t_1(y), y) \in M_+$ for $y \in M_{\text{left}}$. This also shows that

$$D_{\text{left}} = \{y \in M_{\text{left}} : \varphi^t(y) \notin E_{\text{left}} \ \forall t \in [0, t_1(y)]\} , \tag{4.62}$$

since for $y \in M_{\text{left}}$, $t \in [0, t_1(y)[$, and arbitrary $\varepsilon > 0$ we have $\xi_1 \big(B_\varepsilon(\varphi^t(y)) \cap M_{\text{left}} \big) \subset \xi_1(M_{\text{left}}) \subset M_+$. But $\varphi^t(y) = (x, 0, \tau + t)$ for $y = (x, 0, \tau) \in M_{\text{left}}$ and $t \in [0, t_1(y)[$ by Lemma 4.2.2(b), i.e., the flow is a straight line in τ-direction. Thus $\varphi^t(y) \notin E_{\text{left}}$ for all $t \in [0, t_1(y)]$ iff $y = \varphi^0(y) \notin E_{\text{left}}$, and this implies

$D_{\text{left}} = M_{\text{left}} \setminus E_{\text{left}}$. Because every trajectory starting in $y_0 \in M_{\text{left}}$ will reach $C_+ \subset C_{I \cup \{\infty\}}$ at time $t = t_1(y_0)$, we have $\{y \in M_{\text{left}} : \varphi^t(y) \notin C_{I \cup \{\infty\}} \; \forall t \in [0, t_1(y)]\} = \emptyset$, but this makes sense here, since $\lambda^3(M_{\text{left}}) = 0$, see the remark after the introduction of $(A10)$ in Sect. 4.1.4. Hence (4.61) holds.

To prove the second claim, define for $k \in \mathbb{Z}$

$$U_k = \Big\{ y \in \mathbb{R}^3 : x \in]-\gamma - 1, \gamma - 1[, \; v \in]-1, 1[,$$
$$\tau \in](k - 1)(2\pi/\eta) + \pi/2\eta, \, k(2\pi/\eta) + \pi/2\eta[,$$
$$\tau < (k - 1)(2\pi/\eta) + 3\pi/2\eta + \sigma_{\text{left}}(x) \Big\},$$

with σ_{left} from Lemma 4.2.2. Thus $U_k \subset \mathbb{R}^3$ is open and $U_k \cap U_l = \emptyset$ for $k \neq l$. Hence

$$T_1 : \quad U = \bigcup_{k \in \mathbb{Z}} U_k \to \mathbb{R}, \quad T_1(x, v, \tau) = (k - 1)(2\pi/\eta) + 3\pi/2\eta + \sigma_{\text{left}}(x) - \tau,$$

for $(x, v, \tau) \in U_k$, is well-defined, and $T_1 \in C^1(U)$. Since it follows from Lemma 4.2.2 that $D_{\text{left}} \subset M_{\text{left}} \subset U$ and $T_1(y) = t_1(y)$ for $y \in U$, we are done. $\qquad \square$

Next we intend to show that $t_1(\cdot) \in C^1(D_+)$ and $t_1(\cdot) \in C^1(D_-)$, and afterwards we will give a characterization of D_+ resp. D_-.

Lemma 4.2.10. *Let $y_0 \in M_+$ be such that $\varphi^t(y_0) \notin E_+$ for $t \in [0, t_1(y_0)]$. Then the following assertions hold.*

(a) *There exist $\delta_1 > 0$, $\delta_2 \in]0, t_1(y_0)/2[$, and $T_1 : B_{\delta_1}(y_0) \to]t_1(y_0) - \delta_2, t_1(y_0) + \delta_2[$ such that $T_1 \in C^2(B_{\delta_1}(y_0))$ and $T_1(y) = t_1(y)$ for all $y \in B_{\delta_1}(y_0) \cap M_+$. In particular, it holds that $t_1(\cdot) \in C^1(D_+)$.*

(b) *For every $\delta_3 > 0$ there exists $\delta_4 > 0$ such that $B_{\delta_4}(\xi_1(y_0)) \cap \{y \in \mathbb{R}^3 : v = 0\} \subset \xi_1(B_{\delta_3}(y_0) \cap M_+)$.*

(c) *For every $\delta_5 > 0$ and $t \in [0, t_1(y_0)[$ there exists $\delta_6 > 0$ with*

$$\xi_1(B_{\delta_6}(\varphi^t(y_0)) \cap M_+) \subset B_{\delta_5}(\xi_1(y_0)) \cap \{y \in \mathbb{R}^3 : v = 0\}.$$

An analogous statement holds for D_-.

Proof: We start with some general observations. Let $y_0 = (x_0, v_0, \tau_0)$ be as in the lemma. Since $s_1(y_0) \leq t_1(y_0)$, cf. (4.45), we have $\varphi(s_1(y_0), y_0) \notin E_+$, and therefore $j(y_0) = 1$, i.e., $s_1(y_0) = t_1(y_0)$, by Lemma 4.2.3(c), cf. Lemma 4.2.4. Hence $\xi_1(y_0) = \varphi(s_1(y_0), y_0) \notin E_+$. Taking also into account Lemma 4.2.3(f), it follows that $y_1 = (x_1, 0, \tau_1) = \xi_1(y_0) \notin C_+$. By Lemma 4.2.3(b) we thus obtain

$$x_1 > b^-(\tau_1), \tag{4.63}$$

and we have $\tau_1 = \tau_0 + t_1(y_0)$. In addition, because of $\varphi^t(y_0) \in \{y \in \mathbb{R}^3 : v > 0\}$ for $t \in]0, t_1(y_0)[$ and $y_0 \in M_+ \setminus E_+$, even

$$\varphi^t(y_0) \notin C_+ \quad \text{for} \quad t \in [0, t_1(y_0)] \,. \tag{4.64}$$

From the continuity of φ and Lemma 4.2.4(b2), cf. (4.52), it follows that for

$$\varphi(t, y_0) = \Phi_{f_+}(t, y_0) = \big(g_1(t, y_0), g_2(t, y_0), \tau_0 + t\big) \quad \text{for} \quad t \in [0, t_1(y_0)] \,, \tag{4.65}$$

with

$$g_1(t, y) = x \cos t + v \sin t + \int_0^t \sin(t - s) \, b^-(\tau + s) \, ds$$

and $g_2(t, y) = \partial_t \, g_1(t, y)$ for $t \in \mathbb{R}$ and $y = (x, v, \tau) \in \mathbb{R}^3$. By continuity of the global flow Φ_{f_+} we find $\varepsilon_1 > 0$ such that

$$\Phi_{f_+}\big([0, t_1(y_0) + \varepsilon_1[\times B_{\varepsilon_1}(y_0)\big) \subset \mathbb{R}^3 \setminus C_+ \,. \tag{4.66}$$

We prove (a) and (b) together and begin with Case 1: $v_0 > 0$. To show (a) in this case, note that here $B_{\varepsilon_2}(y_0) \subset \{y \in \mathbb{R}^3 : v > 0\} \subset M_+$ for some $\varepsilon_2 > 0$, w.l.o.g. $\varepsilon_2 < t_1(y_0)/2$. Now we consider the map $g_2 :]t_1(y_0) - \varepsilon_2, t_1(y_0) + \varepsilon_2[\times B_{\varepsilon_2}(y_0) \to \mathbb{R}$. As a consequence of

$$\big(g_1(t_1(y_0), y_0), g_2(t_1(y_0), y_0), \tau_0 + t_1(y_0)\big) = \Phi_{f_+}(t_1(y_0), y_0)$$
$$= \varphi(t_1(y_0), y_0) = y_1 \in \{y \in \mathbb{R}^3 : v = 0\},$$

cf. Lemma 4.2.3(b), we obtain $g_2(t_1(y_0), y_0) = 0$. In addition,

$$\partial_t \, g_2(t, y) = -x \cos t - v \sin t + b^-(\tau + t) - \int_0^t \sin(t - s) b^-(\tau + s) \, ds$$
$$= -g_1(t, y) + b^-(\tau + t),$$

and hence $\partial_t \, g_2(t_1(y_0), y_0) = -g_1(t_1(y_0), y_0) + b^-(\tau_0 + t_1(y_0)) = -x_1 + b^-(\tau_1) \neq 0$ by (4.63). Hence, by the implicit function theorem there exist $0 < \varepsilon_3, \varepsilon_4 \leq \varepsilon_2$ and a function $T_1 : B_{\varepsilon_3}(y_0) \to]t_1(y_0) - \varepsilon_4, t_1(y_0) + \varepsilon_4[$ such that $g_2(T_1(y), y) = 0$ for $y \in B_{\varepsilon_3}(y_0)$, and with the property

$$(t, y) \in]t_1(y_0) - \varepsilon_4, t_1(y_0) + \varepsilon_4[\times B_{\varepsilon_3}(y_0) \,, \quad g_2(t, y) = 0 \quad \Longrightarrow \quad t = T_1(y) \,. \tag{4.67}$$

In addition we may assume $T_1 \in C^2(B_{\varepsilon_3}(y_0))$, and by shrinking ε_3 if necessary, also that $\varepsilon_3, \varepsilon_4 < \varepsilon_1$. By keeping ε_4 fixed and reducing ε_3 further, we may moreover suppose that

$$(t, y) \in [0, t_1(y_0) - \varepsilon_4] \times B_{\varepsilon_3}(y_0) \quad \Longrightarrow \quad g_2(t, y) \neq 0, \tag{4.68}$$

since otherwise $g_2(t_*, y_0) = 0$ for some $t_* \in [0, t_1(y_0) - \varepsilon_4]$. This is a contradiction, because $v_0 > 0$, and thus $g_2(t, y_0) > 0$ for $t \in [0, s_1(y_0)[= [0, t_1(y_0)[$ by Lemma 4.2.3(b). Hence we also have (4.68). Now we claim that $T_1(\bar{y}_0) = t_1(\bar{y}_0)$ for all $\bar{y}_0 \in B_{\varepsilon_3}(y_0) = B_{\varepsilon_3}(y_0) \cap M_+$. For this, fix such $\bar{y}_0 = (\bar{x}_0, \bar{v}_0, \bar{\tau}_0)$. By choice of ε_2 we find $g_2(0, \bar{y}_0) = \bar{v}_0 > 0$, and therefore, by Lemma 4.2.3, $s_1(\bar{y}_0)$ is the unique first positive zero of $g_2(\cdot, \bar{y}_0)$. Thus $s_1(\bar{y}_0) > t_1(y_0) - \varepsilon_4$ by (4.68). Because $g_2(T_1(\bar{y}_0), \bar{y}_0) = 0$, we also must have $0 < s_1(\bar{y}_0) \leq T_1(\bar{y}_0) < t_1(y_0) + \varepsilon_4$. Therefore $s_1(\bar{y}_0) \in]t_1(y_0) - \varepsilon_4, t_1(y_0) + \varepsilon_4[$, and it follows from (4.67) that $T_1(\bar{y}_0) = s_1(\bar{y}_0)$. Next, (4.66) implies

$$\varphi([0, s_1(\bar{y}_0)] \times \{\bar{y}_0\}) = \Phi_{f_+}([0, s_1(\bar{y}_0)] \times \{\bar{y}_0\}) \subset \mathbb{R}^3 \setminus C_+ .$$

If $\bar{y}_1 = \varphi(s_1(\bar{y}_0), \bar{y}_0) \in M_+$, then $\bar{y}_1 \in E_+ \subset C_+$ by Lemma 4.2.3(c), a contradiction. Therefore $\bar{y}_1 \notin M_+$, and thus $j(\bar{y}_0) = 1$ by Lemma 4.2.4, i.e., $T_1(\bar{y}_0) = s_1(\bar{y}_0) = t_1(\bar{y}_0)$. So (a) holds in the first case $v_0 > 0$.

To prove (b) for $v_0 > 0$, we want to apply a suitable open mapping theorem, cf. DEIMLING [62, Cor. 15.2]. For this we note that $\xi_1 : B_{\varepsilon_3}(y_0) \rightarrow \mathbb{R}^3$ is C^2 and of the form

$$\xi_1(y) = \varphi(t_1(y), y) = \Phi_{f_+}(t_1(y), y) = \big(g_1(t_1(y), y), 0, \tau + t_1(y)\big)$$
$$\equiv \big(g_1(t_1(y), y), \tau + t_1(y)\big)$$

for $y = (x, v, \tau) \in B_{\varepsilon_3}(y_0)$, i.e., ξ_1 is in fact two-dimensional. Hence, by the cited result, we only have to check that $\xi_1'(y_0) \in \mathcal{L}(\mathbb{R}^3, \mathbb{R}^2)$ has rank two, since $\xi_1'(\cdot)$ is continuous at y_0, and since it is sufficient to show (b) for $\delta_3 > 0$ being sufficiently small. By the implicit function theorem,

$$\nabla t_1(y_0) = -(\partial_t\, g_2(t_1(y_0), y_0))^{-1}\, \partial_y\, g_2(t_1(y_0), y_0)$$
$$= (x_1 - b^-(\tau_1))^{-1}\big(-\sin(t_1(y_0)), \cos(t_1(y_0)), \partial_\tau\, g_2(t_1(y_0), y_0)\big)$$

In addition,

$$\partial_y\big(g_1(t_1(y), y)\big) = \partial_t\, g_1(t_1(y), y)\, \nabla t_1(y_0) + \partial_y\, g_1(t_1(y), y)$$
$$= g_2(t_1(y), y) + \partial_y\, g_1(t_1(y), y)$$
$$= \partial_y\, g_1(t_1(y), y)$$
$$= \big(\cos(t_1(y_0)), \sin(t_1(y_0)), \partial_\tau\, g_1(t_1(y_0), y_0)\big)$$

for $y \in B_{\varepsilon_3}(y_0)$, and therefore

$$\xi_1'(y_0) = \begin{pmatrix} \cos(t_1(y_0)) & \sin(t_1(y_0)) & \partial_\tau\, g_1(t_1(y_0), y_0) \\[2mm] -\dfrac{\sin(t_1(y_0))}{(x_1 - b^-(\tau_1))} & \dfrac{\cos(t_1(y_0))}{(x_1 - b^-(\tau_1))} & 1 + \dfrac{\partial_\tau\, g_2(t_1(y_0), y_0)}{(x_1 - b^-(\tau_1))} \end{pmatrix}$$

The left 2×2-matrix has determinant $(x_1 - b^-(\tau_1))^{-1} \neq 0$, cf. (4.63), and thus we have proved (a) and (b) in case that $v_0 > 0$.

Next we turn to Case 2: $v_0 = 0$. By (4.64) in particular $y_0 \notin C_+$. Therefore there exists $\varepsilon_5 > 0$ such that $t_1(y) \geq s_1(y) > \varepsilon_5$ for $y \in B_{\varepsilon_5}(y_0) \cap M_+$, by Lemma 4.2.3(g). Let $\bar{y}_0 = (\bar{x}_0, \bar{v}_0, \bar{\tau}_0) = \varphi(\varepsilon_5, y_0) = \Phi_{f_+}(\varepsilon_5, y_0)$. Then $\bar{v}_0 > 0$ and $\bar{y}_0 \in M_+$ by Lemma 4.2.3(b). Moreover, $t_1(\bar{y}_0) = t_1(y_0) - \varepsilon_5$ by Lemma 4.2.5, and by assumption this implies $\varphi(t, \bar{y}_0) = \varphi(t + \varepsilon_5, y_0) \notin E_+$ for

$t \in [0, t_1(\bar{y}_0)]$. Hence the first case applies to $\bar{y}_0$, and we obtain the existence of $\varepsilon_3 > 0$, $\varepsilon_4 \in]0, t_1(\bar{y}_0)/2[$, and of a C^2-function $\bar{T}_1 : B_{\varepsilon_3}(\bar{y}_0) \to]t_1(\bar{y}_0) - \varepsilon_4, t_1(\bar{y}_0) + \varepsilon_4[$ such that $B_{\varepsilon_3}(\bar{y}_0) \subset \{y \in \mathbb{R}^3 : v > 0\} \subset M_+$ and $\bar{T}_1(y) = t_1(y)$ for $y \in B_{\varepsilon_3}(\bar{y}_0)$. Since $\Phi_{f_+}(\varepsilon_5, \cdot) : \mathbb{R}^3 \to \mathbb{R}^3$ is a diffeomorphism, we find $\varepsilon_6 \in]0, \varepsilon_5[$ with $B_{\varepsilon_6}(y_0) \subset \Phi_{f_+}(-\varepsilon_5, B_{\varepsilon_3}(\bar{y}_0))$, thus $\Phi_{f_+}(\varepsilon_5, B_{\varepsilon_6}(y_0)) \subset B_{\varepsilon_3}(\bar{y}_0)$. Define $T_1 : B_{\varepsilon_6}(y_0) \to \mathbb{R}$ through $T_1(y) = \varepsilon_5 + \bar{T}_1(\Phi_{f_+}(\varepsilon_5, y))$. Then T_1 is well-defined and of class C^2, and we have $T_1(y) = t_1(y)$ for $y \in B_{\varepsilon_6}(y_0) \cap M_+$. To see this, fix $y \in B_{\varepsilon_6}(y_0) \cap M_+$. By choice of ε_5 hence $t_1(y) \geq s_1(y) > \varepsilon_5$, and consequently Lemma 4.2.4 (b2) resp. Lemma 4.2.5 yield $\varphi(\varepsilon_5, y) = \Phi_{f_+}(\varepsilon_5, y)$ resp. $t_1(\varphi(\varepsilon_5, y)) = t_1(y) - \varepsilon_5$. Thus $T_1(y) = \varepsilon_5 + \bar{T}_1(\Phi_{f_+}(\varepsilon_5, y)) = \varepsilon_5 + t_1(\varphi(\varepsilon_5, y)) = t_1(y)$, as was to be shown for (a) in case that $v_0 = 0$.

Finally, to prove (b) for $v_0 = 0$, we first note that $\xi_1(\bar{y}_0) = \xi_1(y_0)$ by Lemma 4.1.1(d), take $i = 1$ and $j = 0$ there. Since $\bar{y}_0 \neq y_0$, we may assume above, by shrinking ε_6, that $|\bar{y}_0 - y_0| \geq 2\varepsilon_6$. We claim that

$$\forall \, \delta_3 \in]0, \varepsilon_6[\quad \exists \, \delta > 0 : \quad B_\delta(\bar{y}_0) \subset \varphi([0, \varepsilon_5] \times (B_{\delta_3}(y_0) \cap M_+)). \tag{4.69}$$

To see this, we remark that $g_2(0, y_0) = v_0 = 0$ and $g_2(\varepsilon_5, y_0) = \bar{v}_0 > 0$, cf. (4.65). Hence for every large $n \in \mathbb{N}$ there exists $\sigma_n \in]0, \varepsilon_5[$ such that $g_2(\sigma_n, y_0) = \delta_3/n$. Assume $|\varphi(\sigma_n, y_0) - y_0| \geq \delta_3$ for large n. Then by continuity of φ we obtain $\varphi(\sigma_*, y_0) \in \{y \in \mathbb{R}^3 : v = 0\}$ for some $\sigma_* \in [0, \varepsilon_5]$, i.e., $g_2(\sigma_*, y_0) = 0$. But $\varphi(\sigma_*, y_0) \neq y_0$, and thus $\sigma_* \neq 0$. This is a contradiction, because $s_1(y_0) = t_1(y_0)$ is the first positive zero of $g_2(\cdot, y_0)$, and $0 < \sigma_* \leq \varepsilon_5 < s_1(y_0)$. Therefore we find $\varepsilon_7 \in]0, \delta_3 \wedge \bar{v}_0/2]$ and $y_1 \in \varphi(]0, \varepsilon_5[\times \{y_0\}) \cap B_{\delta_3}(y_0) \cap \{y \in \mathbb{R}^3 : v = \varepsilon_7\}$. Next, there exists $c_1 > 0$ such that $|\Phi_{f_+}(-t, \bar{y}) - \Phi_{f_+}(-t, \bar{y}_0)| \leq c_1 |\bar{y} - \bar{y}_0|$ for $t \in [0, \varepsilon_5]$ and $\bar{y} \in B_1(\bar{y}_0)$. By the properties of y_1 we thus obviously we may choose $\delta > 0$ so small that for $\bar{y} \in B_\delta(\bar{y}_0)$ there exists $t(\bar{y}) \in]0, \varepsilon_5[$ with $\Phi_{f_+}(-t(\bar{y}), \bar{y}) \in B_{\delta_3}(y_0) \cap \{y \in \mathbb{R}^3 : v = \varepsilon_7\}$. We find $\eta \in]0, \varepsilon_5[$ such that for all $\delta \in]0, \eta[$: $\bar{y} \in B_\delta(\bar{y}_0) \Rightarrow t(\bar{y}) \leq \varepsilon_5 - \eta$, since in the opposite case there would exist a sequence $\bar{y}_n \to \bar{y}_0$ with $t(\bar{y}_n) \to \varepsilon_5$ and $\Phi_{f_+}(-t(\bar{y}_n), \bar{y}_n) \in \{y \in \mathbb{R}^3 : v = \varepsilon_7\}$, yielding the contradiction $y_0 = \Phi_{f_+}(-\varepsilon_5, \bar{y}_0) \in \{y \in \mathbb{R}^3 : v = \varepsilon_7\}$. Moreover, for sufficiently small $\delta > 0$ we additionally may assume

$$\Phi_{f_+}\big(-[0, \varepsilon_5 - \eta] \times B_\delta(\bar{y}_0)\big) \subset \{y \in \mathbb{R}^3 : v > 0\} \tag{4.70}$$

Indeed, w.l.o.g. $\Phi_{f_+}(0, B_\delta(\bar{y}_0)) = B_\delta(\bar{y}_0) \subset \{y \in \mathbb{R}^3 : v > 0\}$ and hence, if (4.70) were wrong, there would exist sequences $\sigma_n \in [0, \varepsilon_5 - \eta]$ and $\bar{y}_n \to \bar{y}_0$ such that $\Phi_{f_+}(-\sigma_n, \bar{y}_n) \in \{y \in \mathbb{R}^3 : v = 0\}$. This implies $\varphi(\varepsilon_5 - \sigma_*, y_0) = \Phi_{f_+}(\varepsilon_5 - \sigma_*, y_0) = \Phi_{f_+}(-\sigma_*, \Phi_{f_+}(\varepsilon_5, y_0)) = \Phi_{f_+}(-\sigma_*, \bar{y}_0) \in \{y \in \mathbb{R}^3 : v = 0\}$ for some $\sigma_* \in [0, \varepsilon_5 - \eta]$, a contradiction, since $\varepsilon_5 - \sigma_* \in]0, s_1(y_0)[$. Thus (4.70) holds for small $\delta > 0$, and for such $\delta > 0$ we can also prove (4.69): for $\bar{y} \in B_\delta(\bar{y}_0)$ let $\hat{y} = \Phi_{f_+}(-t(\bar{y}), \bar{y}) \in B_{\delta_3}(y_0) \cap \{y \in \mathbb{R}^3 : v = \varepsilon_7\} \subset B_{\delta_3}(y_0) \cap M_+$. Then $\bar{y} = \varphi(t(\bar{y}), \hat{y})$, because by the above reasoning $t(\bar{y}) \leq \varepsilon_5 - \eta$, and hence

$\Phi_{f_+}(-[0, t(\bar{y})], \bar{y}) \subset \{y \in \mathbb{R}^3 : v > 0\}$ by (4.70). This means $\Phi_{f_+}(t, \hat{y}) \in \{y \in \mathbb{R}^3 : v > 0\}$ for $t \in [0, t(\bar{y})]$. By definition, $s_1(\hat{y})$ is the unique first time with $\Phi_{f_+}(s_1(\hat{y}), \hat{y}) \in \{y \in \mathbb{R}^3 : v = 0\}$. Thus we obtain $t(\bar{y}) < s_1(\hat{y}) \leq t_1(\hat{y})$, and hence $\varphi(t(\bar{y}), \hat{y}) = \Phi_{f_+}(t(\bar{y}), \hat{y}) = \bar{y}$, and this completes the proof of (4.69).

Now we can continue to show (b) in case that $v_0 = 0$ as follows. Fix $\delta_3 \in]0, \varepsilon_6[$ and choose $\delta > 0$ according to (4.69). Additionally we may assume that $B_\delta(\bar{y}_0) \subset \{y \in \mathbb{R}^3 : v > 0\} \subset M_+$. By the first case of (b) applied to $\bar{y}_0$ as above, we find $\delta_4 > 0$ with $B_{\delta_4}(\xi_1(\bar{y}_0)) \cap \{y \in \mathbb{R}^3 : v = 0\} \subset \xi_1(B_\delta(\bar{y}_0) \cap M_+) = \xi_1(B_\delta(\bar{y}_0))$. Therefore by (4.69)

$$B_{\delta_4}(\xi_1(y_0)) \cap \{y \in \mathbb{R}^3 : v = 0\} \subset \xi_1\left(\varphi([0, \varepsilon_5] \times (B_{\delta_3}(y_0) \cap M_+))\right)$$
$$\subset \xi_1(B_{\delta_3}(y_0) \cap M_+). \qquad (4.71)$$

To see the last inclusion, fix $t \in [0, \varepsilon_5]$ and $y \in B_{\delta_3}(y_0) \cap M_+$. Since $\delta_3 < \varepsilon_6 < \varepsilon_5$, we obtain $t_1(y) \geq s_1(y) > \varepsilon_5$ by choice of ε_5, and thus $t \in [0, t_1(y)[$. Since it follows from Lemma 4.1.1(d) that $\xi_1(\varphi(t, y)) = \xi_1(y)$, we have shown (4.71). Hence the proof of (a) and (b) is complete.

Concerning (c), we may restrict ourselves to $t = 0$, since for $\bar{y}_0 = \varphi^{t_0}(y_0)$ with $t_0 \in [0, t_1(y_0)[$ we have $\xi_1(\bar{y}_0) = \xi_1(y_0)$ resp. $t_1(\bar{y}_0) = t_1(y_0) - t_0$ by Lemma 4.1.1(d) resp. Lemma 4.2.5, and also $\varphi^t(\bar{y}_0) \in \mathbb{R}^3 \setminus C_+ \subset \mathbb{R}^3 \setminus E_+$ for $t \in [0, t_1(\bar{y}_0)]$ by (4.64). So we consider only $t = 0$. By (a) we find $\delta_1 > 0$, $\delta_2 \in]0, t_1(y_0)/2[$, and a C^2-function $T_1 : B_{\delta_1}(y_0) \to]t_1(y_0) - \delta_2, t_1(y_0) + \delta_2[$ with $T_1(y) = t_1(y)$ for $y \in B_{\delta_1}(y_0) \cap M_+$. Define $\Xi_1 : B_{\delta_1}(y_0) \to \mathbb{R}^3$ through $\Xi_1(y) = \Phi_{f_+}(T_1(y), y)$. If $y \in B_{\delta_1}(y_0) \cap M_+$, then $\Xi_1(y) = \Phi_{f_+}(t_1(y), y) = \varphi(t_1(y), y) = \xi_1(y)$. In particular, $\Xi_1(y_0) = \xi_1(y_0)$, and hence, by the continuity of Ξ_1, for given $\delta_5 > 0$ there exists $\delta_6 \in]0, \delta_1[$ such that $\Xi_1(B_{\delta_6}(y_0)) \subset B_{\delta_5}(\xi_1(y_0))$. Since $\xi_1(M_+) \subset \{y \in \mathbb{R}^3 : v = 0\}$ by Lemma 4.2.4(c), thus (c) follows, and the proof of the lemma is finished. $\square$

By means of Lemma 4.2.9 and Lemma 4.2.10(a) we obtain

Lemma 4.2.11. *Assumption (A8) from Sect. 4.1.1 is satisfied.*

Thus is remains to show that also $(A10)$ from Sect. 4.1.4 holds.

Lemma 4.2.12. *We have*

$$D_+ = \{y \in M_+ : \varphi^t(y) \notin E_+ \ \forall t \in [0, t_1(y)], \text{ and } \xi_1(y) \notin C_{I \cup \{\infty\}}\}$$
$$\supset \{y \in M_+ : \varphi^t(y) \notin C_{I \cup \{\infty\}} \ \forall t \in [0, t_1(y)]\},$$

and an analogous statement for D_-.

Proof: To show the equality, we remark that for both $y_0 \in D_+$ and y_0 being an element of the right-hand side, the assumptions of Lemma 4.2.10 are satisfied. "$\subseteq$": If $y_0 \in D_+$, then $\xi_1(y_0) \notin C_+$ by (4.64). Choose $l \in$

{left, ∞, right, $-$} such that $\xi_1(y_0) \in M_l$. Then $\xi_1(B_\varepsilon(y_0) \cap M_+) \subset M_l$ for some $\varepsilon > 0$ by assumption. Consequently, Lemma 4.2.10(b) implies that $B_{\delta_4}(\xi_1(y_0)) \cap \{y \in \mathbb{R}^3 : v = 0\} \subset M_l$ for some $\delta_4 > 0$. Thus by definition of C_{left}, C_{right}, and C_+ we obtain $\xi_1(y_0) \notin C_{\text{left}} \cup C_{\text{right}} \cup C_-$, since in the opposite case every neighborhood of $\xi_1(y_0)$ in $\{y \in \mathbb{R}^3 : v = 0\}$ would intersect at least two of the manifolds. "$\supseteq$": Let y_0 be an element of the right-hand side and choose $l \in \{\text{left}, \infty, \text{right}, -\}$ with $\xi_1(y_0) \in M_l$. Because $\xi_1(M_+) \subset \{y \in \mathbb{R}^3 : v = 0\}$, cf. Lemma 4.2.4(c), we find $\delta > 0$ such that $B_\delta(\xi_1(y_0)) \cap \{y \in \mathbb{R}^3 : v = 0\} \subset M_l$. Hence, for fixed $t \in [0, t_1(y_0)[$, we may apply Lemma 4.2.10(c) to find $\delta_6 > 0$ such that $\xi_1(B_{\delta_6}(\varphi^t(y_0)) \cap M_+) \subset M_l$, as was to be shown for $D_+ = \{y \in M_+ : \ldots\}$. The missing inclusion "$\supseteq$" follows from the definition of $C_{I \cup \{\infty\}}$, cf. (4.60), since $E_+ \subset C_{I \cup \{\infty\}}$ and $\xi_1(y) = \varphi(t_1(y), y)$ for $y \notin M_\infty$. $\square$

Hence Lemma 4.2.8, Lemma 4.2.9, and Lemma 4.2.12 imply that as well (A10) from Sect. 4.1.4 holds, with $C_{I \cup \{\infty\}}$ from (4.60). Thus we can summarize this, (4.54), Lemma 4.2.6, Lemma 4.2.8, Lemma 4.2.11, and Lemma 4.2.7 to

Theorem 4.2.2. *If $\gamma \neq 1$, then assumptions $(A1) - (A10)$ from Sect. 4.1.1 and Sect. 4.1.4 are satisfied for the pendulum with dry friction, described through (1.2) in Sect. 4.2.*

Although we have shown this explicitly only in the case $\gamma \in]0, 1[$, cf. (4.35), it may be seen by a similar reasoning that Thm. 4.2.2 also holds for $\gamma \in]1, \infty[$, with the main difference being that then $M_\infty = \emptyset$. The case $\gamma = 1$ is considered to be exceptional, since then M_∞ degenerates to a line.

4.2.3 Dynamics of the pendulum with dry friction

In this section we collect and extend some results obtained earlier concerning the existence and stability of periodic solutions to

$$\ddot{x}(\tau) + x(\tau) + \text{Sgn}\,\dot{x}(\tau) \ni \gamma \sin(\eta\tau) \quad \text{a.e.}, \tag{1.2}$$

with $\gamma, \eta \in [0, \infty[$, and we are also interested in the boundedness of solutions. Therefore the reader sometimes might find it helpful to refer backward to Chap. 3.

We start by recalling Thm. 3.2.7.

Lemma 4.2.13. *(a) If $\eta \neq 1$, then (1.2) has a $(2\pi/\eta)$-periodic solution. This solution is unique for $\gamma > 1$, and denoting it x_*, then*

$$\left(x(\tau) - x_*(\tau)\right)^2 + \left(\dot{x}(\tau) - \dot{x}_*(\tau)\right)^2 \to 0 \quad as \quad \tau \to \infty$$

for every other solution x of (1.2). For $\gamma \in [0,1]$, the $(2\pi/\eta)$-periodic so-
lutions are exactly the equilibria $x_0 \in [\gamma - 1, -\gamma + 1]$. In this case, for any
solution x of (1.2) there is $c = c_x \in [\gamma - 1, -\gamma + 1]$ satisfying

$$(x(\tau) - c)^2 + \dot{x}^2(\tau) \to 0 \quad as \quad \tau \to \infty. \tag{4.72}$$

(b) For $\eta = 1$ the following holds.
  (b1) If $\gamma \in [0,1]$, then the equilibria $x_0 \in [\gamma - 1, -\gamma + 1]$ are the only
    2π-periodic solutions of (1.2). Moreover, (4.72) holds.
  (b2) If $\gamma \in]1, 4/\pi[$, then there exists a unique 2π-periodic solution x_γ :
    $[0, 2\pi] \to \mathbb{R}$ of (1.2). This x_γ is globally asymptotically stable for (1.2),
    i.e., for every $\xi_0 = (x_0, v_0) \in \mathbb{R}^2$ and $\tau_0 \in [0, 2\pi]$, in the notation of
    Thm. 4.2.1,

$$\left(x(\tau_0 + t; \xi_0, \tau_0) - x_\gamma(\tau_0 + t)\right)^2 + \left(\dot{x}(\tau_0 + t; \xi_0, \tau_0) - \dot{x}_\gamma(\tau_0 + t)\right)^2 \to 0 \tag{4.73}$$

as $t \to \infty$. In addition, there are $\tau_1^\gamma \in]0, \pi/2[$ and $\tau_2^\gamma \in]\pi/2, \pi[$ such that
x_γ is given through

$$
\begin{cases}
x_\gamma(\tau) \equiv -1 + \gamma \sin \tau_1^\gamma = b^-(\tau_1^\gamma) & for \quad \tau \in [0, \tau_1^\gamma], \\
\dot{x}_\gamma(\tau) > 0 & for \quad \tau \in]\tau_1^\gamma, \tau_2^\gamma[, \\
x_\gamma(\tau) \equiv 1 - \gamma \sin \tau_1^\gamma = -b^-(\tau_1^\gamma) & for \quad \tau \in [\tau_2^\gamma, \pi], \\
x_\gamma(\tau) = -x_\gamma(\tau + \pi) & for \quad \tau \in \mathbb{R}
\end{cases}
$$

Here τ_1^γ and τ_2^γ are the unique solutions of

$$\cotan\tau_2^\gamma = -\frac{\tau_2^\gamma - \tau_1^\gamma - \sin \tau_1^\gamma \cos \tau_1^\gamma}{\sin^2 \tau_1^\gamma}, \quad 2/\gamma = \frac{1}{2}\frac{\sin^2 \tau_1^\gamma}{\sin \tau_2^\gamma} + \frac{1}{2}\sin \tau_2^\gamma + \sin \tau_1^\gamma,$$

or equivalently

$$\tau_2^\gamma - \tau_1^\gamma = \frac{\sin \tau_1^\gamma}{\sin \tau_2^\gamma}\sin(\tau_2^\gamma - \tau_1^\gamma), \quad 4\sin \tau_2^\gamma = \gamma\left(\sin \tau_1^\gamma + \sin \tau_2^\gamma\right)^2.$$

  (b3) If $\gamma = 4/\pi$, then there are infinitely many 2π-periodic solutions of
    (1.2), and they may be calculated explicitly.
  (b4) If $\gamma \in]4/\pi, \infty[$, then (1.2) has no 2π-periodic solution.

Proof: All assertions besides (4.73) have been verified in Thm. 3.2.7, where
we used t to denote time, instead of τ as we do here. Concerning (4.73),
we note that by Thm. 3.2.4 also $\ddot{x}(\tau) + x(\tau) + \mathrm{Sgn}\,\dot{x}(\tau) \ni \gamma \sin(\tau_0 + \tau)$
a.e. has a unique 2π-periodic solution x_{γ, τ_0} for every $\tau_0 \in [0, 2\pi]$, whence by
forward uniqueness $x_{\gamma, \tau_0} = x_\gamma(\tau_0 + \cdot)$. Because $x(\tau_0 + \cdot; \xi_0, \tau_0)$ is a solution of
$\ddot{x} + x + \mathrm{Sgn}\,\dot{x} \ni \gamma \sin(\tau_0 + \tau)$ a.e., (4.73) is just (3.42) in Thm. 3.2.5 applied
to the problem with forcing $\gamma \sin(\tau_0 + \cdot)$. $\qquad\square$

Next we summarize what has been found in Chap. 3 concerning bound-edness resp. unboundedness of solutions of (1.2). For $\xi_0 = (x_0, v_0) \in \mathbb{R}^2$ and $\tau_0 \in \mathbb{R}$ define

$$\mathcal{E}(\tau; \xi_0, \tau_0) = x^2(\tau; \xi_0, \tau_0) + \dot{x}^2(\tau; \xi_0, \tau_0) \quad \text{for} \quad \tau \in [\tau_0, \infty[; \quad (4.74)$$

see also Thm. 4.2.1 for the notation.

Definition 4.2.1. (a) We say that the solutions of (1.2) are uniformly bounded w.r. to initial values and phases, if for every $R > 0$ there exists a constant $c > 0$ such that

$$\sup \left\{ \mathcal{E}(\tau; \xi_0, \tau_0) : |\xi_0| \le R, \tau_0 \in [0, 2\pi/\eta], \tau \in [\tau_0, \infty[\right\} \le c. \quad (4.75)$$

(b) We call the solutions of (1.2) uniformly unbounded w.r. to initial values and phases, if for every $R > 0$

$$\inf_{|\xi_0| \le R, \tau_0 \in [0, 2\pi/\eta]} \mathcal{E}(\tau; \xi_0, \tau_0) \to \infty \quad \text{as} \quad \tau \to \infty.$$

The following theorem gives an answer to the question for which parameter values the solutions of (1.2) are bounded resp. unbounded.

Theorem 4.2.3. (a) If either $\gamma \in [0, \infty[$ and $\eta \ne 1$, or $\gamma \in [0, 4/\pi]$ and $\eta = 1$, then the solutions of (1.2) are uniformly bounded w.r. to initial values and phases.
(b) If $\gamma \in]4/\pi, \infty[$ and $\eta = 1$, then the solutions of (1.2) are uniformly unbounded w.r. to initial values and phases.

Proof: Ad (a): This follows from Cor. 3.2.1 and Thm. 3.1.2. Ad (b): Cf. Ex. 3.1.1.   □

We recall that the result from (a) already has been obtained before in Rem. 3.1.1(b), although non-uniformly in initial values and phases, and not including $\gamma = 4/\pi$. Note moreover that Thm. 4.2.3(b) implies Lemma 4.2.13(b4). We also call attention to the fact that it is equivalent to have "$\tau_0 \in [0, 2\pi/\eta]$" and "$\tau_0 \in [0, \infty[$" in (4.75), cf. (3.16) on p. 25, since the forcing is $(2\pi/\eta)$-periodic, and thus $\varphi^t(\xi_0, \tau_0 + k(2\pi/\eta)) = \varphi^t(\xi_0, \tau_0) + (0, 0, k(2\pi/\eta))$ for $k \in \mathbb{N}_0$; see Thm. 4.2.1.
We also state a consequence of Thm. 4.2.3(a).

Corollary 4.2.3. If either $\gamma \in [0, \infty[$ and $\eta \ne 1$, or $\gamma \in [0, 4/\pi]$ and $\eta = 1$, then for every $R > 0$ the semiflow φ from Thm. 4.2.1 is Lipschitz continuous on $[0, \infty[\times (\overline{B}_R(0) \times [0, 2\pi/\eta])$.

Proof: Since the uniform boundedness of $\mathcal{E}$ from (4.75) implies via the inclusion (1.2) also the uniform boundedness of the corresponding second derivatives, for $R > 0$ we find $c_1 > 0$ such that for all $|\xi_0| \le R$, $\tau_0 \in [0, 2\pi/\eta]$, and $\tau, \bar{\tau} \in [\tau_0, \infty[$

$$|x(\tau; \xi_0, \tau_0) - x(\bar{\tau}; \xi_0, \tau_0)| + |\dot{x}(\tau; \xi_0, \tau_0) - \dot{x}(\bar{\tau}; \xi_0, \tau_0)| \leq c_1 |\tau - \bar{\tau}|. \quad (4.76)$$

As φ is Lipschitz continuous on $[0, 2\pi/\eta] \times (\overline{B}_R(0) \times [0, 2\pi/\eta])$ by Thm. 4.2.1, there exists $c_2 > 0$ such that for (σ, ξ, τ) and $(\bar{\sigma}, \bar{\xi}, \bar{\tau})$ in this set

$$|\varphi^\sigma(\xi, \tau) - \varphi^{\bar{\sigma}}(\bar{\xi}, \bar{\tau})| \leq c_2 (|\sigma - \bar{\sigma}| + |\xi - \bar{\xi}| + |\tau - \bar{\tau}|). \quad (4.77)$$

Let $\Delta = \Delta_{\varepsilon_\gamma \sin(\eta \cdot)}$ be defined as in (3.12) and fix $t, s \in [0, \infty[$, $\xi_0, \xi_1 \in \overline{B}_R(0)$, and $\tau_0, \tau_1 \in [0, 2\pi/\eta]$ with $\tau_1 \geq \tau_0$. Then by (4.33), Hölder's inequality, and (4.76)

$$|\varphi^t(\xi_0, \tau_0) - \varphi^s(\xi_1, \tau_1)|$$

$$\leq \left\{ \left(x(\tau_0 + t; \xi_0, \tau_0) - x(\tau_1 + s; \xi_1, \tau_1) \right)^2 \right.$$

$$\left. + \left(\dot{x}(\tau_0 + t; \xi_0, \tau_0) - \dot{x}(\tau_1 + s; \xi_1, \tau_1) \right)^2 \right\}^{1/2} + |\tau_0 - \tau_1| + |t - s| \quad (4.78)$$

$$\leq \left\{ \left(x(\tau_0 + t; \xi_0, \tau_0) - x(\tau_1 + t; \xi_0, \tau_0) \right)^2 \right.$$

$$\left. + \left(\dot{x}(\tau_0 + t; \xi_0, \tau_0) - \dot{x}(\tau_1 + t; \xi_0, \tau_0) \right)^2 \right\}^{1/2} + \Delta^{1/2}(\tau_1 + t; \xi_0, \tau_0, \xi_1, \tau_0)$$

$$+ \Delta^{1/2}(\tau_1 + t; \xi_1, \tau_0, \xi_1, \tau_1) + |\tau_0 - \tau_1| + |t - s|$$

$$+ \left\{ \left(x(\tau_1 + t; \xi_1, \tau_1) - x(\tau_1 + s; \xi_1, \tau_1) \right)^2 \right.$$

$$\left. + \left(\dot{x}(\tau_1 + t; \xi_1, \tau_1) - \dot{x}(\tau_1 + s; \xi_1, \tau_1) \right)^2 \right\}^{1/2}$$

$$\leq \Delta^{1/2}(\tau_1 + t; \xi_0, \tau_0, \xi_1, \tau_0) + \Delta^{1/2}(\tau_1 + t; \xi_1, \tau_0, \xi_1, \tau_1)$$

$$+ (1 + c_1)(|\tau_0 - \tau_1| + |t - s|). \quad (4.79)$$

By Ex. 3.1.3 on p. 24 we know that for the pendulum with dry friction the assumptions of Lemma 3.1.2 are satisfied. Thus part (a) of this lemma implies $\Delta^{1/2}(\tau_1 + t; \xi_0, \tau_0, \xi_1, \tau_0) \leq |\xi_0 - \xi_1|$, whereas part (b), (4.33), and (4.77) yield

$$\Delta(\tau_1 + t; \xi_1, \tau_0, \xi_1, \tau_1) \leq \Delta(\tau_1; \xi_1, \tau_0, \xi_1, \tau_1)$$

$$= \left(x(\tau_1; \xi_1, \tau_0) - x(\tau_1; \xi_1, \tau_1) \right)^2 + \left(\dot{x}(\tau_1; \xi_1, \tau_0) - \dot{x}(\tau_1; \xi_1, \tau_1) \right)^2$$

$$= \left| \varphi^{\tau_1 - \tau_0}(\xi_1, \tau_0) - \varphi^0(\xi_1, \tau_1) \right|^2_{\mathbb{R}^3} \leq 4c_2^2 |\tau_0 - \tau_1|^2$$

Inserting both estimates into (4.78) gives the claim.   □

Next we have a closer look on some special cases. In Sect. 4.2.3(i) we consider parameters $\gamma \in [0, 1]$, $\eta \in [0, \infty[$, calling this the "small amplitude case", whereas in Sect. 4.2.3(ii) we investigate the "resonant case" $\gamma \in [0, \infty[$, $\eta = 1$.

4.2.3 (i) The small amplitude case $\gamma \in [0,1]$, $\eta \in [0,\infty[$

Our objective is here to prove that for these parameters $\mathcal{S}_\gamma = [\gamma - 1, 1 - \gamma]$, cf. (4.36), the set of stationary solutions of (1.2), is uniformly asymptotically stable w.r. to initial values and phases. This will imply that the only ergodic measures for the periodic flow φ_{per} from Sect. 4.1.5 are the Dirac measures on $\mathcal{S}_\gamma$, or, to be more precise, their counterparts in the periodic setting; see Cor. 4.2.6 below.

Definition 4.2.2. *The set $\mathcal{S}_\gamma$ is called uniformly asymptotically stable w.r. to initial values and phases, if*

$$\forall R > 0 \ \forall \varepsilon > 0 \ \exists \tau_* > 0 \ \forall \tau \in [\tau_*, \infty[\ \forall |\xi_0| \leq R \ \forall \tau_0 \in [0, 2\pi/\eta]:$$

$$\mathrm{dist}_{\mathbf{R}^2}\Big(\big(x(\tau; \xi_0, \tau_0), \dot{x}(\tau; \xi_0, \tau_0)\big), \mathcal{S}_\gamma \times \{0\} \Big) \leq \varepsilon. \tag{4.80}$$

Theorem 4.2.4. *For $\gamma \in [0,1]$ and $\eta \in]0, \infty[$, $\mathcal{S}_\gamma$ is uniformly asymptotically stable w.r. to initial values and phases.*

Proof: We will assume w.l.o.g. that $\gamma > 0$. Step 1. Fix $R > 0$, $\varepsilon > 0$, $\xi_0 \in \mathbf{R}^2$ with $|\xi_0| \leq R$, and $\tau_0 \in [0, 2\pi/\eta]$. We show that there exists $\tau_*(\xi_0, \tau_0) > 0$ such that for $\tau \in [\tau_*(\xi_0, \tau_0), \infty[$ we have $\big(x(\tau; \xi_0, \tau_0), \dot{x}(\tau; \xi_0, \tau_0)\big) \in U_\varepsilon$, where $U_\varepsilon \subset \mathbf{R}^2$ is any ε-neighborhood of $\mathcal{S}_\gamma \times \{0\}$ in $\mathbf{R}^2$. Let $x(\tau) = x(\tau; \xi_0, \tau_0)$ and $\mathcal{E}(\tau) = \mathcal{E}(\tau; \xi_0, \tau_0)$, cf. (4.74). The first part of the argument (i.e., $\dot{x}(\tau) \to 0$ as $\tau \to \infty$) now is similar to DEIMLING [65, Example 14.2]. With a suitable $w(\tau) \in \mathrm{Sgn}\,\dot{x}(\tau)$ a.e. it follows a.e. on $[\tau_0, \infty[$ that

$$\dot{\mathcal{E}}(\tau) = 2\,\dot{x}(\tau)\,[x(\tau) + \ddot{x}(\tau)] = 2\,\dot{x}(\tau)\,[\gamma \sin(\eta\tau) - w(\tau)] \leq -2(1-\gamma)|\dot{x}(\tau)| \leq 0, \tag{4.81}$$

and therefore $\mathcal{E}(\tau) \leq \mathcal{E}(\tau_0)$ for $\tau \in [\tau_0, \infty[$, and $\mathcal{E}(\tau) \to \mathcal{E}_\infty$ for some $\mathcal{E}_\infty \geq 0$ as $\tau \to \infty$. We claim that $\dot{x}(\tau) \to 0$ as $\tau \to \infty$. If not, then we find $\delta_0 > 0$ and $\tau_{n+1} \geq \tau_n + 1 \geq \tau_0$ such that w.l.o.g. $\dot{x}(\tau_n) \geq 2\delta_0$ for $n \in \mathbb{N}$. Because of the boundedness of $\mathcal{E}$ it follows from the inclusion that $\mathrm{esssup}_{\tau \in [\tau_0, \infty[}|\ddot{x}(\tau)| \leq c_0$ for some finite constant c_0, w.l.o.g. $c_0 > 0$. With $\delta_1 = \delta_0/c_0 \wedge 1 > 0$ we obtain for $n \in \mathbb{N}$ and $\tau \in [\tau_n, \tau_n + \delta_1]$ that $\dot{x}(\tau) = \dot{x}(\tau_n) + \int_{\tau_n}^\tau \ddot{x}(\sigma)d\sigma \geq 2\delta_0 - (\tau - \tau_n)c_0 \geq \delta_0$. Hence (4.81) yields by the property of w

$$\mathcal{E}(\tau_{n+1}) - \mathcal{E}(\tau_n) = 2\int_{\tau_n}^{\tau_{n+1}} \dot{x}(\tau)\,[\gamma \sin(\eta\tau) - w(\tau)]\,d\tau$$

$$\leq 2\int_{\tau_n}^{\tau_n + \delta_1} \dot{x}(\tau)\,[\gamma \sin(\eta\tau) - w(\tau)]\,d\tau$$

$$= 2\int_{\tau_n}^{\tau_n + \delta_1} \dot{x}(\tau)\,[\gamma \sin(\eta\tau) - 1]\,d\tau$$

$$\leq -2\delta_0\left(\delta_1 - \gamma \int_{\tau_n}^{\tau_n + \delta_1} \sin(\eta\tau)\,d\tau \right) \tag{4.82}$$

$$\leq -2\delta_0\delta_1\,(1 - \gamma) \leq 0.$$

Since the left-hand side tends to zero as $n \to \infty$, this gives a contradiction for $\gamma \in [0,1[$. On the other hand, (4.82) implies for $\gamma = 1$ that $\lim_{n\to\infty} \int_{\tau_n}^{\tau_n+\delta_1} \sin(\eta\tau)\, d\tau = \delta_1$, which is also contradictory, because with $\tau_n = k_n(2\pi/\eta) + r_n$, $k_n \in \mathbb{N}_0$, $r_n \in [0, 2\pi/\eta[$, we then would have w.l.o.g. $r_n \to r_\infty \in [0, 2\pi/\eta]$, and thus $0 \leftarrow \delta_1 - \int_{\tau_n}^{\tau_n+\delta_1} \sin(\eta\tau)\, d\tau = \int_{r_n}^{r_n+\delta_1}[1 - \sin(\eta\tau)]\, d\tau \to \int_{r_\infty}^{r_\infty+\delta_1}[1 - \sin(\eta\tau)]\, d\tau$, a contradiction. Consequently, $\dot{x}(\tau) \to 0$ as $\tau \to \infty$, and from this also $x^2(\tau) = \mathcal{E}(\tau) - \dot{x}^2(\tau) \to \mathcal{E}_\infty$. Since we are done if $\mathcal{E}_\infty = 0$, we can assume w.l.o.g. that $(x(\tau), \dot{x}(\tau)) \to (x_\infty, 0)$ as $\tau \to \infty$ for some $x_\infty < 0$. If the claim were wrong, then $(x_\infty, 0) \notin U_\varepsilon$, and thus $x_\infty \le \gamma - 1 - 2\varepsilon_0$ for some $\varepsilon_0 > 0$. Hence also $x(\tau) \le \gamma - 1 - \varepsilon_0$ for $\tau \in [\tau_0 + \tau_1, \infty[$, with a suitable $\tau_1 \ge 0$. Let $y_1 = (x(\tau_0 + \tau_1), \dot{x}(\tau_0 + \tau_1), \tau_0 + \tau_1) = \varphi^{\tau_1}(y_0)$ with $y_0 = (\xi_0, \tau_0)$. In terms of the semiflow, $\varphi^\tau(y_0) \in \{y \in \mathbb{R}^3 : x \le \gamma - 1 - \varepsilon_0\}$ for $\tau \in [\tau_1, \infty[$, and thus in particular $y_1 \in M_+ \cup M_{\mathrm{left}} \cup M_-$. Therefore we find $\tau_2 \ge \tau_1$ such that $y_2 = \varphi^{\tau_2}(y_0) \in M_+$, by Lemma 4.2.2(b) or Lemma 4.2.4(c) for M_-. Let $\tau_i = t_{i-2}(y_2)$ and $y_i = \varphi(\tau_i, y_0)$ for $i \ge 3$. Then $\tau_{i+1} = t_1(y_i) + \tau_i$ and $y_{i+1} = \varphi(t_1(y_i), y_i)$ for $i \ge 2$. Since under the present hypotheses only switchings from M_+ to M_{left} and vice versa are possible, it follows that $y_3 \in M_{\mathrm{left}}$, $y_4 \in M_+$, $y_5 \in M_{\mathrm{left}}$, ... Therefore $y_{2k} \in \{y \in \mathbb{R}^3 : v = 0, x = b^-(\tau), \tau_{\mathrm{mod}} \in [0, \pi/2\eta[\cup]3\pi/2\eta, 2\pi/\eta[, x \le \gamma - 1 - \varepsilon_0\} \subset M_+$ for $k \ge 2$ by assumption and by (4.43) in Lemma 4.2.2(b). Corresponding to ε_0 from above choose $\varepsilon_1 > 0$ and $\delta_1 > 0$ with the property described in Lemma 4.2.3(e). Then $\varphi(\varepsilon_1 + \tau_{2k}, y_0) = \varphi(\varepsilon_1, y_{2k}) \in \{y \in \mathbb{R}^3 : v \ge \delta_1\}$, meaning that $\dot{x}(\tau_0 + \varepsilon_1 + \tau_{2k}) \ge \delta_1$ for $k \ge 2$. Since $y_i \in \{y \in \mathbb{R}^3 : x \le \gamma - 1 - \varepsilon_0\} \subset \mathbb{R}^3 \setminus M_\infty$, we have $t_i(y_2) < \infty$ for $i \in \mathbb{N}_0$, thus $t_i(y_2) \to \infty$ as $i \to \infty$ by Lemma 4.2.6, and hence also $\tau_{2k} \to \infty$ as $k \to \infty$. This contradicts $\dot{x}(\tau) \to 0$ as $\tau \to \infty$. Therefore we must have $(x_\infty, 0) \in U_\varepsilon$, yielding the claim of the first step.

Step 2. By compactness of $\overline{B}_R(0) \times [0, 2\pi/\eta]$ it is sufficient to prove

$$\forall R > 0 \ \forall \varepsilon > 0 \ \forall |\xi_0| \le R \ \forall \bar{\tau}_0 \in [0, 2\pi/\eta] \ \exists \delta > 0 \ \exists \tau_* > 0 :$$
$$\xi_0 \in \overline{B}_R(0) \cap \overline{B}_\delta(\bar{\xi}_0), \ \tau_0 \in [0, 2\pi/\eta] \cap [\bar{\tau}_0 - \delta, \bar{\tau}_0 + \delta] \implies$$
$$\left(x(\tau; \xi_0, \tau_0), \dot{x}(\tau; \xi_0, \tau_0)\right) \in U_\varepsilon \quad \text{for} \quad \tau \in [\tau_*, \infty[, \tag{4.83}$$

where $U_\varepsilon \subset \mathbb{R}^2$ is a ε-neighborhood of $\mathcal{S}_\gamma \times \{0\}$. Let L denote the Lipschitz constant of φ on $[0, \infty[\times (\overline{B}_R(0) \times [0, 2\pi/\eta])$, cf. Cor. 4.2.3. For $\bar{y}_0 = (\bar{\xi}_0, \bar{\tau}_0)$ as above choose $\bar{\tau}_* = \tau_*(\bar{\xi}_0, \bar{\tau}_0) \ge 2\pi/\eta$ according to the first step, with $U_{\varepsilon/2}$ there. Moreover, fix $\varepsilon_1 > 0$ such that $U_{\varepsilon/2} + \overline{B}_{\varepsilon_1}(0) \subset U_\varepsilon$. Define $\delta = \varepsilon_1/(3L)$ and $\tau_* = \bar{\tau}_*$, and let $y_0 = (\xi_0, \tau_0)$ be as in (4.83). Then for $\tau \in [\tau_*, \infty[$

$$\left| \begin{pmatrix} x(\tau;y_0) \\ \dot{x}(\tau;y_0) \end{pmatrix} - \begin{pmatrix} x(\tau;\bar{y}_0) \\ \dot{x}(\tau;\bar{y}_0) \end{pmatrix} \right|_{\mathbb{R}^2} = |\varphi(\tau - \tau_0, y_0) - \varphi(\tau - \bar{\tau}_0, \bar{y}_0)|_{\mathbb{R}^3}$$

$$\leq L\left(|\tau_0 - \bar{\tau}_0| + |y_0 - \bar{y}_0|_{\mathbb{R}^3}\right)$$

$$\leq (1 + \sqrt{2})L\,\delta \leq \varepsilon_1,$$

and hence $\left(x(\tau;\xi_0,\tau_0), \dot{x}(\tau;\xi_0,\tau_0)\right) \in U_\varepsilon$ for $\tau \in [\tau_*, \infty[$. $\square$

Corollary 4.2.4. *Let φ_{per} be the periodic flow on $\mathbb{R}^3_{2\pi/\eta} = \mathbb{R}^2 \oplus S^1_{2\pi/\eta}$ corresponding to φ, see Sect. 4.1.5. Then*

$$\forall R > 0 \ \ \forall \varepsilon > 0 \ \ \exists t_* > 0 \ \ \forall t \in [t_*, \infty[\ \ \forall |\xi_0| \leq R \ \ \forall \tau_0 \in [0, 2\pi/\eta] :$$

$$\mathrm{dist}_{\mathbb{R}^3_{2\pi/\eta}} \left(\varphi^t_{\mathrm{per}}(\xi_0, e^{i\tau_0\eta}), S_\gamma \times \{0\} \times S^1_{2\pi/\eta} \right) \leq \varepsilon.$$

Proof: By (4.31) and (4.33),

$$\varphi^t_{\mathrm{per}}(\xi_0, e^{i\tau_0\eta}) = \left(\varphi_1(t, \xi_0, \tau_0), \varphi_2(t, \xi_0, \tau_0), e^{i[\tau_0 + t]\eta} \right)$$

$$= \left(x(\tau_0 + t; \xi_0, \tau_0), \dot{x}(\tau_0 + t; \xi_0, \tau_0), e^{i[\tau_0 + t]\eta} \right).$$

Hence we may define $t_* = \tau_*$, with τ_* from (4.80). $\square$

In the small amplitude case it is also possible to explicitly find forward invariant neighborhoods of the globally attracting set $S_\gamma \times \{0\}$. An example is given in the following lemma. There we let, for $\varepsilon > 0$ small, $V_\varepsilon = V_\varepsilon^+ \cup (-V_\varepsilon^+) \subset \mathbb{R}^2$, with

$$V_\varepsilon^+ = \left([\gamma - 1, 1 - \gamma + \varepsilon] \times [0, \varepsilon] \cap \overline{B}_{1-\gamma+\varepsilon}(0) \right)$$

$$\bigcup \left([\gamma - 1 - \varepsilon, \gamma - 1] \times [0, \varepsilon] \cap \overline{B}_\varepsilon((\gamma - 1, 0)) \right); \qquad (4.84)$$

see Fig. 4.2 below, where we assumed $\gamma < 1$.

Lemma 4.2.14. *Let $\gamma < 1$ and V_ε be defined by (4.84). Then $S_\gamma \times \{0\} \subset V_\varepsilon$, and V_ε is forward invariant for (1.2). To be more precise, if $\tau_0 \in [0, \infty[$ and $\xi_0 = (x_0, v_0) \in V_\varepsilon$, then also $\left(x(\tau_0 + t; \xi_0, \tau_0), \dot{x}(\tau_0 + t; \xi_0, \tau_0)\right) \in V_\varepsilon$ for $t \in [0, \infty[$.*

Proof: Since V_ε is closed convex, and since for every $a > 0$ the multi-valued mapping $F : [0, a] \times V_\varepsilon \to 2^{\mathbb{R}^2} \setminus \{\emptyset\}$ given through

$$F(\tau, (x, v)) = \{v\} \times \left(\gamma \sin(\eta[\tau_0 + \tau]) - x - \mathrm{Sgn}\,v \right)$$

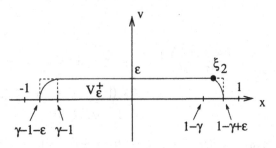

Fig. 4.2. The forward invariant set V_ε

has closed convex values, is ε-δ-usc, linearly bounded, and $F(\tau, \cdot)$ satisfies a dissipativity condition, cf. (2.13), we only have to verify that

$$F(\tau, \xi_0) \cap T_{V_\varepsilon}(\xi_0) \neq \emptyset \quad \text{for} \quad \tau \in [0, a], \quad \xi_0 \in \partial V_\varepsilon, \qquad (4.85)$$

where $T_{V_\varepsilon}(\xi_0) = \overline{\{\lambda(\xi - \xi_0) : \lambda \geq 0, \xi \in V_\varepsilon\}}$ is the associated tangent cone; recall Lemma 2.1.2. Indeed, if (4.85) holds, then it follows from DEIMLING [65, Thm. 5.1] that $(x_{\tau_0}(\tau), \dot{x}_{\tau_0}(\tau)) \in V_\varepsilon$ for $\tau \in [0, a]$, where x_{τ_0} uniquely solves

$$\ddot{x} + x + \text{Sgn}\,\dot{x} \ni \gamma \sin(\eta[\tau_0 + \tau]) \quad \text{a.e. in } [0, a], \quad (x_{\tau_0}(0), \dot{x}_{\tau_0}(0)) = \xi_0.$$

From this we obtain $x(\tau_0 + t; \xi_0, \tau_0) = x_{\tau_0}(t)$ for $t \in [0, a]$, and hence the claim, since $a > 0$ was arbitrary.

To verify (4.85), we note that in case that there exists a unique outer unit normal $n(\xi_0)$ at ξ_0 to V_ε, then $\langle n(\xi_0), \xi \rangle \leq 0$ implies $\xi \in T_{V_\varepsilon}(\xi_0)$. Because ∂V_ε is smooth at $(\gamma - 1 - \varepsilon, 0)$ and $(1 - \gamma + \varepsilon, 0)$, it suffices to check (4.85) for fixed $\xi_0 = (x_0, v_0) \in \partial V_\varepsilon$ with $v_0 \geq 0$. In this case, $\xi_1 = (v_0, -x_0 - 1 + \gamma \sin\theta) \in F(\tau, \xi_0)$, with $\theta = \eta(\tau_0 + \tau)$, and we will show that also $\xi_1 \in T_{V_\varepsilon}(\xi_0)$. Let $\xi_2 = (x_2, \varepsilon)$ denote the intersection point of $\overline{B}_{1-\gamma+\varepsilon}(0)$ with the line $v = \varepsilon$ in $\{(x, v) \in \mathbb{R}^2 : x > 0\}$. Case 1: $x_0 \in [\gamma - 1, x_2[$. Then $n(\xi_0) = (0, 1)$, and hence $\langle n(\xi_0), \xi_1 \rangle = -x_0 - 1 + \gamma \sin\theta \leq 0$. Case 2: $x_0 \in [\gamma - 1 - \varepsilon, \gamma - 1]$, i.e. we are on the left quarter circle. Here $n(\xi_0) = \varepsilon^{-1}(1 - \gamma + x_0, v_0)$ and $\varepsilon \langle n(\xi_0), \xi_1 \rangle = \gamma v_0 (\sin\theta - 1) \leq 0$. Case 3: $x_0 \in]x_2, 1 - \gamma + \varepsilon]$. Again $n(\xi_0) = (1 - \gamma + \varepsilon)^{-1}(x_0, v_0)$ uniquely exists, and thus $(1 - \gamma + \varepsilon)\langle n(\xi_0), \xi_1 \rangle = v_0 (\gamma \sin\theta - 1) \leq 0$. Case 4: $x_0 = x_2$, i.e. $\xi_0 = \xi_2$. Now there is no unique $n(\xi_0)$, but it may be shown that $T_{V_\varepsilon}(\xi_0) = \{\xi \in \mathbb{R}^2 : \langle (0, 1), \xi \rangle \leq 0 \text{ and } \langle \xi_2, \xi \rangle \leq 0\}$. Because $\langle \xi_2, \xi_1 \rangle = \varepsilon(-1 + \gamma \sin\theta) \leq 0$, we find $\xi_1 \in T_{V_\varepsilon}(\xi_0)$. Hence we obtain (4.85) in all cases.   □

4.2.3 (ii) The resonant case $\gamma \in [0, \infty[,\ \eta = 1$

Since we are interested in the bifurcations when γ varies, we first summarize what has been obtained so far.

(a) For $\gamma \in [0, 1]$, the fixed points S_γ are uniformly asymptotically stable w.r. to initial values and phases, cf. Thm. 4.2.4.

(b) For $\gamma \in]4/\pi, \infty[$, the solutions of (1.2) are uniformly unbounded w.r. to initial values and phases, cf. Thm. 4.2.3(b).

So we are left to investigate more closely the case $\gamma \in]1, 4/\pi[$, where we already have some information from Lemma 4.2.13(b2). For fixed $\gamma \in]1, 4/\pi[$ we let

$$gr(x_\gamma) = \{(x_\gamma(\tau), \dot{x}_\gamma(\tau), e^{i\tau}) : \tau \in [0, 2\pi]\} \subset \mathbb{R}^3_{2\pi},$$

cf. Sect. 4.1.5, denote the graph in $\mathbb{R}^3_{2\pi}$ of the unique periodic solution x_γ.

Our aim is to prove the following uniform asymptotic stability of $gr(x_\gamma)$ which will be used in Sect. 4.2.4(iii) below to show that

$$\mathcal{P}_{\text{erg}}(\varphi_{\text{per}}) = \mathcal{P}_{\text{inv}}(\varphi_{\text{per}}) = \{\mu_\gamma\}$$

contains exactly one probability measure μ_γ determined by x_γ.

Lemma 4.2.15. *Let φ_{per} be the periodic flow on $\mathbb{R}^3_{2\pi} = \mathbb{R}^2 \oplus S^1_{2\pi}$ corresponding to φ, see Sect. 4.1.5. Then*

$$\forall R > 0\ \forall \varepsilon > 0\ \exists t_* > 0\ \forall t \in [t_*, \infty[\ \ \forall |\xi_0| \le R\ \forall \tau_0 \in [0, 2\pi] :$$

$$\text{dist}_{\mathbb{R}^3_{2\pi}}\left(\varphi^t_{\text{per}}(\xi_0, e^{i\tau_0}), gr(x_\gamma)\right) \le \varepsilon.$$

Proof: We can proceed similarly to the proof of Thm. 4.2.4, cf. the second step there. Due to the compactness of $\overline{B}_R(0) \times [0, 2\pi]$ it is again sufficient to validate a local result, i.e.,

$$\forall R > 0\ \forall \varepsilon > 0\ \forall |\bar{\xi}_0| \le R\ \forall \bar{\tau}_0 \in [0, 2\pi]\ \exists \delta > 0\ \exists \tau_* > 0 :$$
$$\bar{\xi}_0 \in \overline{B}_R(0) \cap \overline{B}_\delta(\bar{\xi}_0),\ \tau_0 \in [0, 2\pi] \cap [\bar{\tau}_0 - \delta, \bar{\tau}_0 + \delta] \implies$$
$$\varphi^t_{\text{per}}(\xi_0, e^{i\tau_0}) \in U_\varepsilon\ \ \text{for}\ \ t \in [\tau_*, \infty[, \tag{4.86}$$

where $U_\varepsilon = \{z \in \mathbb{R}^3_{2\pi} : d_{\mathbb{R}^3_{2\pi}}(z, \bar{z}) < \varepsilon\ \text{for some}\ \bar{z} \in gr(x_\gamma)\}$ is the ε-neighborhood of $gr(x_\gamma)$ in $\mathbb{R}^3_{2\pi}$, with $d_{\mathbb{R}^3_{2\pi}}$ being the canonical metric on $\mathbb{R}^3_{2\pi}$. To prove (4.86), fix $\bar{y}_0 = (\bar{\xi}_0, \bar{\tau}_0)$ as in the statement. Then by (4.73) in Lemma 4.2.13 there exits τ_* such that for $t \in [\tau_*, \infty[$

$$\left(x(\bar{\tau}_0 + t; \bar{y}_0) - x_\gamma(\bar{\tau}_0 + t)\right)^2 + \left(\dot{x}(\bar{\tau}_0 + t; \bar{y}_0) - \dot{x}_\gamma(\bar{\tau}_0 + t)\right)^2 < \varepsilon^2/4. \tag{4.87}$$

Moreover, because x_γ is uniformly bounded, also $\ddot{x}_\gamma$ is uniformly bounded by (1.2). Thus for some $c_1 > 0$ and all $\sigma, \bar{\sigma} \in [0, \infty[$

$$|x_\gamma(\sigma) - x_\gamma(\bar{\sigma})| + |\dot{x}_\gamma(\sigma) - \dot{x}_\gamma(\bar{\sigma})| \le c_1 |\sigma - \bar{\sigma}|. \tag{4.88}$$

Let L be the Lipschitz constant of φ on $[0, \infty[\times (\overline{B}_R(0) \times [0, 2\pi])$, cf. Cor. 4.2.3. Define $\delta = \varepsilon/2(c_1 + 2)$ and let $y_0 = (\xi_0, \tau_0)$ be as in the statement of (4.86). Thus for $t \in [\tau_*, \infty[$ by (4.87) and (4.88)

$$d_{\mathbb{R}^3_{2\pi}}\left(\varphi^t_{\mathrm{per}}(\xi_0, e^{i\tau_0}), (x_\gamma(\tau_0 + t), \dot{x}_\gamma(\tau_0 + t), e^{i[\tau_0 + t]})\right)$$

$$= \left| \begin{pmatrix} x(\tau_0 + t; y_0) \\ \dot{x}(\tau_0 + t; y_0) \end{pmatrix} - \begin{pmatrix} x_\gamma(\tau_0 + t) \\ \dot{x}_\gamma(\tau_0 + t) \end{pmatrix} \right|_{\mathbb{R}^2}$$

$$\le \left| \begin{pmatrix} x(\tau_0 + t; y_0) \\ \dot{x}(\tau_0 + t; y_0) \end{pmatrix} - \begin{pmatrix} x(\bar{\tau}_0 + t; \bar{y}_0) \\ \dot{x}(\bar{\tau}_0 + t; \bar{y}_0) \end{pmatrix} \right|_{\mathbb{R}^2}$$

$$+ \left| \begin{pmatrix} x(\bar{\tau}_0 + t; \bar{y}_0) \\ \dot{x}(\bar{\tau}_0 + t; \bar{y}_0) \end{pmatrix} - \begin{pmatrix} x_\gamma(\bar{\tau}_0 + t) \\ \dot{x}_\gamma(\bar{\tau}_0 + t) \end{pmatrix} \right|_{\mathbb{R}^2}$$

$$+ \left| \begin{pmatrix} x_\gamma(\bar{\tau}_0 + t) \\ \dot{x}_\gamma(\bar{\tau}_0 + t) \end{pmatrix} - \begin{pmatrix} x_\gamma(\tau_0 + t) \\ \dot{x}_\gamma(\tau_0 + t) \end{pmatrix} \right|_{\mathbb{R}^2}$$

$$\le |\varphi^t(y_0) - \varphi^t(\bar{y}_0)|_{\mathbb{R}^3} + \varepsilon/2 + c_1 |\tau_0 - \bar{\tau}_0| \le (\sqrt{2} + c_1)\delta + \varepsilon/2 < \varepsilon.$$

Hence we have shown (4.86).   □

4.2.4 Lyapunov exponents for the pendulum with dry friction

Let us first sum up what we have got so far. In Sect. 4.2.2 we have shown that assumptions $(A1)$-$(A10)$ from Sects. 4.1.1 and 4.1.4 hold for the pendulum with dry friction, cf. Thm. 4.2.2, if $\gamma \ne 1$. Hence Thm. 4.1.1 yields a canonical cocycle $T : [0, \infty[\times \mathbb{R}^3 \to \mathcal{L}(\mathbb{R}^3)$ corresponding to the semiflow $\varphi : [0, \infty[\times \mathbb{R}^3 \to \mathbb{R}^3$, cf. Sect. 4.2.1. This cocycle in fact lives on a set $G \subset \mathbb{R}^3$ of "good" initial values, cf. Cor. 4.1.3, which has full Lebesgue measure by Thm. 4.1.2. Since (1.2) is $\omega = 2\pi/\eta$-periodic, we also have a canonical periodic flow φ_{per} and cocycle T_{per} on $\mathbb{R}^3_{2\pi/\eta}$ by Sect. 4.1.5, and it is clear that the MET Thm. 4.3.1 also holds in the setting of $\mathbb{R}^3_{2\pi/\eta}$. Moreover, by Cor. 4.2.5 below, the integrability conditions (4.108) in the MET are satisfied.

In this section, we first show in part 4.2.4(i) that for the pendulum with dry friction the upper Lyapunov exponents, cf. (4.1),

$$\lambda^+(y_0, y) = \limsup_{t \to \infty} \left(\frac{1}{t} \ln |T(t, y_0)y|\right), \quad y_0, y \in \mathbb{R}^3, \tag{4.89}$$

are non-positive; see Thm. 4.2.5. In particular, this implies that no positive Lyapunov exponents are possible. Afterwards we turn to some special cases. First, in Sect. 4.2.4(ii), we investigate the "small amplitude case" $\gamma \in [0, 1[$, $\eta \in [0, \infty[$, from Sect. 4.2.3(i) in greater detail, while Sect. 4.2.4(iii) is devoted to the study of the "resonant case" $\gamma \in [0, \infty[$, $\gamma \neq 1$, $\eta = 1$, from Sect. 4.2.3(ii). Note that in comparison to Sects. 4.2.3(i) and 4.2.3(ii) we have to exclude the value $\gamma = 1$ here, since we did not construct a cocycle for this value of γ.

4.2.4 (i) Non-positivity of the Lyapunov exponents

To estimate the upper Lyapunov exponents for (1.2) we investigate closer the form of the cocycle $T : [0, \infty[\times \mathbb{R}^3 \to \mathcal{L}(\mathbb{R}^3)$. We start with some preliminary results.

Let

$$p(\tau) = \gamma \sin(\eta \tau), \quad \tau \in \mathbb{R}, \tag{4.90}$$

denote the forcing, i.e., $b^\pm(\tau) = p(\tau) \pm 1$, cf. (4.37). Define

$$P(t, \bar{t}) = -\sin(t - \bar{t})p(\bar{t}) + \int_{\bar{t}}^{t} \cos(t - s)p(s)\,ds, \quad t, \bar{t} \in \mathbb{R}. \tag{4.91}$$

Therefore

$$P'(t, \bar{t}) = \partial_t P(t, \bar{t}) = -\cos(t - \bar{t})p(\bar{t}) + p(t) - \int_{\bar{t}}^{t} \sin(t - s)p(s)\,ds \text{ for } t, \bar{t} \in \mathbb{R}.$$

We first state some elementary properties of P and P'.

Lemma 4.2.16. *For $0 \leq \bar{t} \leq t$*

$$P(t, \bar{t}) = -\sin(t - \bar{t})b^\pm(\bar{t}) + \int_{\bar{t}}^{t} \cos(t - s)b^\pm(s)\,ds$$

and

$$P'(t, \bar{t}) = -\cos(t - \bar{t})b^\pm(\bar{t}) + b^\pm(t) - \int_{\bar{t}}^{t} \sin(t - s)b^\pm(s)\,ds.$$

Proof: The first equality follows from inserting $b^\pm(\bar{t}) = p(\bar{t}) \pm 1$ and $b^\pm(s) = p(s) \pm 1$, the second by differentiating the first. $\qquad \square$

Lemma 4.2.17. *There exists a constant c_* such that for all $0 \leq \bar{t} \leq t$*

$$|P(t, \bar{t})| \leq c_*(t - \bar{t}) \quad and \quad |P'(t, \bar{t})| \leq c_*(t - \bar{t}).$$

Proof: Since $1 - \cos(x) \leq x$ for $x \in [0, \infty[$, we have e.g.

$$|P'(t, \bar{t})| = \left| [1 - \cos(t - \bar{t})]p(\bar{t}) + [p(t) - p(\bar{t})] - \int_{\bar{t}}^{t} \sin(t - s)p(s)\, ds \right|$$

$$\leq |p|_\infty (t - \bar{t}) + |p'|_\infty (t - \bar{t}) + |p|_\infty (t - \bar{t})$$

$$= (2|p|_\infty + |p'|_\infty)(t - \bar{t}).$$

We thus may define $c_* = 2|p|_\infty + |p'|_\infty$ to obtain both claimed estimates, using $|\sin(x)| \leq x$, $x \in [0, \infty[$, for the first. $\qquad\square$

Next we introduce the notation

$$A(t, \bar{t}, \alpha) = \begin{pmatrix} \cos(t - \bar{t}) & \alpha \sin(t - \bar{t}) & P(t, \bar{t}) \\ -\sin(t - \bar{t}) & \alpha \cos(t - \bar{t}) & P'(t, \bar{t}) \\ 0 & 0 & 1 \end{pmatrix}, \quad t, \bar{t}, \alpha \in \mathbb{R}, \quad (4.92)$$

and in particular

$$A(\alpha = 0) = A(t, t, 0) = \begin{pmatrix} 1 & 0 & 0 \\ 0 & 0 & 0 \\ 0 & 0 & 1 \end{pmatrix}, \quad t \in \mathbb{R}.$$

By direct calculation we obtain

Lemma 4.2.18. *We have $A(t, \bar{t}, 1) \circ A(\bar{t}, s, 1) = A(t, s, 1)$ for $t, \bar{t}, s \in \mathbb{R}$.*

Let $T : [0, \infty[\times \mathbb{R}^3 \to \mathcal{L}(\mathbb{R}^3)$ denote the cocycle for the pendulum with dry friction, constructed according to Thm. 4.1.1 with G from Cor. 4.1.3. The next two lemmas describe the form of $T(\cdot, y)$ for $y \in G = G_{<\infty} \cup G_\infty = G_{<\infty} \cup \bigcup_{i \in \mathbb{N}_0} G_\infty^i$.

Lemma 4.2.19. *Let $y_0 \in G_\infty^0 = D_\infty$. Then $T(t, y_0) = A(\alpha = 0)$ for $t \in [0, \infty[$. In particular, we have $\|T(t, y_0)\|_3 = 1$ for $t \in [0, \infty[$.*

Proof: By the 3rd line of the definition of T in Thm. 4.1.1 we have $T(t, y_0) = \partial_y \varphi^t(y_0)$ for $t \in [0, \infty[$, and $\varphi^t(y) = (x, 0, \tau + t)$ if $y = (x, v, \tau)$ by Lemma 4.2.1. Thus the claim follows. $\qquad\square$

Next we turn to

Lemma 4.2.20. *For $y_0 = (x_0, v_0, \tau_0) \in G \setminus G_\infty^0$ fixed let $t_i = t_i(y_0)$ be the switching times along the trajectory generated by y_0, i.e., $i \in \mathbb{N}_0$ for $y_0 \in G_{<\infty}$, and $i \in \{0, \ldots, i_0 + 1\}$ for $y_0 \in G_\infty^{i_0+1}$ with $i_0 \in \mathbb{N}_0$. Moreover, let $y_{i+1} = \xi_{i+1}(y_0) = \xi_1(y_i) = \varphi(t_{i+1}, y_0) = \varphi(t_1(y_i), y_i) = (x_{i+1}, v_{i+1}, \tau_{i+1})$, as well as*

$$\alpha_{i+1}^+ = \left[1 - \frac{2}{x_{i+1} - b^-(\tau_{i+1})} \right] \quad \text{and} \quad \alpha_{i+1}^- = \left[1 - \frac{2}{b^+(\tau_{i+1}) - x_{i+1}} \right], \quad (4.93)$$

with $i \in \mathbb{N}_0$ for $y_0 \in G_{<\infty}$, and $i \in \{0, \ldots, i_0\}$ for $y_0 \in G_\infty^{i_0+1}$.

There are the following possibilities of switchings at time t_{i+1}:

(a) from $M_+ \to M_-$, i.e., $y_i \in M_+$ and $y_{i+1} \in M_-$. Then for $t \in [t_{i+1}, t_{i+2}[$

$$T(t, y_0) = A(\tau_0 + t, \tau_0 + t_{i+1}, \alpha^+_{i+1}) \circ \left(\lim_{s \to t^-_{i+1}} T(s, y_0) \right).$$

In this case, additionally $\alpha^+_{i+1} \in [0, 1[$.

(b) switching $M_+ \to M_{\text{right}}$, i.e., $y_i \in M_+$ and $y_{i+1} \in M_{\text{right}}$. Then for $t \in [t_{i+1}, t_{i+2}[$

$$T(t, y_0) = A(\alpha = 0) \circ \left(\lim_{s \to t^-_{i+1}} T(s, y_0) \right).$$

(c) switching $M_+ \to M_{\text{left}}$, i.e., $y_i \in M_+$ and $y_{i+1} \in M_{\text{left}}$. Then for $t \in [t_{i+1}, t_{i+2}[$

$$T(t, y_0) = A(\alpha = 0) \circ \left(\lim_{s \to t^-_{i+1}} T(s, y_0) \right).$$

(d) switching $M_+ \to M_\infty$, i.e., $y_i \in M_+$ and $y_{i+1} \in M_\infty$. Then for $t \in [t_{i+1}, t_{i+2}[$

$$T(t, y_0) = A(\alpha = 0) \circ \left(\lim_{s \to t^-_{i+1}} T(s, y_0) \right).$$

(e) switching $M_- \to M_+$, i.e., $y_i \in M_-$ and $y_{i+1} \in M_+$. Then for $t \in [t_{i+1}, t_{i+2}[$

$$T(t, y_0) = A(\tau_0 + t, \tau_0 + t_{i+1}, \alpha^-_{i+1}) \circ \left(\lim_{s \to t^-_{i+1}} T(s, y_0) \right).$$

In this case, additionally $\alpha^-_{i+1} \in [0, 1[$.

(f) switching $M_- \to M_{\text{left}}$, i.e., $y_i \in M_-$ and $y_{i+1} \in M_{\text{left}}$. Then for $t \in [t_{i+1}, t_{i+2}[$

$$T(t, y_0) = A(\alpha = 0) \circ \left(\lim_{s \to t^-_{i+1}} T(s, y_0) \right).$$

(g) switching $M_- \to M_{\text{right}}$, i.e., $y_i \in M_-$ and $y_{i+1} \in M_{\text{right}}$. Then for $t \in [t_{i+1}, t_{i+2}[$

$$T(t, y_0) = A(\alpha = 0) \circ \left(\lim_{s \to t^-_{i+1}} T(s, y_0) \right).$$

(h) switching $M_- \to M_\infty$, i.e., $y_i \in M_-$ and $y_{i+1} \in M_\infty$. Then for $t \in [t_{i+1}, t_{i+2}[$

$$T(t, y_0) = A(\alpha = 0) \circ \left(\lim_{s \to t^-_{i+1}} T(s, y_0) \right).$$

(i) switching $M_{\text{left}} \to M_+$, i.e., $y_i \in M_{\text{left}}$ and $y_{i+1} \in M_+$. Then for $t \in [t_{i+1}, t_{i+2}[$

$$T(t, y_0) = A(\tau_0 + t, \tau_0 + t_{i+1}, 1) \circ \left(\lim_{s \to t_{i+1}^-} T(s, y_0) \right).$$

(j) switching $M_{\text{right}} \to M_-$, i.e., $y_i \in M_{\text{right}}$ and $y_{i+1} \in M_-$. Then for $t \in [t_{i+1}, t_{i+2}[$

$$T(t, y_0) = A(\tau_0 + t, \tau_0 + t_{i+1}, 1) \circ \left(\lim_{s \to t_{i+1}^-} T(s, y_0) \right).$$

For the transition matrices $A_{i+1}(y_0)$ of the linearization, cf. Lemma 4.1.6 resp. Lemma 4.1.7 and Cor. 4.1.6, we obtain

$$(a): \qquad A_{i+1}(y_0) = \begin{pmatrix} 1 & 0 & 0 \\ 0 & \alpha_{i+1}^+ & 0 \\ 0 & 0 & 1 \end{pmatrix},$$

$$(b),(c),(d),(f),(g),(h): \quad A_{i+1}(y_0) = A(\alpha = 0),$$

$$(e): \qquad A_{i+1}(y_0) = \begin{pmatrix} 1 & 0 & 0 \\ 0 & \alpha_{i+1}^- & 0 \\ 0 & 0 & 1 \end{pmatrix},$$

$$(i),(j): \qquad A_{i+1}(y_0) = \text{id}_{\mathbb{R}^3}. \tag{4.94}$$

In particular, $\|A_{i+1}(y_0)\|_3 = 1$ for $y_0 \in G \setminus G_\infty^0$.

Proof: Ad (a): By Rem. 4.1.1 we have $T(t, y_0) = \partial_y \varphi^t(y_0)$ for $t \in]t_{i+1}, t_{i+2}[$, and this may be calculated by means of (4.17) in Cor. 4.1.6. In the notation of this corollary, $f_l = f_+$, $h_l = h_+$, and $f_p = f_-$, cf. (4.39) and (4.59) for the definitions of the functions. Since $\lim_{s \to t_{i+1}^-} \partial_y \varphi^s(y_0) = \lim_{s \to t_{i+1}^-} T(s, y_0)$ by definition of T, we obtain from (4.17) for $t \in]t_{i+1}, t_{i+2}[$

$$T(t, y_0) = \left(\frac{1}{\langle \nabla h_+, f_+ \rangle (y_{i+1})} \left\{ \left[f_- (\Phi_{f_-}(t - t_{i+1}, y_{i+1})) \right. \right. \right.$$
$$\left. - \partial_y \Phi_{f_-}(t - t_{i+1}, y_{i+1}) f_+(y_{i+1}) \right]_\kappa \cdot \left[\nabla h_+(y_{i+1}) \right]_\nu \Big\}_{1 \le \kappa, \nu \le 3}$$
$$+ \partial_y \Phi_{f_-}(t - t_{i+1}, y_{i+1}) \Big) \circ \left(\lim_{s \to t_{i+1}^-} T(s, y_0) \right). \tag{4.95}$$

The right-continuity of $T(\cdot, y_0)$ at t_{i+1}, cf. Lemma 4.1.6, shows that (4.95) in fact holds for $t \in [t_{i+1}, t_{i+2}[$. To derive (a) from (4.95), we first remark that $\langle h_+, f_+ \rangle (y_{i+1}) = -x_{i+1} + b^-(\tau_{i+1})$. Moreover, by the analogue of (4.52) in Lemma 4.2.4, for $t \in \mathbb{R}$ and $y = (x, v, \tau) \in \mathbb{R}^3$

$$\Phi_{f_-}(t,y) = \begin{pmatrix} x\cos t + v\sin t + \int_0^t \sin(t-s)\,b^+(\tau+s)\,ds \\ -x\sin t + v\cos t + \int_0^t \cos(t-s)\,b^+(\tau+s)\,ds \\ \tau + t \end{pmatrix}, \qquad (4.96)$$

and thus by changing variables $\sigma = \tau + s$ in the integrals, and by Lemma 4.2.16 and (4.92),

$$\partial_y\,\Phi_{f_-}(t,y) = \begin{pmatrix} \cos t & \sin t & -\sin(t)\,b^+(\tau) \\ & & +\int_\tau^{\tau+t}\cos(\tau+t-s)\,b^+(s)\,ds \\ -\sin t & \cos t & -\cos(t)\,b^+(\tau) + b^+(\tau+t) \\ & & -\int_\tau^{\tau+t}\sin(\tau+t-s)\,b^+(s)\,ds \\ 0 & 0 & 1 \end{pmatrix}$$

$$= \begin{pmatrix} \cos t & \sin t & P(\tau+t,\tau) \\ -\sin t & \cos t & P'(\tau+t,\tau) \\ 0 & 0 & 1 \end{pmatrix} = A(\tau+t,\tau,1).$$

$$(4.97)$$

Since $y_{i+1} = \xi_1(y_i) \in \{y \in \mathbb{R}^3 : v = 0\}$ by Lemma 4.2.4(c), we have $v_{i+1} = 0$. Therefore we obtain from Lemma 4.2.16 and $b^\pm(\tau) = p(\tau) \pm 1$, cf. (4.37),

$$f_-\big(\Phi_{f_-}(t-t_{i+1},y_{i+1})\big) - \partial_y\,\Phi_{f_-}(t-t_{i+1},y_{i+1})\,f_+(y_{i+1})$$

$$= \begin{pmatrix} -x_{i+1}\sin(t-t_{i+1}) + \int_0^{t-t_{i+1}}\cos(t-t_{i+1}-s)\,b^+(\tau_{i+1}+s)\,ds \\ -x_{i+1}\cos(t-t_{i+1}) - \int_0^{t-t_{i+1}}\sin(t-t_{i+1}-s)\,b^+(\tau_{i+1}+s)\,ds \\ +b^+(\tau_{i+1}+t-t_{i+1}) \\ 1 \end{pmatrix}$$

$$- \begin{pmatrix} [-x_{i+1}+b^-(\tau_{i+1})]\sin(t-t_{i+1}) + P(\tau_{i+1}+t-t_{i+1},\tau_{i+1}) \\ [-x_{i+1}+b^-(\tau_{i+1})]\cos(t-t_{i+1}) + P'(\tau_{i+1}+t-t_{i+1},\tau_{i+1}) \\ 1 \end{pmatrix}$$

$$= \begin{pmatrix} [b^+(\tau_{i+1})-b^-(\tau_{i+1})]\sin(t-t_{i+1}) \\ [b^+(\tau_{i+1})-b^-(\tau_{i+1})]\cos(t-t_{i+1}) \\ 0 \end{pmatrix} = 2\begin{pmatrix} \sin(t-t_{i+1}) \\ \cos(t-t_{i+1}) \\ 0 \end{pmatrix}.$$

Because $\nabla h_+(y) = (0,1,0)^{tr}$, we can insert into (4.95) and find by (4.92) and (4.93) that

$$T(t,y_0)$$

$$= \Bigg(\frac{2}{-x_{i+1}+b^-(\tau_{i+1})}\begin{pmatrix} 0 & \sin(t-t_{i+1}) & 0 \\ 0 & \cos(t-t_{i+1}) & 0 \\ 0 & 0 & 0 \end{pmatrix} + A\big(\tau_{i+1}+t-t_{i+1},\tau_{i+1},1\big)\Bigg)$$

$$\circ \left(\lim_{s\to t_{i+1}^-} T(s,y_0)\right)$$

$$= A(\tau_{i+1} + t - t_{i+1}, \tau_{i+1}, \alpha_{i+1}^+) \circ \left(\lim_{s \to t_{i+1}^-} T(s, y_0) \right).$$

Since $\varphi^t(y) = (*, *, \tau + t)$ for every $y = (x, v, \tau) \in \mathbb{R}^3$ and $t \in [0, t_1(y)[$, we have $\tau_1 = \tau_0 + t_1, \tau_2 = \tau_1 + t_1(y_1), \dots, \tau_{i+1} = \tau_i + t_1(y_i)$, and thus by definition of the t_i in (4.5), $\tau_{i+1} = \tau_0 + t_{i+1}$. This gives the claimed form of $T(t, y_0)$. In addition, note that $y_{i+1} \in \{y \in \mathbb{R}^3 : v = 0\} \cap M_-$ yields $x_{i+1} \geq b^+(\tau_{i+1})$ by definition of M_-, cf. (4.38), and thus $x_{i+1} \geq b^+(\tau_{i+1}) - b^-(\tau_{i+1}) + b^-(\tau_{i+1}) = 2 + b^-(\tau_{i+1})$. It follows that $\alpha_{i+1}^+ \in [0, 1[$.

Ad (b): We have to replace f_- through f_{right} in (4.95). Here $\Phi_{f_{\text{right}}}(t, y) = (x, v, \tau + t)$ for $t \in [0, \infty[$ and $y = (x, v, \tau) \in \mathbb{R}^3$, cf. (4.39). Hence $\partial_y \Phi_{f_{\text{right}}}(t, y) = \text{id}_{\mathbb{R}^3}$, and $v_{i+1} = 0$ yields $f_{\text{right}}(\Phi_{f_{\text{right}}}(t - t_{i+1}, y_{i+1})) - f_+(y_{i+1}) = (0, x_{i+1} - b^-(\tau_{i+1}), 0)^{\text{tr}}$. Therefore we obtain from

$$\langle h_+, f_+ \rangle(y_{i+1}) = -x_{i+1} + b^-(\tau_{i+1})$$

that

$$T(t, y_0) = \left((-1) \begin{pmatrix} 0 & 0 & 0 \\ 0 & 1 & 0 \\ 0 & 0 & 0 \end{pmatrix} + \text{id}_{\mathbb{R}^3} \right) \circ \left(\lim_{s \to t_{i+1}^-} T(s, y_0) \right)$$

$$= A(\alpha = 0) \circ \left(\lim_{s \to t_{i+1}^-} T(s, y_0) \right).$$

Since $f_{\text{left}} = f_\infty = f_{\text{right}}$, this additionally yields (c) and (d), while (e) is analogous to (a), and (f), (g), and (h) are obtained in the same way as (b).

Ad (i): Here $f_l = f_{\text{left}}$ and $f_p = f_+$ in (4.17). By (4.52) in Lemma 4.2.4(b2) we know that $\Phi_{f_+}(t, y)$ is obtained by replacing $b^+(\tau + s)$ through $b^-(\tau + s)$ in (4.96). Thus Lemma 4.2.16 shows that also $\partial_y \Phi_{f_+}(t, y) = A(\tau + t, \tau, 1)$ for $t \in \mathbb{R}$ and $y = (x, v, \tau) \in \mathbb{R}^3$, since by this lemma it plays no role whether one has b^+ or b^- in (4.97). Hence by Lemma 4.2.16

$$f_+(\Phi_{f_+}(t - t_{i+1}, y_{i+1})) - \partial_y \Phi_{f_+}(t - t_{i+1}, y_{i+1}) f_{\text{left}}(y_{i+1})$$

$$= \begin{pmatrix} -x_{i+1} \sin(t - t_{i+1}) + \int_0^{t-t_{i+1}} \cos(t - t_{i+1} - s)\, b^-(\tau_{i+1} + s)\, ds \\ -x_{i+1} \cos(t - t_{i+1}) - \int_0^{t-t_{i+1}} \sin(t - t_{i+1} - s)\, b^-(\tau_{i+1} + s)\, ds \\ +b^-(\tau_{i+1} + t - t_{i+1}) \\ 1 \end{pmatrix}$$

$$- \begin{pmatrix} P(\tau_{i+1} + t - t_{i+1}, \tau_{i+1}) \\ P'(\tau_{i+1} + t - t_{i+1}, \tau_{i+1}) \\ 1 \end{pmatrix}$$

$$= [b^-(\tau_{i+1}) - x_{i+1}] \begin{pmatrix} \sin(t - t_{i+1}) \\ \cos(t - t_{i+1}) \\ 0 \end{pmatrix}.$$

But $y_{i+1} = \xi_1(y_i)$ has $x_{i+1} = b^-(\tau_{i+1})$ by (4.43) in Lemma 4.2.2(b). It follows from (4.17), cf. (4.95), that

$$T(t, y_0) = \partial_y \Phi_{f_+}(t - t_{i+1}, y_{i+1}) \circ \left(\lim_{s \to t_{i+1}^-} T(s, y_0) \right)$$

$$= A(\tau_{i+1} + t - t_{i+1}, \tau_{i+1}, 1) \circ \left(\lim_{s \to t_{i+1}^-} T(s, y_0) \right),$$

and thus we obtain (i), cf. (a). In the same way, (j) can be shown.

Concerning the form of the $A_{i+1}(y_0)$, we remark that $A_{i+1}(y_0)$ is obtained as the limit $t \to t_{i+1}^+$ of the respective matrices above, see Lemma 4.1.6, Lemma 4.1.7, and Cor. 4.1.6. For example, in case (a) we find

$$A_{i+1}(y_0) = \lim_{t \to t_{i+1}^+} A(\tau_0 + t, \tau_0 + t_{i+1}, \alpha_{i+1}^+) = A(\tau_0 + t_{i+1}, \tau_0 + t_{i+1}, \alpha_{i+1}^+)$$

$$= A(\tau_{i+1}, \tau_{i+1}, \alpha_{i+1}^+),$$

and since $P(t, t) = P'(t, t) = 0$ for $t \in \mathbb{R}$, this gives the claimed matrices. So the proof of the lemma is complete.   □

Corollary 4.2.5. *For every probability measure μ the integrability conditions (4.108) in the MET Thm. 4.3.1 are satisfied.*

Proof: By definition of the f_k in (4.39) and of $b^\pm$ in (4.37) we have $|f_k'|_\infty \leq c$ for some constant $c > 0$ and $k \in \{+, -, \text{left}, \text{right}, \infty\}$. Consequently the claim follows from $\|A_{i+1}(y_0)\|_3 = 1$ for $y_0 \in G \setminus G_\infty^0$ and the corresponding $i's$, cf. Lemma 4.2.20, by applying Cor. 4.1.8.   □

Now we are going to estimate the upper Lyapunov exponents for the pendulum with dry friction; see (4.89) for the definition.

Theorem 4.2.5. *For all $y_0, y \in \mathbb{R}^3$*

$$\lambda^+(y_0, y) \leq 0. \tag{4.98}$$

In particular, the pendulum with dry friction has no positive Lyapunov exponent.

Proof: By definition of T, cf. Thm. 4.1.1, and by Lemma 4.2.19 we can restrict ourselves to consideration of $y_0 = (x_0, v_0, \tau_0) \in G \setminus G_\infty^0$. We retain the notation from Lemma 4.2.20. <u>Case 1</u>: $y_0 \in G_\infty^{i_0+1}$ for some $i_0 \in \mathbb{N}_0$. Then $y_{i_0+1} \in D_\infty \subset M_\infty$, $t_{i_0+1} < \infty$, and $t_{i_0+2} = \infty$. We are thus in one of the cases (d) or (h) in Lemma 4.2.20, and thus it follows by taking $i = i_0$ that for $t \in [t_{i_0+1}, \infty[$

$$\|T(t, y_0)\|_3 = \left\| A(\alpha = 0) \circ \left(\lim_{s \to t_{i_0+1}^-} T(s, y_0) \right) \right\|_3 \le c \| A(\alpha = 0) \|_3 = c$$

for some constant $c > 0$. This yields (4.98). Case 2: $y_0 \in G_{<\infty}$. Then $t_i < \infty$, $y_i \in \mathbb{R}^3 \setminus M_\infty$ for all $i \in \mathbb{N}_0$, and also $t_i \to \infty$ as $i \to \infty$ by Lemma 4.2.6. For $i \in \mathbb{N}_0$ define functions $\vartheta_{i+1} : [t_{i+1}, t_{i+2}] \to [t_{i+1}, t_{i+2}]$ and numbers $\beta_{i+1} \in [0, 1]$ as follows:

$$\vartheta_{i+1}(t) = t, \quad \beta_{i+1} = \alpha_{i+1}^+ \quad \text{for} \quad y_i \in M_+, \ y_{i+1} \in M_-,$$

$$\vartheta_{i+1}(t) = t_{i+1}, \quad \beta_{i+1} = 0 \quad \text{for} \quad [y_i \in M_+, \ y_{i+1} \in M_{\text{right}}], \ \text{or}$$
$$[y_i \in M_+, \ y_{i+1} \in M_{\text{left}}], \ \text{or}$$
$$[y_i \in M_-, \ y_{i+1} \in M_{\text{left}}], \ \text{or}$$
$$[y_i \in M_-, \ y_{i+1} \in M_{\text{right}}],$$

$$\vartheta_{i+1}(t) = t, \quad \beta_{i+1} = \alpha_{i+1}^- \quad \text{for} \quad y_i \in M_-, \ y_{i+1} \in M_+,$$

$$\vartheta_{i+1}(t) = t, \quad \beta_{i+1} = 1 \quad \text{for} \quad [y_i \in M_{\text{left}}, \ y_{i+1} \in M_+], \ \text{or}$$
$$[y_i \in M_{\text{right}}, \ y_{i+1} \in M_-].$$

Since $A(\alpha = 0) = A(t, t, 0)$ for $t \in \mathbb{R}$, it follows from Lemma 4.2.20(a), (b), (c), (e), (f), (g), (i), and (j) that for $i \in \mathbb{N}_0$ and $t \in [t_{i+1}, t_{i+2}[$

$$T(t, y_0) = A\big(\tau_0 + \vartheta_{i+1}(t), \tau_0 + t_{i+1}, \beta_{i+1}\big) \circ \left(\lim_{s \to t_{i+1}^-} T(s, y_0) \right). \quad (4.99)$$

Let $T_1^- = \lim_{t \to t_1^-} T(t, y_0)$. Then (4.99) may be iterated to yield for $k \in \mathbb{N}_0$ and $t \in [t_{k+1}, t_{k+2}[$

$$T(t, y_0) = A\big(\tau_0 + \vartheta_{k+1}(t), \tau_0 + t_{k+1}, \beta_{k+1}\big)$$
$$\circ \left(\prod_{i=0}^{k-1} A\big(\tau_0 + \vartheta_{i+1}(t_{i+2}), \tau_0 + t_{i+1}, \beta_{i+1}\big) \right) \circ T_1^-$$
$$= T_k(t, y_0) \circ T_1^-. \quad (4.100)$$

Define $e_1 = (1, 0, 0)^{\text{tr}}$, $e_2 = (0, 1, 0)^{\text{tr}}$, and $e_3 = (0, 0, 1)^{\text{tr}}$. We claim that there exists a constant $c_{**} > 0$ such that for $k \in \mathbb{N}_0$, $t \in [t_{k+1}, t_{k+2}[$, and $j = 1, 2, 3$

$$|T_k(t, y_0)e_j| \le 1 + c_{**} t. \quad (4.101)$$

Indeed, let $c_{**} = \sqrt{2}c_*$, with c_* from Lemma 4.2.17. We are in the situation of Lemma 4.3.4: For $i = 0, \ldots, k - 1$ define $\theta_i = \vartheta_{i+1}(t_{i+2}) - t_{i+1}$, $\alpha_i = \beta_{i+1}$, $P_i = P(\tau_0 + \vartheta_{i+1}(t_{i+2}), \tau_0 + t_{i+1})$ and $P_i' = P'(\tau_0 + \vartheta_{i+1}(t_{i+2}), \tau_0 + t_{i+1})$. Moreover, let $\theta_k = \vartheta_{k+1}(t) - t_{k+1}$, $\alpha_k = \beta_{k+1}$, $P_k = P(\tau_0 + \vartheta_{k+1}(t), \tau_0 + t_{k+1})$ and $P_k' = P'(\tau_0 + \vartheta_{k+1}(t), \tau_0 + t_{k+1})$. Then $A\big(\tau_0 + \vartheta_{i+1}(t_{i+2}), \tau_0 + t_{i+1}, \beta_{i+1}\big) = B(\theta_i, \alpha_i, P_i, P_i') = B_i$ for $i = 0, \ldots, k-1$, and $A\big(\tau_0 + \vartheta_{k+1}(t), \tau_0 + t_{k+1}, \beta_{k+1}\big) = B(\theta_k, \alpha_k, P_k, P'_k) = B_k$, cf. (4.92) and the definition of the B's in (4.110). Since $T_k(t, y_0) = \prod_{i=0}^{k} B_i$ and since $\alpha_i \in [0, 1]$ we first conclude from Lemma 4.3.4(d) and (e) that

$$|T_k(t, y_0)e_1| \le 1 \quad \text{and} \quad |T_k(t, y_0)e_2| \le 1. \tag{4.102}$$

Because $\vartheta_{i+1}(\bar{t}) \le \bar{t} \le t_{i+2}$ for $\bar{t} \in [t_{i+1}, t_{i+2}]$ by definition of ϑ_{i+1}, Lemma 4.2.17 implies that for $i = 0, \ldots, k-1$

$$\left| \begin{pmatrix} P_i \\ P'_i \end{pmatrix} \right|^2_{\mathbb{R}^2} = (P_i)^2 + (P'_i)^2 \le 2c_*^2 \left(\vartheta_{i+1}(t_{i+2}) - t_{i+1} \right)^2 \le 2c_*^2 \left(t_{i+2} - t_{i+1} \right)^2.$$

In the same way we obtain from $\vartheta_{k+1}(t) \le t$ that

$$\left| \begin{pmatrix} P_k \\ P'_k \end{pmatrix} \right|^2_{\mathbb{R}^2} \le 2c_*^2 \left(t - t_{k+1} \right)^2.$$

Hence by Lemma 4.3.4(f)

$$|T_k(t, y_0)e_3| \le 1 + \sum_{i=0}^{k} \left| \begin{pmatrix} P_i \\ P'_i \end{pmatrix} \right|_{\mathbb{R}^2} \le 1 + \sqrt{2}c_* \left(t - t_1 \right) \le 1 + c_{**}t.$$

Together with (4.102) this shows (4.101).

To prove the theorem, fix $y \in \mathbb{R}^3$ and $t \in [t_1, \infty[$, and choose $k \in \mathbb{N}_0$ with $t \in [t_{k+1}, t_{k+2}[$. Then (4.100) and (4.101) imply for some constant $c(y) > 0$

$$|T(t, y_0)y| = |T_k(t, y_0)(T_1^- y)| \le c(y) \left(1 + c_{**}t \right),$$

and thus $\lambda^+(y_0, y) \le 0$. $\square$

4.2.4 (ii) The small amplitude case $\gamma \in [0, 1[, \eta \in [0, \infty[$

We intend to show that here the Lyapunov exponents equal $-\infty, 0, 0$. Since we will derive that now the only ergodic measures are just the δ_{x_0} with $x_0 \in S_\gamma = [\gamma - 1, 1 - \gamma]$, considered on $\mathbb{R}^3_{2\pi/\eta}$, cf. Sect. 4.1.5, this in fact amounts to an application of the MET Thm. 4.3.1 to Dirac measures. For notation used in this section we refer to Sects. 4.1.5 and 4.3.1. The following lemma does not rely on the cocycle, but only on Sect. 4.2.3(i); whence the value $\gamma = 1$ is included.

Lemma 4.2.21. *For $\gamma \in [0, 1]$ and $\eta \in [0, \infty[$ we have*

$$\mathcal{P}_{\text{inv}}(\varphi_{\text{per}}) = \left\{ \mu \in \mathcal{P}(\mathbb{R}^3_{2\pi/\eta}) : \mu = \mu_1 \otimes \delta_0 \otimes \sigma, \ \mu_1 \in \mathcal{P}(\mathbb{R}), \ \mu_1(S_\gamma) = 1 \right\},$$

where σ denotes arclength on $S^1_{2\pi/\eta} = S^1$.

Proof: Let $\alpha : [0, 2\pi/\eta] \ni \tau \to e^{i\tau\eta} \in S^1_{2\pi/\eta}$ denote the canonical map. We consider α also on $\mathbb{R}$, and for simplicity we write $[\tau_1, \tau_2]$ instead of $\alpha([\tau_1, \tau_2])$, i.e., $\sigma([\tau_1, \tau_2]) = \tau_2 - \tau_1$ if $\tau_2 - \tau_1 \in [0, 2\pi/\eta]$. Since $\mathcal{B}(\mathbb{R}^3_{2\pi/\eta})$ is generated by sets of the form $A_1 \times A_2 \times [\tau_1, \tau_2]$ with $A_1, A_2 \in \mathcal{B}(\mathbb{R})$ and $\tau_1, \tau_2 \in \mathbb{R}$, $\tau_2 - \tau_1 \in [0, 2\pi/\eta]$, equality of measures only has to be checked for sets of this type. "$\subseteq$": For $\varepsilon > 0$ define $V_\varepsilon = \{(\xi, e^{i\tau\eta}) \in \mathbb{R}^3_{2\pi/\eta} : \mathrm{dist}_{\mathbb{R}^3_{2\pi/\eta}}((\xi, e^{i\tau\eta}), S_\gamma \times \{0\} \times S^1_{2\pi/\eta}) \leq \varepsilon\}$. Fix $\mu \in \mathcal{P}_{\mathrm{inv}}(\varphi_{\mathrm{per}})$ and $R > 0$. It follows from Cor. 4.2.4 that $\overline{B}_R(0) \times S^1_{2\pi/\eta} \subset (\varphi^t_{\mathrm{per}})^{-1}(V_\varepsilon)$ for sufficiently large t, and thus $\mu(\overline{B}_R(0) \times S^1_{2\pi/\eta}) \leq \mu((\varphi^t_{\mathrm{per}})^{-1}(V_\varepsilon)) = \mu(V_\varepsilon)$. With $R \to \infty$ we arrive at $\mu(V_\varepsilon) = 1$ for every $\varepsilon > 0$, hence $\varepsilon \to 0^+$ implies $\mu(S_\gamma \times \{0\} \times S^1_{2\pi/\eta}) = 1$. This in turn yields for $t \in [0, \infty[$ and $A \in \mathcal{B}(\mathbb{R}^2)$, because the $x_0 \in S_\gamma$ are stationary solutions,

$$\mu(A \times [\tau_1 + t, \tau_2 + t])$$
$$= \mu\Big((\varphi^t_{\mathrm{per}})^{-1}(A \times [\tau_1 + t, \tau_2 + t]) \cap (S_\gamma \times \{0\} \times S^1_{2\pi/\eta})\Big)$$
$$= \mu(\{(x, 0, e^{i\tau\eta}) \in S_\gamma \times \{0\} \times S^1_{2\pi/\eta} : (x, 0, e^{i[\tau+t]\eta}) = \varphi^t_{\mathrm{per}}(x, 0, e^{i\tau\eta})$$
$$\in A \times [\tau_1 + t, \tau_2 + t])\})$$
$$= \mu(\{(x, 0, e^{i\tau\eta}) \in S_\gamma \times \{0\} \times S^1_{2\pi/\eta} : (x, 0, e^{i\tau\eta}) \in A \times [\tau_1, \tau_2]\})$$
$$= \mu(A \times [\tau_1, \tau_2]). \tag{4.103}$$

Let $\bar{\mu}(A) = \mu(A \times S^1_{2\pi/\eta})$ denote the marginal distribution of μ on $\mathcal{B}(\mathbb{R}^2)$, and for fixed A with $\bar{\mu}(A) > 0$ define $\bar{\mu}_A([\tau_1, \tau_2]) = \mu(A \times [\tau_1, \tau_2])/\bar{\mu}(A)$ on $\mathcal{B}(S^1_{2\pi/\eta})$. Then (4.103) implies $\bar{\mu}_A([\tau_1, \tau_2]) = \bar{\mu}_A([\tau_1 + t, \tau_2 + t])$ for $t \in [0, \infty[$, and thus $\bar{\mu}_A = \sigma$, i.e., $\mu = \bar{\mu} \otimes \sigma$. Since $\mu(S_\gamma \times \{0\} \times S^1_{2\pi/\eta}) = 1$, we also have $\bar{\mu}(S_\gamma \times \{0\}) = 1$ and $\bar{\mu}(A) = \bar{\mu}(A \cap (S_\gamma \times \{0\}))$ for $A \in \mathcal{B}(\mathbb{R}^2)$. Hence with $\mu_1(A_1) = \bar{\mu}(A_1 \times \mathbb{R})$ for $A_1 \in \mathcal{B}(\mathbb{R})$ we obtain $\mu_1 \in \mathcal{P}(\mathbb{R})$ and $\bar{\mu} = \mu_1 \otimes \delta_0$, and therefore $\mu = \mu_1 \otimes \delta_0 \otimes \sigma$. "$\supseteq$": This follows directly from the fact that all $x_0 \in S_\gamma$ are equilibria. $\qquad\square$

Corollary 4.2.6. *For $\gamma \in [0, 1]$ and $\eta \in [0, \infty[$ we have $\mathcal{P}_{\mathrm{erg}}(\varphi_{\mathrm{per}}) = \{\delta_{x_0} \otimes \delta_0 \otimes \sigma : x_0 \in S_\gamma\}$, with σ denoting arclength on $S^1_{2\pi/\eta} = S^1$.*

Proof: Since $\{\mu_1 \in \mathcal{P}(\mathbb{R}) : \mu_1(S_\gamma) = 1\}$ may be identified with $\mathcal{P}(S_\gamma) = \mathcal{P}([\gamma - 1, 1 - \gamma])$ and since $\mathcal{P}_{\mathrm{erg}}(\varphi_{\mathrm{per}}) = \mathrm{ext}(\mathcal{P}_{\mathrm{inv}}(\varphi_{\mathrm{per}}))$, cf. Lemma 4.3.1, we clearly obtain from Lemma 4.2.21 that

$$\mathcal{P}_{\mathrm{erg}}(\varphi_{\mathrm{per}}) = \{\mu_1 \otimes \delta_0 \otimes \sigma : \mu_1 \in \mathrm{ext}(\mathcal{P}(S_\gamma))\}.$$

Hence the claim follows from Ex. 4.3.1. $\qquad\square$

Now we are going to apply the MET to the ergodic measures found in Cor. 4.2.6. Since the stationary solutions $x_0 = \gamma - 1$ and $x_0 = 1 - \gamma$ lie on the boundary of M_∞, in particular $y_0 = (x_0, 0, \tau_0) \notin G$ for these both x_0. Hence the cocycle $T(t, y_0)$ does not give information, and thus $x_0 = \gamma - 1$ resp. $x_0 = 1 - \gamma$ will not be taken into account.

Theorem 4.2.6. In the "small amplitude case" $\gamma \in [0, 1[$, $\eta \in [0, \infty[$, the Lyapunov exponents equal $-\infty, 0, 0$. To be more precise, we obtain these values when the MET Thm. 4.3.1 is applied in the periodic setting to one of the ergodic measures $\delta_{x_0} \otimes \delta_0 \otimes \sigma$ from Cor. 4.2.6, with $x_0 \in]\gamma - 1, 1 - \gamma[$.

Proof: Since the integrability conditions hold, cf. Cor. 4.2.5 and Sect. 4.1.5, in particular (4.32), we find $\Gamma_{2\pi/\eta} \subset \mathbb{R}^3_{2\pi/\eta}$ with $(\delta_{x_0} \otimes \delta_0 \otimes \sigma)(\Gamma_{2\pi/\eta}) = 1$ such that the conclusions of the MET are satisfied for $(\xi_0, e^{i\tau_0\eta}) \in \Gamma_{2\pi/\eta}$. In particular, there must be one $(x_0, 0, e^{i\tau_0\eta}) \in \Gamma_{2\pi/\eta}$. Let $y_0 = (x_0, 0, \tau_0) \in \mathbb{R}^3$. Then $y_0 \in D_\infty = G^0_\infty \subset G$ by definition, cf. Cor. 4.1.3. Hence $T(t, y_0) = A(\alpha = 0)$ by Lemma 4.2.19. Letting e_j, $j = 1, 2, 3$, denote the unit vectors in $\mathbb{R}^3$, we obtain $|T(t, y_0)e_1| = 1 = |T(t, y_0)e_3|$ and $|T(t, y_0)e_2| = 0$. Taking into account the criterion (4.109) in the MET and (4.32), we hence have the exponents $-\infty, 0, 0$. □

4.2.4 (iii) The resonant case $\gamma \in [0, \infty[$, $\gamma \neq 1$, $\eta = 1$

We already know from Lemma 4.2.21 resp. Cor. 4.2.6 what are the invariant resp. ergodic probability measures for φ_{per}, when $\gamma \in [0, 1]$. The next lemma shows that there are no such invariant measures, if $\gamma \in]4/\pi, \infty[$. Hence the MET Thm. 4.3.1 may not be applied, i.e., there are no reasonably defined Lyapunov exponents in this case.

Lemma 4.2.22. For $\gamma \in]4/\pi, \infty[$ and $\eta = 1$ we have $\mathcal{P}_{\text{inv}}(\varphi_{\text{per}}) = \emptyset$.

Proof: Suppose that $\mu \in \mathcal{P}_{\text{inv}}(\varphi_{\text{per}})$ and let $A_R = \overline{B}_R(0) \times S^1_{2\pi} \subset \mathbb{R}^3_{2\pi}$ for $R > 0$. We claim that

$$\forall R_1, R_2 > 0 \ \exists t \in [0, \infty[: \quad A_{R_1} \subset (\varphi^t_{\text{per}})^{-1}(A^c_{R_2}). \tag{4.104}$$

Indeed, from Thm. 4.2.3(b) it follows that for the parameter values under consideration the solutions of (1.2) are uniformly unbounded w.r. to initial values and phases, cf. Definition 4.2.1(b). Thus we find $\tau_* \geq 2\pi$ such that for $\tau \in [\tau_*, \infty[$

$$\inf_{|\xi_0| \leq R_1, \tau_0 \in [0, 2\pi]} \mathcal{E}(\tau; \xi_0, \tau_0) \geq 2R_2^2. \tag{4.105}$$

If (4.104) were false, then $\varphi^t_{\text{per}}(\xi_0, e^{i\tau_0}) \in A_{R_2}$ for some $|\xi_0| \leq R_1$ and $\tau_0 \in [0, 2\pi]$, and thus by definition of φ_{per}, cf. (4.31) and (4.33), and by (4.105) we obtain for $t \in [\tau_* - \tau_0, \infty[$

$$R_2^2 \geq x^2(\tau_0 + t; \xi_0, \tau_0) + \dot{x}^2(\tau_0 + t; \xi_0, \tau_0) = \mathcal{E}(\tau_0 + t; \xi_0, \tau_0) \geq 2R_2^2.$$

Hence (4.104) holds, and since $A_R \uparrow \mathbb{R}_{2\pi}^3$ as $R \uparrow \infty$, this gives a contradiction by invariance of μ. $\square$

Thus it remains to investigate $\mathcal{P}_{\mathrm{inv}}(\varphi_{\mathrm{per}})$ for $\gamma \in]1, 4/\pi[$ in the resonant case. Recall from Lemma 4.2.13(b2) that here we have a unique 2π-periodic solution x_γ with uniformly asymptotically stable graph $\mathrm{gr}(x_\gamma) \subset \mathbb{R}_{2\pi}^3$, in the sense of Lemma 4.2.15.

Theorem 4.2.7. *For $\gamma \in]1, 4/\pi[$ and $\eta = 1$ we have*

$$\mathcal{P}_{\mathrm{inv}}(\varphi_{\mathrm{per}}) = \mathcal{P}_{\mathrm{erg}}(\varphi_{\mathrm{per}}) = \{\mu_\gamma\},$$

where for $A \in \mathcal{B}(\mathbb{R}_{2\pi}^3)$

$$\mu_\gamma(A) = \frac{1}{2\pi} \int_0^{2\pi} \delta_{(x_\gamma(\tau), \dot{x}_\gamma(\tau))}(A_\tau)\, d\tau$$

$$= \frac{1}{2\pi} \lambda^1 \Big(\{\tau \in [0, 2\pi] : \varphi_{\mathrm{per}}^\tau(\xi_\gamma, (1,0)) \in A\} \Big). \qquad (4.106)$$

Here $A_\tau = \{\xi \in \mathbb{R}^2 : (\xi, e^{i\tau}) \in A\} \subset \mathbb{R}^2$ and $\xi_\gamma = (x_\gamma(0), \dot{x}_\gamma(0))$, i.e., we have the representation $\varphi_{\mathrm{per}}^\tau(\xi_\gamma, (1,0)) = (x_\gamma(\tau), \dot{x}_\gamma(\tau), e^{i\tau})$ for $\tau \in [0, \infty[$.

Proof: By Lemma 4.3.1 we only have to show that $\mathcal{P}_{\mathrm{inv}}(\varphi_{\mathrm{per}}) = \{\mu_\gamma\}$. For simplicity we write $\delta_{[\tau]} = \delta_{(x_\gamma(\tau), \dot{x}_\gamma(\tau))}$. "$\supseteq$": Let $t \in [0, \infty[$, $t = 2k\pi + t_{\mathrm{mod}}$, with $k \in \mathbb{N}_0$ and $t_{\mathrm{mod}} \in [0, 2\pi[$. From the definitions, and since x_γ is (2π)-periodic, it follows that $\delta_{[\tau]}\Big(((\varphi_{\mathrm{per}}^t)^{-1}A)_\tau\Big) = 1$ iff $\delta_{[\tau+t]}(A_{\tau+t}) = 1$ iff $\delta_{[\tau+t_{\mathrm{mod}}]}(A_{\tau+t_{\mathrm{mod}}}) = 1$, and hence

$$\big(\varphi_{\mathrm{per}}(\mu_\gamma)\big)(A) = \frac{1}{2\pi} \int_0^{2\pi} \delta_{[\tau+t_{\mathrm{mod}}]}(A_{\tau+t_{\mathrm{mod}}})\, d\tau$$

$$= \frac{1}{2\pi} \int_{t_{\mathrm{mod}}}^{2\pi} \delta_{[\tau]}(A_\tau)\, d\tau + \frac{1}{2\pi} \int_{2\pi}^{2\pi+t_{\mathrm{mod}}} \delta_{[\tau]}(A_\tau)\, d\tau$$

$$= \mu_\gamma(A),$$

where the last equality follows again from the periodicity of x_γ. Therefore $\mu_\gamma \in \mathcal{P}_{\mathrm{inv}}(\varphi_{\mathrm{per}})$, and we also note that $\mu_\gamma(\mathrm{gr}(x_\gamma)) = 1$, cf. the notation in Sect. 4.2.3(ii). "$\subseteq$": Fix $\mu \in \mathcal{P}_{\mathrm{inv}}(\varphi_{\mathrm{per}})$. First we will show with an argument analogous to one in the proof of Lemma 4.2.21 that $\mu(\mathrm{gr}(x_\gamma)) = 1$. For $\varepsilon > 0$ define $V_\varepsilon = \{(\xi, e^{i\tau}) \in \mathbb{R}_{2\pi}^3 : \mathrm{dist}_{\mathbb{R}_{2\pi}^3}((\xi, e^{i\tau}), \mathrm{gr}(x_\gamma)) \leq \varepsilon\}$. If $R > 0$ and $\varepsilon > 0$ are fixed, then by Lemma 4.2.15 there exists $t \in [0, \infty[$ such that $\varphi_{\mathrm{per}}^t(\overline{B}_R(0) \times S_{2\pi}^1) \subset V_\varepsilon$. Hence $\mu(\overline{B}_R(0) \times S_{2\pi}^1) \leq \mu(V_\varepsilon)$ by the invariance of μ, and with $R \to \infty$ and $\varepsilon \to 0^+$ this implies $\mu(\mathrm{gr}(x_\gamma)) = 1$. Define the random variable $X : [0, \infty[\to \mathrm{gr}(x_\gamma)$, $X(\tau) = (x_\gamma(\tau), \dot{x}_\gamma(\tau), e^{i\tau})$. Then $X(\tau + 2k\pi) = X(\tau)$ for $k \in \mathbb{N}_0$, and $X(\tau) = X(\sigma)$ iff $\tau = \sigma$, if $\tau, \sigma \in [0, 2\pi[$. For $0 \leq \tau_1 \leq \tau_2$ let $\|[\tau_1, \tau_2]\| = X([\tau_1, \tau_2])$. The periodicity of x_γ

implies $\text{gr}(x_\gamma) \cap (\varphi_{\text{per}}^t)^{-1}(|[\tau_1 + t, \tau_2 + t]|) = |[\tau_1, \tau_2]|$ for $t \in [0, \infty[$. Thus $\mu(|[\tau_1 + t, \tau_2 + t]|) = \mu(|[\tau_1, \tau_2]|)$ by the invariance of μ. Taking $\tau_1 = \tau_2$, this in particular yields $\mu(\{X(\tau)\}) = 0$ for every $\tau \in [0, 2\pi[$, and hence $\mu(|[\tau_1, \tau_2]|) = \mu(X([\tau_1, \tau_2[))$ for $0 \le \tau_1 < \tau_2$. Fix $n \in \mathbb{N}$. Then because of

$$\text{gr}(x_\gamma) = X([0, 2\pi[) = \bigcup_{l=0}^{n-1} X([2\pi l/n, 2\pi(l+1)/n[)$$

we obtain from the translation invariance that $\mu(X([0, 2\pi/n[)) = 1/n$, and thus by translation invariance $\mu(X([\tau, \tau + 2\pi/n])) = \mu(X([\tau, \tau + 2\pi/n[)) = 1/n$ for $\tau \in [0, \infty[$ and $n \in \mathbb{N}$. Hence approximation clearly gives

$$\mu(X([\tau_1, \tau_2[)) = (\tau_2 - \tau_1)/(2\pi), \quad 0 \le \tau_1 \le \tau_2 < 2\pi.$$

Note that by (4.106) for $0 \le \tau_1 \le \tau_2 < 2\pi$ also $\mu_\gamma(X([\tau_1, \tau_2[)) = (2\pi)^{-1}\lambda^1([\tau_1, \tau_2[) = (\tau_2 - \tau_1)/(2\pi)$. To exploit this, we remark that $X : [0, 2\pi[\to \text{gr}(x_\gamma)$ is bijective with measurable inverse $X^{-1} : (\xi, e^{i\tau}) \mapsto \tau$. Hence $X(\mathcal{B}([0, 2\pi[)) = \mathcal{B}(\text{gr}(x_\gamma))$. Let $\mathcal{A} = \{[\tau_1, \tau_2[: 0 \le \tau_1 \le \tau_2 < 2\pi\}$. Then $\mathcal{B}([0, 2\pi[) = \sigma(\mathcal{A})$, the generated σ-algebra, and thus

$$\sigma(X(\mathcal{A})) = \sigma((X^{-1})^{-1}(\mathcal{A})) = (X^{-1})^{-1}(\sigma(\mathcal{A})) = X(\mathcal{B}([0, 2\pi[)) = \mathcal{B}(\text{gr}(x_\gamma)),$$

i.e., $\mathcal{B}(\text{gr}(x_\gamma))$ is generated by sets of the form $X([\tau_1, \tau_2[)$ with $0 \le \tau_1 \le \tau_2 < 2\pi$. From this we conclude that $\mu = \mu_\gamma$ on $\mathcal{B}(\text{gr}(x_\gamma))$, and therefore finally $\mu = \mu_\gamma$, because both measures are supported on $\text{gr}(x_\gamma)$. $\qquad\square$

Hence for $\gamma \in]1, 4/\pi[$ we can apply the MET Thm. 4.3.1 only to the single ergodic measure μ_γ from Thm. 4.2.7. Before doing so, we have to extend the cocycle T_{per} to $\text{gr}(x_\gamma)$, or equivalently, T to

$$G_\gamma = \{\varphi^\tau(\xi_\gamma, 0) = (x_\gamma(\tau), \dot{x}_\gamma(\tau), \tau) : \tau \in [0, \infty[\} \subset \mathbb{R}^3.$$

To explain the reason for this, we first note that although in Sect. 4.2.2 we described the relevant sets M_k, E_k, and D_k only in case that $\gamma \in [0, 1[$, it should be clear what are the analogues in the present situation of $\gamma > 1$. In particular, here $M_\infty = \emptyset$, and thus $G = G_{<\infty}$ and $t_i(y) < \infty$ for all $i \in \mathbb{N}_0$ and $y \in \mathbb{R}^3$, since trajectories cannot end in M_∞. Due to the very special form of x_γ, cf. Lemma 4.2.13(b2) we have

$$
\begin{aligned}
y_0 &= (\xi_\gamma, 0) = (b^-(\tau_1^\gamma), 0, 0) \in M_{\text{left}}, \\
t_1(y_0) &= \tau_1^\gamma, & \xi_1(y_0) &= (b^-(\tau_1^\gamma), 0, \tau_1^\gamma) \in M_+, \\
t_2(y_0) &= \tau_2^\gamma, & \xi_2(y_0) &= (-b^-(\tau_1^\gamma), 0, \tau_2^\gamma) \in M_{\text{right}}, \\
t_3(y_0) &= \pi + \tau_1^\gamma, & \xi_3(y_0) &= (-b^-(\tau_1^\gamma), 0, \pi + \tau_1^\gamma) \in M_-, \\
t_4(y_0) &= \pi + \tau_2^\gamma, & \xi_4(y_0) &= (b^-(\tau_1^\gamma), 0, \pi + \tau_2^\gamma) \in M_{\text{left}}, \\
t_5(y_0) &= 2\pi + \tau_1^\gamma, & \xi_5(y_0) &= (b^-(\tau_1^\gamma), 0, 2\pi + \tau_1^\gamma) \in M_+, \dots, \quad \text{etc.}
\end{aligned}
$$

Hence $\xi_1(y_0) \in E_+$, $\xi_3(y_0) \in E_-$, $\xi_5(y_0) \in E_+$, ..., etc. Consequently, y_0 $G_{<\infty} = G$ here, cf. Lemma 4.1.3, and therefore by the periodicity of x_γ an the forward invariance of G even $G_\gamma \subset \mathbb{R}^3 \setminus G$. Now application of the ME will yield a set $\Gamma_\gamma \subset \mathbb{R}^3_{2\pi}$ with $\mu_\gamma(\Gamma_\gamma) = 1$ such that for every $(\xi, e^{i\tau}) \in I$ the statements of the MET hold. Since μ_γ is supported on $\mathrm{gr}(x_\gamma)$, we thus ca take advantage only for $(\xi, e^{i\tau}) \in \mathrm{gr}(x_\gamma)$. But by definition in Thm. 4.1. $T(t, y) = 0$ for the corresponding $y = (\xi, \tau) \in G_\gamma \subset \mathbb{R}^3 \setminus G$, and we s $T(t, y) = 0$ only to have T defined on the whole of $\mathbb{R}^3$. Thus we of cour: cannot expect that the corresponding Lyapunov exponents $-\infty$, $-\infty$, $-\infty$ a related to the dynamical behaviour.

Although this seems to be a problem, we can remedy in a simple wa since the trajectory of x_γ in $\mathbb{R}^3$, i.e., G_γ, is a forward invariant set. Th allows us to extend the definition of T from G to $G \dot\cup G_\gamma$ and so to find th cocycle being appropriate for an application of the MET.

Lemma 4.2.23. *If $\gamma \in]1, 4/\pi[$, $\eta = 1$, we obtain a cocycle $T : [0, \infty[\times \mathbb{R}^3 -$ $\mathbb{R}^3$ by letting $T(t, y)$ be as before in Thm. 4.1.1 if $y \in \mathbb{R}^3 \setminus G_\gamma$, whereas f $t \in [0, \infty[$ and $y = \varphi^s(y_0) = \varphi^s(\xi_\gamma, 0) \in G_\gamma$ we adopt the following definitio cf. the notation in Sect. 4.2.4. Choose $i \in \mathbb{N}_0$ such that $s \in [t_i(y_0), t_{i+1}(y_0$ and $k \in \mathbb{N}_0$ as well as $l \in \{0, 1, 2, 3\}$ such that $i = 4k + l$.*
Case 1: $l = 0$ or $l = 2$. Let

$$
\begin{aligned}
t \in [0, t_1(y)[: \quad & T(t, y) = A(\alpha = 0), \\
t \in [t_1(y), t_2(y)[: \quad & T(t, y) = A(s + t, t_{i+1}(y_0), 1) \circ A(\alpha = 0), \\
t \in [t_2(y), t_3(y)[: \quad & T(t, y) = A(\alpha = 0) \circ A(t_{i+2}(y_0), t_{i+1}(y_0), 1) \\
& \qquad \circ A(\alpha = 0), \\
t \in [t_3(y), t_4(y)[: \quad & T(t, y) = A(s + t, t_{i+3}(y_0), 1) \circ A(\alpha = 0) \\
& \qquad \circ A(t_{i+2}(y_0), t_{i+1}(y_0), 1) \circ A(\alpha = 0),
\end{aligned}
$$

..., etc.

Case 2: $l = 1$ or $l = 3$. Let

$$
\begin{aligned}
t \in [0, t_1(y)[: \quad & T(t, y) = A(s + t, s, 1), \\
t \in [t_1(y), t_2(y)[: \quad & T(t, y) = A(\alpha = 0) \circ A(t_{i+1}(y_0), s, 1), \\
t \in [t_2(y), t_3(y)[: \quad & T(t, y) = A(s + t, t_{i+2}(y_0), 1) \circ A(\alpha = 0) \\
& \qquad \circ A(t_{i+1}(y_0), s, 1), \\
t \in [t_3(y), t_4(y)[: \quad & T(t, y) = A(\alpha = 0) \circ A(t_{i+3}(y_0), t_{i+2}(y_0), 1) \\
& \qquad \circ A(\alpha = 0) \circ A(t_{i+1}(y_0), s, 1),
\end{aligned}
$$

..., etc.

Proof: Since G_γ is forward invariant, we have to check $T(t + s, y)$: $T(t, \varphi^s(y))T(s, y)$ only for $y \in G_\gamma$. But this is straightforward from the de inition, taking into account that $A(t, \bar{t}, 1) \circ A(\bar{t}, s, 1) = A(t, s, 1)$, cf. Lemm 4.2.18. Note how T is constructed on G_γ: if e.g.

then $l = 1$ above by the form of the trajectory of x_γ, and hence we let $T(t,y) = \partial_y \Phi_{f_+}(t,y) = A(s+t,s,1)$, cf. (4.97), for $t \in [0,t_1(y)[= [0,t_{i+1}(y_0) - s[$. At time $t = t_1(y)$ we must have a switching from M_+ to M_{left}, and therefore we define in accordance with Lemma 4.2.20(c), $T(t,y) = A(\alpha = 0) \circ \left(\lim_{\varrho \to t_1(y)-} T(\varrho, y) \right) = A(\alpha = 0) \circ A(t_{i+1}(y_0), s, 1)$ for $t \in [t_1(y), t_2(y)[= [t_{i+1}(y_0) - s, t_{i+2}(y_0) - s[$, and so on. □

Now we can apply the MET to the extended cocycle, or to be more precise, to its periodic analogue T_{per}, cf. Sect. 4.1.5. Of course in the present case of a uniformly stable periodic solution, the values of T on $\mathbb{R}^3 \setminus G_\gamma$ are of no importance.

Theorem 4.2.8. *If $\gamma \in]1, 4/\pi[$, $\eta = 1$, the Lyapunov exponents equal $-\infty$, $\pi^{-1} \ln |\cos(\tau_2^\gamma - \tau_1^\gamma)|$, 0, i.e., these values are obtained from the MET Thm. 4.3.1 applied with μ_γ from Thm. 4.2.7 and the extended cocycle from Lemma 4.2.23. Here τ_1^γ and τ_2^γ are from Lemma 4.2.13(b2).*

Proof: Again the integrability conditions are satisfied, cf. Cor. 4.2.5 and (4.32). Therefore we find $\Gamma_\gamma \subset \mathbb{R}_{2\pi}^3$ with $\mu_\gamma(\Gamma_\gamma) = 1$ such that for every $(\xi_0, e^{i\tau_0}) \in \Gamma_\gamma$ the results stated in the MET can be used. Since $\mu_\gamma(\text{gr}(x_\gamma)) = 1$, in particular $\Gamma_\gamma \cap \text{gr}(x_\gamma) \neq \emptyset$. Because Γ_γ is forward invariant w.r. to φ_{per}, thus by periodicity $(\xi_\gamma, (1,0)) = \varphi_{\text{per}}^0(\xi_\gamma, (1,0)) \in \Gamma_\gamma$. Hence even $\text{gr}(x_\gamma) \subset \Gamma_\gamma$, and this corresponds to the fact that all the $y \in G_\gamma \subset \mathbb{R}^3$ are admissible to calculate $T(t,y)$ with. In particular, we may take $y = y_0 = (\xi_\gamma, 0)$. From Lemma 4.2.23 we explicitly know the form of $T(t, y_0)$, and with $t_i = t_i(y_0)$ we have $t_{2i} - t_{2i-1} = \tau_2^\gamma - \tau_1^\gamma$ for $i \in \mathbb{N}$, cf. the remarks before Lemma 4.2.23. We claim that for $k \in \mathbb{N}_0$ and $t \in [t_{4k}, t_{4k+1}[$

$$T(t, y_0) = \begin{pmatrix} \cos(\tau_2^\gamma - \tau_1^\gamma)^{2k} & 0 & \alpha_k \\ 0 & 0 & 0 \\ 0 & 0 & 1 \end{pmatrix}, \quad \text{with}$$

$$\alpha_k = \sum_{j=1}^{2k} \cos(\tau_2^\gamma - \tau_1^\gamma)^{2k-j} P(t_{2j}, t_{2j-1}). \tag{4.107}$$

Indeed, we have $\alpha_0 = 0$ and $T(t, y_0) = A(\alpha = 0)$ for $t \in [0, t_1[$. If (4.107) already holds for some $k \in \mathbb{N}_0$, then by definition for $t \in [t_{4(k+1)}, t_{4(k+1)+1}[$, because of $\xi_{4k}(y_0) \in M_{\text{left}}$, $\xi_{4k+1}(y_0) \in M_+$, $\xi_{4k+2}(y_0) \in M_{\text{right}}$, $\xi_{4k+3}(y_0) \in M_-$, and $\xi_{4k+4}(y_0) \in M_{\text{left}}$,

$$T(t, y_0) = A(\alpha = 0) \circ A(t_{4k+4}, t_{4k+3}, 1) \circ A(\alpha = 0) \circ A(t_{4k+2}, t_{4k+1}, 1)$$
$$\circ \left(\lim_{s \to t_{4k+1}^-} T(s, y_0) \right).$$

By multiplication of the matrices and using

$$\cos(t_{4k+4} - t_{4k+3})\cos(t_{4k+2} - t_{4k+1})\cos(\tau_2^\gamma - \tau_1^\gamma)^{2k} = \cos(\tau_2^\gamma - \tau_1^\gamma)^{2(k+1)}$$

as well as the fact that

$$\cos(t_{4k+4} - t_{4k+3})\cos(t_{4k+2} - t_{4k+1})\alpha_k$$
$$+ \cos(t_{4k+2} - t_{4k+1})P(t_{4k+2}, t_{4k+1}) + P(t_{4k+4}, t_{4k+3}) = \alpha_{k+1},$$

this gives (4.107) for $k + 1$. With e_j, $j = 1, 2, 3$, being the unit vectors in $\mathbb{R}^3$, we conclude from the criterion (4.109) in the MET Thm. 4.3.1 and (4.32) in Sect. 4.1.5 that the Lyapunov exponents equal $\{\rho^{(1)}, \rho^{(2)}, \rho^{(3)}\}$, where

$$\rho^{(j)} = \lim_{k \to \infty} \left(\frac{1}{t_{4k}}\right) \ln |T(t_{4k}, y_0)e_j|,$$

since $t_{4k} \to \infty$ as $k \to \infty$. Because of $t_{4k} = (2k - 1)\pi + \tau_2^\gamma$ for $k \in \mathbb{N}$, we thus obtain from (4.107) that $\rho^{(2)} = -\infty$ and $\rho^{(1)} = \pi^{-1} \ln |\cos(\tau_2^\gamma - \tau_1^\gamma)| \in [-\infty, 0[$, the latter, since $\tau_2^\gamma - \tau_1^\gamma \in]0, \pi[$ by Lemma 4.2.13(b2). Finally, it follows from Lemma 4.2.17 that

$$|\alpha_k| \leq \sum_{j=1}^{2k} |P(t_{2j}, t_{2j-1})| \leq c_* \sum_{j=1}^{2k} (t_{2j} - t_{2j-1}) \leq c_* t_{4k},$$

and therefore $|T(t_{4k}, y_0)e_3|^2 = 1 + \alpha_k^2 \leq 1 + c_*^2 t_{4k}^2$ yields

$$0 \leq \left(\frac{1}{t_{4k}}\right) \ln |T(t_{4k}, y_0)e_3| \leq \left(\frac{1}{2t_{4k}}\right) \ln(1 + c_*^2 t_{4k}^2) \to 0 \quad \text{as} \quad k \to \infty,$$

hence $\rho^{(3)} = 0$. $\qquad\qquad\qquad\qquad\qquad\qquad\qquad\qquad\qquad\qquad\qquad \Box$

To summarize Thm. 4.2.6, Thm. 4.2.8, and Lemma 4.2.22, we obtain the following bifurcation picture w.r. to $\gamma \in [0, \infty[$ in the resonant case $\eta = 1$: for $\gamma \in [0, 1[$ the exponents are $-\infty, 0, 0$, whereas they equal $-\infty, \pi^{-1} \ln |\cos(\tau_2^\gamma - \tau_1^\gamma)|, 0$ for $\gamma \in]1, 4/\pi[$. If $\gamma \in]4/\pi, \infty[$, the MET does not apply.

Remark 4.2.1. The exponents are continuous at $\gamma = 1$. To see this, let $\lambda_\gamma^{(2)} = \pi^{-1} \ln |\cos(\tau_2^\gamma - \tau_1^\gamma)|$ with $\tau_1^\gamma \in]0, \pi/2[$ and $\tau_2^\gamma \in]\pi/2, \pi[$ as in Lemma 4.2.13(b2); from there we moreover recall that

$$\cot\tau_2^\gamma = -\frac{\tau_2^\gamma - \tau_1^\gamma - \sin\tau_1^\gamma \cos\tau_1^\gamma}{\sin^2 \tau_1^\gamma}, \quad 2/\gamma = \frac{1}{2}\frac{\sin^2 \tau_1^\gamma}{\sin\tau_2^\gamma} + \frac{1}{2}\sin\tau_2^\gamma + \sin\tau_1^\gamma,$$

or

$$\tau_2^\gamma - \tau_1^\gamma = \frac{\sin\tau_1^\gamma}{\sin\tau_2^\gamma}\sin(\tau_2^\gamma - \tau_1^\gamma), \quad 4\sin\tau_2^\gamma = \gamma (\sin\tau_1^\gamma + \sin\tau_2^\gamma)^2.$$

This shows that $\tau_1^\gamma \uparrow \pi/2$ and $\tau_2^\gamma \downarrow \pi/2$ as $\gamma \to 1^+$. Hence $\lambda_\gamma^{(2)} \to 0$ as $\gamma \to 1^+$. In addition, we note that $\tau_1^\gamma \downarrow 0$ and $\tau_2^\gamma \uparrow \pi$ for $\gamma \to (4/\pi)^-$, and therefore also $\lambda_\gamma^{(2)} \to 0$ as $\gamma \to (4/\pi)^-$. $\diamond$

4.3 Appendix

In this appendix we collect some definitions and results that have been previously used.

4.3.1 On ergodic theory

We introduce some standard notation. For a given measurable semiflow $\varphi : [0,\infty[\times M \to M$ on a complete separable metric manifold M we let $\mathcal{P}_{\text{inv}} = \mathcal{P}_{\text{inv}}(\varphi) = \{\mu : \forall t \in [0,\infty[: \varphi^t(\mu) = \mu\}$ denote the set of all $(\varphi^t)_{t \geq 0}$-invariant probability measures on $(M, \mathcal{B}(M))$, where $\mathcal{B}(M)$ is the Borel σ-algebra and $(\varphi^t(\mu))(A) = \mu((\varphi^t)^{-1}(A))$. In general, $\mathcal{P}_{\text{inv}}$ can be empty, cf. Lemma 4.2.22. Observe that $\mathcal{P}_{\text{inv}}$ is a convex set, i.e., if $\nu, \bar{\nu} \in \mathcal{P}_{\text{inv}}$, then also $\lambda\nu + (1-\lambda)\bar{\nu} \in \mathcal{P}_{\text{inv}}$ for $\lambda \in [0,1]$. Hence it makes sense to consider the extreme points $\text{ext}(\mathcal{P}_{\text{inv}})$ of $\mathcal{P}_{\text{inv}}$, consisting of those $\mu \in \mathcal{P}_{\text{inv}}$ such that if $\mu = \lambda\nu + (1-\lambda)\bar{\nu}$ for some $\nu, \bar{\nu} \in \mathcal{P}_{\text{inv}}$ and $\lambda \in [0,1]$, then necessarily $\lambda = 0$ or $\lambda = 1$. This means the extreme points are the "end points" of line segments in $\mathcal{P}_{\text{inv}}$.

Next, we call $(\varphi^t)_{t \geq 0}$-invariant a set $A \in \mathcal{B}(M)$, if $(\varphi^t)^{-1}(A) = A$ for all $t \in [0,\infty[$, and the set of ergodic invariant probability measures is

$$\mathcal{P}_{\text{erg}} = \mathcal{P}_{\text{erg}}(\varphi)$$
$$= \{\, \mu \in \mathcal{P}_{\text{inv}} : A \in \mathcal{B}(M) \text{ is } (\varphi^t)_{t \geq 0} - \text{invariant} \ \Rightarrow \ \mu(A) \in \{0,1\} \,\}.$$

Since we did not find a suitable reference, we also include a proof of

Lemma 4.3.1. $\mathcal{P}_{\text{erg}} = \text{ext}(\mathcal{P}_{\text{inv}})$ *is the set of extreme points of the convex set* $\mathcal{P}_{\text{inv}}$.

Proof: Let $\mathcal{I} = \{A \in \mathcal{B}(M) : A \text{ is } (\varphi^t)_{t \geq 0} - \text{invariant}\}$ denote the σ-algebra of invariant sets. Then $\mu \in \mathcal{P}_{\text{erg}}$ iff $\mu \in \mathcal{P}_{\text{inv}}$ and μ is trivial on $\mathcal{I}$. "$\supseteq$": If $\mu(A) \in]0,1[$ for some $\mu \in \text{ext}(\mathcal{P}_{\text{inv}})$ and $A \in \mathcal{I}$, then by the invariance of A and μ, $\nu(B) = \mu(B \cap A)/\mu(A)$ defines a probability measure $\nu \in \mathcal{P}_{\text{inv}}$, and the same is true for $\bar{\nu}(B) = \mu(B \cap A^c)/\mu(A^c)$. Because of $\mu = \mu(A)\nu + (1-\mu(A))\bar{\nu}$, μ cannot be extreme. "$\subseteq$": We intend to apply GEORGII [91, Cor. 7.4]. To see that this is possible, define for every $t \in [0,\infty[$ a probability kernel π_t from $\mathcal{B}(M)$ to $\mathcal{B}(M)$ through $\pi_t(A, y) = 1_A(\varphi^t(y))$. Then $\mu\pi_t = \varphi^t(\mu)$ for $\mu \in \mathcal{P}(M)$, and thus $\mathcal{P}_\Pi = \{\mu \in \mathcal{P}(M) : \forall t \in [0,\infty[: \mu\pi_t = \mu\} = \mathcal{P}_{\text{inv}}$.

Therefore, if we want to show that a fixed $\mu \in \mathcal{P}_{\text{erg}} \subset \mathcal{P}_{\text{inv}}$ is ergodic, we have to verify that μ is trivial on $\mathcal{I}_\Pi(\mu) = \{A \in \mathcal{B}(M) : \forall t \in [0, \infty[:$ $1_A(\varphi^t(y)) = 1_A \ \mu-\text{a.e.}\,\}$. For that, fix $A \in \mathcal{I}_\Pi(\mu)$. Then μ-a.e. for all $i \in \mathbb{N}_0$ we have $1_A(\varphi^i(\cdot)) = 1_A$ by assumption. Consequently, the Birkhoff Ergodic Theorem, cf. POLLICOTT [181, Thm. 1.2], implies $1_A(\cdot) = \mu(A)$ μ-a.e., and thus $\mu(A) \in \{0, 1\}$ as required. Hence GEORGII [91, Cor. 7.4] applies to yield $\mu \in \text{ext}(\mathcal{P}_{\text{inv}})$. $\square$

Example 4.3.1. We have $\text{ext}(\mathcal{P}([a, b])) = \{\delta_x : x \in [a, b]\}$.

Proof: Clearly, "$\supseteq$" holds. "$\subseteq$": Fix $\mu \in \text{ext}(\mathcal{P}([a, b]))$. It is known that $\mathcal{P} = \mathcal{P}([a, b])$ is convex and a compact Hausdorff space in the weak topology. Moreover, $\Delta : [a, b] \to \mathcal{P}$, $x \mapsto \delta_x$, is $\mathcal{B}([0, a]) - \mathcal{B}(\mathcal{P})$-measurable, and hence $\bar{\mu} = \Delta(\mu)$ is a probability measure on $(\mathcal{P}, \mathcal{B}(\mathcal{P}))$, given through $\bar{\mu}(\mathcal{M}) = \mu(\{x \in [a, b] : \delta_x \in \mathcal{M}\})$ for $\mathcal{M} \in \mathcal{B}(\mathcal{P})$. From the transformation rule we obtain

$$\int_{\mathcal{P}} \nu \, \bar{\mu}(d\nu) = \int_{\mathcal{P}} \nu \, \Delta(\mu)(d\nu) = \int_{[a,b]} \delta_x \, \mu(dx) = \mu,$$

i.e., $\bar{\mu}$ has barycenter μ. Because also $\int_{\mathcal{P}} \nu \, \delta_\mu(d\nu) = \mu$, and since μ is extreme, we conclude from ALFSEN [6, Cor. I.2.4] that $\bar{\mu} = \delta_\mu$. By applying this to $\mathcal{M} = \{\delta_x : x \in [a, b]\}$ we find $1 = \delta_\mu(\mathcal{M})$, and consequently $\mu \in \mathcal{M}$. $\square$

Define $\ln^+(s) = \max\{\ln(s), 0\}$ for $s \in [0, \infty[$.

Theorem 4.3.1 (Oseledets' Multiplicative Ergodic Theorem). *Let* $\varphi : [0, \infty[\times\mathbb{R}^d \to \mathbb{R}^d$ *be a measurable semiflow, and* $\mu \in \mathcal{P}_{\text{inv}}$. *Moreover, assume that* $T : [0, \infty[\times\mathbb{R}^d \to \mathcal{L}(\mathbb{R}^d)$ *is a measurable cocycle for* $(\varphi^t)_{t \geq 0}$, *i.e.,* T *is measurable and* $T(t+s, y) = T(t, \varphi^s(y)) \, T(s, y)$ *for* $t, s \in [0, \infty[$ *and* $y \in \mathbb{R}^d$.

Then, if the integrability conditions

$$\sup_{\theta \in [0,1]} \ln^+ |T(\theta, \cdot)| \in L^1(\mathbb{R}^d, \mu) \quad and \quad \sup_{\theta \in [0,1]} \ln^+ |T(1 - \theta, \varphi^\theta(\cdot))| \in L^1(\mathbb{R}^d, \mu)$$

(4.108)

are satisfied, there exists $\Gamma \subset \mathbb{R}^d$ *with* $\mu(\Gamma) = 1$ *such that* $\varphi^t(\Gamma) \subset \Gamma$ *for* $t \in [0, \infty[$, *and the following assertions hold for all* $y \in \Gamma$:

(a) $\lim_{t \to \infty} \left(T(t, y)^* T(t, y)\right)^{1/2t} = \Lambda_y$ *exists.*

(b) Let $\exp \lambda_y^{(1)} < \ldots < \exp \lambda_y^{(k)}$ *be the eigenvalues of* Λ_y *(where* $k = k(y)$, *the* $\lambda_y^{(i)}$ *are real and* $\lambda_y^{(1)}$ *may be* $-\infty$), *and* $U_y^{(1)}, \ldots, U_y^{(k)}$ *the corresponding eigenspaces. Let* $m_y^{(i)} = \dim U_y^{(i)}$. *Then the functions* $y \mapsto \lambda_y^{(i)}, m_y^{(i)}$ *are* $(\varphi^t)_{t \geq 0}$-*invariant. With* $V_y^{(0)} = \{0\}$ *and* $V_y^{(i)} = U_y^{(1)} + \ldots + U_y^{(i)}$ *one has for* $i = 1, \ldots, k$

$$\lim_{t \to \infty} \frac{1}{t} \ln |T(t, y)\bar{y}| = \lambda_y^{(i)} \quad when \quad \bar{y} \in V_y^{(i)} \setminus V_y^{(i-1)}.$$

If in addition $\mu \in \mathcal{P}_{\text{erg}}$, then $y \mapsto k(y)$ resp. $y \mapsto \lambda_y^{(i)}$ are constant on Γ, and denoted by k resp. $\lambda^{(i)}$. In this case, if there are linearly independent $\bar{y}_1, \ldots, \bar{y}_d \in \mathbb{R}^d$ and sequences $(t_n^j)_{n \in \mathbb{N}}$ for $j = 1, \ldots, d$ with $t_n^j \to \infty$ as $n \to \infty$ such that $\lim_{n \to \infty} \left(\frac{1}{t_n^j} \right) \ln |T(t_n^j, y) \bar{y}_j| = \rho^{(j)} \in [-\infty, \infty[$ for $j = 1, \ldots, d$, then

$$\{\lambda^{(i)} : 1 \leq i \leq k\} = \{\rho^{(j)} : 1 \leq j \leq d\}. \tag{4.109}$$

Proof: See RUELLE [196, Thm. B3, p. 304]. The ergodic part follows like in the discrete case, cf. LEDRAPPIER [126, Thm. 3.1], while the added criterion is straightforward.   □

4.3.2 Some technical lemmas

Lemma 4.3.2. Let $\gamma, \eta > 0$, $\tau_*, x_* \in \mathbb{R}$, and $v_* > 0$. Then for the solution of $\ddot{x} + x = b^-(\tau) = \gamma \sin(\eta \tau) - 1$ in $[\tau_*, \infty[$, $x(\tau_*) = x_*$, $\dot{x}(\tau_*) = v_*$, there exists a time $\tau_{**} > \tau_*$ such that $\dot{x}(\tau_{**}) < 0$.

Proof: We assume that $\dot{x}(\tau) \geq 0$ in $]\tau_*, \infty[$. If we suppose that $x(\tilde{\tau}) + 1 = \delta > 0$ for some $\tilde{\tau} \in [\tau_*, \infty[$, then for $\tau \in [\tilde{\tau}, \infty[$ we have $\ddot{x}(\tau) + \delta \leq \ddot{x}(\tau) + x(\tau) + 1 = \gamma \sin(\eta \tau)$, and thus $0 \leq \dot{x}(\tau) \leq \dot{x}(\tilde{\tau}) + 2\gamma/\eta - \delta(\tau - \tilde{\tau})$ in $[\tilde{\tau}, \infty[$, a contradiction. Consequently, $x(\tau) + 1 \leq 0$ in $[\tau_*, \infty[$, and hence $\ddot{x}(\tau) \geq \ddot{x}(\tau) + x(\tau) + 1 = \gamma \sin(\eta \tau)$ in $[\tau_*, \infty[$. Choose $\delta \in]0, \pi/\eta[$ and the sequence $\tau_k \to \infty$ with $\tau_1 = \tau_*$, $\tau_{k+1} = \tau_k + 2\pi/\eta$, and $|\cos(\eta \tau_1) - \cos(\eta \tau)| \leq \eta v_*/2\gamma$ for $\tau \in [\tau_k - \delta, \tau_k + \delta]$. Thus for $\tau \in [\tau_k - \delta, \tau_k + \delta]$ we obtain the estimate $\dot{x}(\tau) = \dot{x}(\tau_1) + \int_{\tau_1}^\tau \ddot{x}(\sigma) d\sigma \geq v_* + \gamma (\cos(\eta \tau_1) - \cos(\eta \tau))/\eta \geq v_*/2$. This implies $x(\tau_k + \delta) \geq x(\tau_k - \delta) + \delta v_*$ for $k \in \mathbb{N}$, and therefore $x_* = x(\tau_1) \leq x(\tau_2 - \delta) \leq x(\tau_2 + \delta) - \delta v_* \leq x(\tau_3 - \delta) - \delta v_* \leq x(\tau_3 + \delta) - 2\delta v_* \leq \ldots$. Inductively we find $x_* + (k-1)\delta v_* \leq x(\tau_k + \delta) \leq -1$ for $k \in \mathbb{N}$, a contradiction.   □

Lemma 4.3.3. Let $d \geq 2$, $f \in C^1(\mathbb{R}^d, \mathbb{R}^d)$, and let Φ_f denote the global flow generated by f, i.e., $x(\cdot) = \Phi_f(\cdot, x_0)$ uniquely solves $\dot{x} = f(x)$ and $x(0) = x_0$ for $x_0 \in \mathbb{R}^d$.

(a) Suppose that $C \subset \mathbb{R}^d$ is a C^1-manifold of dimension $\leq d - 2$ and global chart $\alpha : \mathbb{R}^{d-2} \to C$, $y \mapsto x = \alpha(y)$. Then

$$\lambda^d \left(\{x \in \mathbb{R}^d : \exists t \in [0, \infty[: \Phi_f(t, x) \in C\} \right) = 0.$$

(b) If $[a, b] \subset \mathbb{R}$, and if $N \subset \mathbb{R}^d$ has $\lambda^d(N) = 0$, then $\lambda^d(\Phi_f([a, b] \times N)) = 0$.

Proof: Ad (a): Let $\varrho :]-\infty, 0[\times \mathbb{R}^{d-2} \to \mathbb{R}^d$, $\varrho(s, y) = \Phi_f(s, \alpha(y))$. Then ϱ is C^1 and

$$\varrho(]-\infty, 0[\times \mathbb{R}^{d-2}) = \{x \in \mathbb{R}^d : \exists t \in]0, \infty[: \Phi_f(t, x) \in C\}.$$

Since $\lambda^d\big(\varrho(]-\infty,0[\times \mathbb{R}^{d-2})\big) = 0$, (a) follows. Ad (b): Because $\Phi_f \in C^1(\mathbb{R} \times \mathbb{R}^d)$, this is an immediate consequence of a result related to Sard's lemma; see J.T. SCHWARTZ [199, Thm. 3.1]. $\qquad\square$

Lemma 4.3.4. *Consider the matrices*

$$B(\theta, \alpha, P, P') = \begin{pmatrix} \cos\theta & \alpha\sin\theta & P \\ -\sin\theta & \alpha\cos\theta & P' \\ 0 & 0 & 1 \end{pmatrix} \in \mathbb{R}^{3\times 3}, \quad and$$

$$\tilde{B}(\theta, \alpha) = \begin{pmatrix} \cos\theta & \alpha\sin\theta \\ -\sin\theta & \alpha\cos\theta \end{pmatrix} \in \mathbb{R}^{2\times 2} \qquad (4.110)$$

for $\theta, \alpha, P, P' \in \mathbb{R}$. Let $\theta_i, \alpha_i, P_i, P_i' \in \mathbb{R}$ for $i \in \mathbb{N}_0$ be given, and define $B_i = B(\theta_i, \alpha_i, P_i, P_i')$ as well as $\tilde{B}_i = \tilde{B}(\theta_i, \alpha_i)$. Let $\prod_{i=0}^{k} B_i = B_k \cdot \ldots \cdot B_0$. Then the following holds for every $k \in \mathbb{N}_0$.

(a) *The first two lines of $\left(\prod_{i=0}^{k} B_i\right)(1,0,0)^{\mathrm{tr}} \in \mathbb{R}^3$ are $\left(\prod_{i=0}^{k} \tilde{B}_i\right)(1,0)^{\mathrm{tr}} \in \mathbb{R}^2$, and there is a zero in the third line.*

(b) *The first two lines of $\left(\prod_{i=0}^{k} B_i\right)(0,1,0)^{\mathrm{tr}} \in \mathbb{R}^3$ are $\left(\prod_{i=0}^{k} \tilde{B}_i\right)(0,1)^{\mathrm{tr}} \in \mathbb{R}^2$, and there is a zero in the third line.*

(c) *The first two lines of $\left(\prod_{i=0}^{k} B_i\right)(0,0,1)^{\mathrm{tr}} \in \mathbb{R}^3$ are*

$$\sum_{i=0}^{k} \left(\prod_{j=k+2-i}^{k} \tilde{B}_j\right) \begin{pmatrix} P_{k-i+1} \\ P'_{k-i+1} \end{pmatrix} \in \mathbb{R}^2,$$

and there is a 1 in the third line. (Here we define empty products to be the unit matrix in $\mathbb{R}^2$.)

In addition, $\|\tilde{B}(\theta,\alpha)\|_2 = 1$ for all $\theta \in \mathbb{R}$ and $\alpha \in [0,1]$. So, assuming above that also $\alpha_i \in [0,1]$ for $i \in \mathbb{N}$, we obtain for every $k \in \mathbb{N}$

(d) $\left|\left(\prod_{i=0}^{k} B_i\right)(1,0,0)^{\mathrm{tr}}\right|_{\mathbb{R}^3} \leq 1,$

(e) $\left|\left(\prod_{i=0}^{k} B_i\right)(0,1,0)^{\mathrm{tr}}\right|_{\mathbb{R}^3} \leq 1, \quad and$

(f) $\left|\left(\prod_{i=0}^{k} B_i\right)(0,0,1)^{\mathrm{tr}}\right|_{\mathbb{R}^3} \leq 1 + \sum_{i=0}^{k} \left|\begin{pmatrix} P_i \\ P_i' \end{pmatrix}\right|_{\mathbb{R}^2}.$

Proof: (a) and (b) are similar, so we consider (a). Since $B_0(1,0,0)^{\mathrm{tr}} = (\cos\theta_0, -\sin\theta_0, 0)^{\mathrm{tr}}$ and $\tilde{B}_0(1,0)^{\mathrm{tr}} = (\cos\theta_0, -\sin\theta_0)^{\mathrm{tr}}$, the case $k = 0$ follows. If the claim already holds for $k \in \mathbb{N}_0$ and $\left(\prod_{i=0}^{k} \tilde{B}_i\right)(1,0)^{\mathrm{tr}} = (u_k, v_k)^{\mathrm{tr}}$, then by induction hypotheses

$$\left(\prod_{i=0}^{k+1} B_i\right)(1,0,0)^{\mathrm{tr}} = B_{k+1}\left(\prod_{i=0}^{k} B_i\right)(1,0,0)^{\mathrm{tr}} = B_{k+1}(u_k, v_k, 0)^{\mathrm{tr}}$$

$$= \begin{pmatrix} u_k \cos\theta_{k+1} + v_k\, \alpha_{k+1}\sin\theta_{k+1} \\ -u_k \sin\theta_{k+1} + v_k\, \alpha_{k+1}\cos\theta_{k+1} \\ 0 \end{pmatrix},$$

and also

$$\left(\prod_{i=0}^{k+1} \tilde{B}_i\right)(1,0)^{\mathrm{tr}} = \tilde{B}_{k+1}\left(\prod_{i=0}^{k} \tilde{B}_i\right)(1,0)^{\mathrm{tr}} = \tilde{B}_{k+1}(u_k, v_k)^{\mathrm{tr}}$$

$$= \begin{pmatrix} u_k \cos\theta_{k+1} + v_k\, \alpha_{k+1}\sin\theta_{k+1} \\ -u_k \sin\theta_{k+1} + v_k\, \alpha_{k+1}\cos\theta_{k+1} \end{pmatrix},$$

and therefore we have shown (a).

Concerning (c), we first note that $B_0(0,0,1)^{\mathrm{tr}} = (P_0, P_0', 1)^{\mathrm{tr}}$, and thus the claim is true for $k = 0$. Assuming it already holds for $k \in \mathbb{N}_0$, we obtain

$$\left(\prod_{i=0}^{k+1} B_i\right)(0,0,1)^{\mathrm{tr}}$$

$$= B_{k+1}\left(\prod_{i=0}^{k} B_i\right)(0,0,1)^{\mathrm{tr}}$$

$$= \begin{pmatrix} \cos\theta_{k+1} & \alpha_{k+1}\sin\theta_{k+1} & P_{k+1} \\ -\sin\theta_{k+1} & \alpha_{k+1}\cos\theta_{k+1} & P_{k+1}' \\ 0 & 0 & 1 \end{pmatrix}$$

$$\begin{pmatrix} \sum_{i=0}^{k}\left(\prod_{j=k+2-i}^{k} \tilde{B}_j\right)\begin{pmatrix} P_{k-i+1} \\ P_{k-i+1}' \end{pmatrix} \\ 1 \end{pmatrix}$$

$$= \begin{pmatrix} \tilde{B}_{k+1}\left[\sum_{i=0}^{k}\left(\prod_{j=k+2-i}^{k} \tilde{B}_j\right)\begin{pmatrix} P_{k-i+1} \\ P_{k-i+1}' \end{pmatrix}\right] + \begin{pmatrix} P_{k+1} \\ P_{k+1}' \end{pmatrix} \\ 1 \end{pmatrix},$$

and hence (c) follows, because

$$\tilde{B}_{k+1}\Big[\sum_{i=0}^{k}\Big(\prod_{j=k+2-i}^{k}\tilde{B}_j\Big)(P_{k-i+1},P'_{k-i+1})^{\mathrm{tr}}\Big]+(P_{k+1},P'_{k+1})^{\mathrm{tr}}$$

$$=\sum_{i=0}^{k+1}\Big(\prod_{j=k+3-i}^{k+1}\tilde{B}_j\Big)(P_{k-i+2},P'_{k-i+2})^{\mathrm{tr}}.$$

Since $\tilde{B}(\theta,\alpha)^{\mathrm{tr}}\tilde{B}(\theta,\alpha)=\begin{pmatrix}1&0\\0&\alpha^2\end{pmatrix}$ has spectral radius $1\vee\alpha^2=1$ for $\alpha\in[0,1]$, we conclude that $\|\tilde{B}(\theta,\alpha)\|_2=1$ for $\theta\in\mathbb{R}$ and $\alpha\in[0,1]$. This yields (d), (e), and (f), using $\sqrt{1+x^2}\le 1+x$, $x\in[0,\infty[$, for the latter.   □

5. On the application of Conley index theory to non-smooth dynamical systems

In this chapter we follow KUNZE/KÜPPER/LI [118] and introduce Conley index theory (also called homotopy index theory) to non-smooth dynamical systems, having in mind that Conley index theory is a powerful tool for the analysis of the qualitative behaviour of smooth systems. One of our main concerns is to prove bifurcation results (of non-equilibrium solutions) for non-smooth dynamical systems, and in particular we are going to obtain a global bifurcation theorem analogous to the one of WARD [218, Thm. 4]; see WARD [219, Thm. 2] for a slightly corrected version.

As a model problem, we can take

$$\ddot{x} + f(\dot{x}) + g(x) \in -\varphi(x, \dot{x}) \operatorname{Sgn} \dot{x} \quad \text{a.e.}, \qquad (5.1)$$

where a suitable bifurcation parameter μ is introduced in one (or several) of the functions f, g, or φ. Here Sgn is the multi-valued sgn-function from (1.3) on p. 3. Note that with $f(v) = 0$, $g(x) = x$, and $\varphi(x, v) = 1$ this becomes (1.1) on p. 90 without external forcing, i.e., $\gamma = 0$.

With regard to bifurcations, systems like (5.1) have two main drawbacks: first, they need not have a well-defined linearization at equilibrium points, and second, they need not define unique solutions, i.e., a global flow or even a semiflow. Since the approach of Ward does not rely on properties of the linearization, in this respect a generalization to non-smooth dynamical systems looks possible. The obstacle that there need not be a flow induced by (5.1) (not even a "multi-valued flow"), can be overcome as follows: the general form of initial-value problem we are going to consider is

$$\dot{y} \in F(y) \quad \text{a.e.}, \quad y(0) = y_0, \qquad (5.2)$$

recall (2.7), with a multi-valued mapping $F : \mathbb{R}^n \to 2^{\mathbb{R}^n} \setminus \{\emptyset\}$ which is assumed to be ε-δ-usc and to have a certain boundedness property (cf. the basic assumptions (I) on p. 147). This implies that there are globally bounded C^1-functions $f : \mathbb{R}^n \to \mathbb{R}^n$ which are "almost" selections of F, i.e., $f(y) \in F(y)$ is "almost" true (up to some error which can be made arbitrarily small). Then $\dot{y} = f(y)$ defines a global flow, and it remains to show that, starting from an isolated invariant set of (5.2), there is a suitable approximation f such that $\dot{y} = f(y)$ has a related isolated invariant set. This program can be carried out, and therefore in fact both above mentioned problems can be resolved.

Our construction of the Conley index for non-smooth dynamical systems is similar to the definition of the (Brouwer or Leray-Schauder) degree of mapping DEG for upper semicontinuous multi-valued mappings, cf. DEIM-LING [65, Ch. 11.4]. Roughly speaking, if f is an "almost" selection of such an F with $0 \notin (\mathrm{id} - F)(\partial\Omega)$, then f will also have $0 \notin (\mathrm{id} - f)(\partial\Omega)$. Thus one can reasonably define $\mathrm{DEG}(\mathrm{id} - F, \Omega, 0) = \deg(\mathrm{id} - f, \Omega, 0)$, with deg being the usual degree for single-valued functions. This new DEG turns out to be well-defined, due to stability of deg w.r. to small perturbations (i.e., homotopy invariance). Analogously, to define the Conley index for non-smooth dynamical systems, if we are given an isolated invariant set I of (5.2) with isolating neighborhood $N = \overline{U}$, then we will define $H(I)$, the generalized Conley index, to be $h(\mathrm{inv}(N); f)$, where $h(\cdot; f)$ is the classical Conley index w.r. to the global flow generated by $\dot{y} = f(y)$, and $\mathrm{inv}(N)$ denotes the corresponding maximal invariant subset of N. Here it will be the stability of the classical Conley index under small perturbations (i.e., the continuation theorem) which ensures that H is well-defined, and that, moreover, H inherits all the useful properties of h. Then the bifurcation theorems we want to prove do follow more or less analogously to the single-valued case.

To put our approach into more context, we note that DANCER [59, Remark 1, p. 6] has a remark on Conley index theory for single-valued equations with non-unique solutions. In MROZEK [151] there was already introduced a Conley type index for multi-valued systems, but the underlying equation was supposed to define a multi-valued flow, cf. MROZEK [151, Sect. 3]. In addition, our approach is more basic, since we are not going to use cohomology methods. Next, a Conley index for discrete multi-valued systems was developed in KACZYNSKI/MROZEK [103], cf. also MROZEK [152], with the main goal of providing a theoretical framework for a computer-assisted proof of chaos in the Lorenz equation; see MISCHAIKOW/MROZEK [142, 143]. Roughly speaking, when considering the discrete dynamics generated by a single-valued mapping f, a multi-valued $F(y)$ comes up in a natural way as the set of possible numerical values for the true value $f(y)$. Finally, what concerns the bifurcation theorems, we often rely on the ideas of WARD in [218] and [219].

For the convenience of readers not familiar with classical Conley index theory, we included a short introduction to the subject in Sect. 5.1. Then in Sect. 5.2 we develop the basic concepts like isolated invariant sets and isolating neighborhoods for multi-valued equations, whereas in Sect. 5.3 we give the definition of the Conley index for non-smooth dynamical systems and prove its fundamental properties. Finally, in Sect. 5.4, we are going to apply these results to obtain bifurcation theorems, and we will illustrate this with an example which is a special case of (5.1).

5.1 A very brief introduction to Conley index theory

The purpose of this section is to provide a short summary of some ideas of Conley index theory, without going into too much details. In part, we follow CONLEY [55, Ch. I.2] or SMOLLER [206, Sect. 22.A]. We will restrict ourselves to the finite dimensional case and refer to RYBAKOWSKI [197] for infinite dimensional systems (which often arise from partial differential equations and require compactness assumptions to be verified). Some additional references concerning Conley index theory are CONLEY/ZEHNDER [56] and MISCHAIKOW [141].

As the name already indicates, the Conley index, as an index, assigns certain values to certain objects. It is not only in this respect that it shares common features with the degree of mapping (or fixed-point index), where in that case the objects are triples (f, Ω, y) with $\Omega \subset \mathbb{R}^n$ (for simplicity) open and bounded, $f : \overline{\Omega} \to \mathbb{R}^n$ continuous, and $y \notin f(\partial \Omega)$. Then the Brouwer degree $\deg(f, \Omega, y)$ is an integer, and deg has several useful properties, some of them we want to recall first; see DEIMLING [62, p. 17]. To begin with, deg enjoys a stability property, since it is invariant w.r. to certain homotopies: if $f = f_\lambda(y) : [a, b] \times \overline{\Omega} \to \mathbb{R}^n$ is a continuous family of mappings such that $y \notin f_\lambda(\partial \Omega)$ for $\lambda \in [a, b]$, then $\deg(f_\lambda, \Omega, y)$ does not depend on $\lambda \in [a, b]$. Additionally, if $\Omega_1, \Omega_2 \subset \Omega$ are disjoint and $y \notin f(\overline{\Omega} \setminus (\Omega_1 \cup \Omega_2))$, then $\deg(f, \Omega, y) = \deg(f, \Omega_1, y) + \deg(f, \Omega_2, y)$. Usually this is called the addition property. The degree also provides a useful criterion to solve the nonlinear equation $f(y) = z$ for $y \in \overline{\Omega}$, since $\deg(f, \Omega, z) \neq 0$ implies $z \in f(\overline{\Omega})$.

The objects for which Conley index is defined are isolated invariant sets, which we are going to introduce now. Consider a dynamical systems $\dot{y} = f(y)$ that generates a (global) flow $\pi_f(\cdot, y)$, i.e., $y(t) = \pi_f(t, y_0)$ is the solution to the equation with initial value $y(0) = y_0 \in \mathbb{R}^n$; recall that in Chap. 4 we have written this as $\Phi_f(\cdot, y)$, but in the present context the notation $\pi_f(\cdot, y)$ is more standard. Sometimes $\pi_f(t, y_0)$ is as well abbreviated as $t \cdot y_0$, so that π_f being a flow means $t \cdot (s \cdot y_0) = (t + s) \cdot y_0$ and $0 \cdot y_0 = y_0$ for $t, s \in \mathbb{R}$ and $y_0 \in \mathbb{R}^n$. First we need to recall the notion of an isolating neighborhood.

Definition 5.1.1. *If $U \subset \mathbb{R}^n$ is open and bounded, then $N = \overline{U}$ is called an isolating neighborhood (of the dynamical system generated by f), if every boundary point $y_0 \in \partial N$ of N leaves N either forward or backward in time, i.e., there exists $t \in \mathbb{R}$ such that $t \cdot y_0 \notin N$.*

An important concept in dynamical systems are invariant sets. Here $I \subset \mathbb{R}^n$ is called invariant, if every trajectory starting in I will remain in I for all forward and backward times, i.e., $y_0 \in I$ implies $\mathbb{R} \cdot y_0 \subset I$. The most basic examples are stationary points (also called fixed points or equilibria), i.e., points y_0 with $f(y_0) = 0$, or periodic orbits. Then, for a given $N \subset \mathbb{R}^n$, one naturally defines $\mathrm{inv}(N) \subset N$ as the maximal invariant subset of N, i.e., those points of N which never leave N.

Lemma 5.1.1. *An $N = \overline{U}$ is an isolating neighborhood if and only if* $\mathrm{inv}(N) \subset U$.

This is readily seen from the definitions. Now we can define isolated invariant sets.

Definition 5.1.2. *An $I \subset \mathbb{R}^n$ is called an isolated invariant set, if $I = \mathrm{inv}(\overline{U}) \subset U$ for some open bounded $U \subset \mathbb{R}^n$.*

Thus $N = \overline{U}$ is an isolating neighborhood in this case, but it may be possible that different isolating neighborhoods $N_1 = \overline{U_1}$ and $N_2 = \overline{U_2}$ define the same isolated invariant set $\mathrm{inv}(N_1) = I = \mathrm{inv}(N_2)$. Note also that all the definitions depend on f resp. on the flow π_f generated by f.

We are going to illustrate these concepts with a simple one-dimensional example, cf. CONLEY [55, Fig. 2, p. 4] or SMOLLER [206, Fig. 22.3, p. 450]. Consider the parameter-dependent

$$\dot{y} = f_\lambda(y) = y(1 - y^2) - \lambda, \quad y \in \mathbb{R}. \tag{5.3}$$

For $\lambda = 0$ there are three equilibria $y_0 = -1, 0, 1$. Moreover, there exists a unique $\lambda_1 > 0$ such that there are exactly two equilibria $a_1 = a(\lambda_1) < -1$ and $r_1 = r(\lambda_1) \in]0, 1[$, and for $\lambda > \lambda_1$ there is left only one equilibrium $a(\lambda)$ which then has $a(\lambda) < a_1 < -1$; see Fig. 5.1.

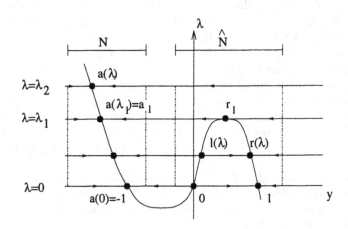

Fig. 5.1. Sketch of the flows for (5.3) on different λ-levels

The corresponding flows are indicated by arrows: for $\lambda = 0$, -1 and 1 are stable while zero is unstable. Moreover, for $\lambda = \lambda_1$, a_1 is stable and r_1 is degenerated (since $f'_{\lambda_1}(r_1) = 0$), and finally $a(\lambda)$ is stable for $\lambda > \lambda_1$. Thus there are plenty of isolating neighborhoods. In particular, each interval having no equilibrium as an endpoint will serve. The isolated invariant sets can also be read off from Fig. 5.1; e.g. for $\lambda = \lambda_1$ there are four isolated invariant sets, namely $\{a_1\}$, $\{r_1\}$, $\{a_1, r_1\}$, and $[a_1, r_1]$, because, as an example, $[a_1, r_1] = $ inv$([a_1 - 1, r_1 + 1]) \subset]a_1 - 1, r_1 + 1[$.

From the definition of an isolating neighborhood it is clear that boundary points of such a set will play a special role. Consider again the above example and the interval N as indicated in Fig. 5.1; N is the same for $\lambda \in [0, \lambda_2]$, and an isolating neighborhood for the isolated invariant set $\{a(\lambda)\}$ consisting of the single fixed point in N. For all parameter values $\lambda \in [0, \lambda_2]$, the boundary points of N behave in the same way, since (in forward time) they will move to interior of N immediately. As a second example, consider $\hat{N}$ for $\lambda \in [0, \lambda_1]$, cf. Fig. 5.1, which again is an isolating neighborhood for the maximal invariant set contained in it. This isolated invariant set $[l(\lambda), r(\lambda)]$ is a proper interval for $\lambda \in [0, \lambda_1[$, it degenerates to the single point $r(\lambda_1)$ for $\lambda = \lambda_1$, while it is empty for $\lambda > \lambda_1$. Here the situation is different, as (in forward time) the right boundary point of $\hat{N}$ will move to the interior of $\hat{N}$ on all levels $\lambda \in [0, \lambda_1]$, whereas the left endpoint will leave $\hat{N}$.

Suppose now that for a given isolated invariant set I there exists a special isolating neighborhood B (called an isolating block) such that each boundary point leaves B immediately either in forward or in backward time, and denote b^+ the set of all boundary points which leave B immediately in forward time. Then it was the idea of Conley to consider the homotopy type of B/b^+ and define an index as $h(I) = [B/b^+]$, the equivalence class of B/b^+ modulo homotopy invariance in the category of pointed spaces. Here B/b^+ is the quotient space of B modulo identification of all points in b^+, so B/b^+ can also be seen as a pointed space, with all elements of b^+ collapsed to a single distinguished point. This definition of h again depends on the underlying flow, i.e., $h(I) = h(I; \pi_f) = h(I; f)$ in case that this flow π_f is generated via a dynamical system with right-hand side f. In addition, to make h well-defined, it of course needs to be shown that indeed an isolating block exists, and, moreover, that the index does not depend on the special choice of isolating block, i.e., if B_1, B_2 are isolated blocks with exit sets b_1^+, b_2^+ and inv$(B_1) = I = $ inv(B_2), then $[B_1/b_1^+] = [B_2/b_2^+]$.

Let us again illustrate this with the above example. First, consider the isolating block $B = [-1/2, 1/2]$ for the isolated invariant set $I = \{0\}$ on the level $\lambda = 0$. Then the exit set is $b^+ = \{-1/2, 1/2\} = \partial B$, so in order to find the homotopy type of B/b^+, both boundary points have to be identified to a new distinguished point, which then has to be attached to the set. By identifying $-1/2$ and $1/2$, B becomes a (pointed) circle, i.e., the (pointed) unit sphere in $\mathbb{R}^2$, and this homotopy type usually is denoted by Σ^1. (More generally, Σ^k is the equivalence class of the pointed k-sphere.) Thus we have $h(\{0\}; f_0) = \Sigma^1$. Next consider the isolating block N for $\lambda \in [0, \lambda_2]$ as shown in Fig. 5.1. Here the exit set n^+ is empty, and therefore $N/n^+ \sim N \cup \{*\} \sim \{p, *\}$ (with $*$ being the distinguished point), since N can be contracted to a single point p. Thus $h(\text{inv}(N); f_\lambda) = h(\{a(\lambda)\}; f_\lambda) = \Sigma^0$ is the (pointed) zero-sphere in $\mathbb{R}$ for $\lambda \in [0, \lambda_2]$. As a second example we will use again $\hat{N}$, with $\mathrm{inv}(\hat{N}) = [l(\lambda), r(\lambda)]$ for $\lambda \in [0, \lambda_1]$, and $\mathrm{inv}(\hat{N}) = \emptyset$ for $\lambda > \lambda_1$. Now for $\lambda \in [0, \lambda_1]$ the exit set $\hat{n}^+$ consists of the left boundary point of $\hat{N}$, thus $\hat{N}/\hat{n}^+ \sim [c, d]/\{c\} \sim \{c\}/\{c\}$ has the homotopy type of $\bar{0}$, the (pointed) one-point space. Since from the very definition moreover $\emptyset/\emptyset$ has homotopy type $\bar{0}$, we find that $h(\mathrm{inv}(\hat{N}); f_\lambda) = \bar{0}$ for $\lambda \in [0, \lambda_2]$.

Although we considered only examples, this illustrates the useful continuation property of the Conley index.

Definition 5.1.3. *A family* (π_λ), $\lambda \in [a, b] \subset \mathbb{R}$, *of flows on* $\mathbb{R}^n$ *will be called continuous in* λ, *if* $\lambda_k \to \lambda$ *and* $(t_k, y_k) \to (t, y)$ *implies* $\pi_{\lambda_k}(t_k, y_k) \to \pi_\lambda(t, y)$.

Theorem 5.1.1 (Continuation Theorem). *For each* $\lambda \in [a, b]$ *let* π_λ *be a flow on* $\mathbb{R}^n$ *such that* (π_λ) *is continuous in* λ *and such that* $I_\lambda = \mathrm{inv}(\overline{U}; \pi_\lambda) \subset U$ *for some open bounded* $U \subset \mathbb{R}^n$, *i.e.,* $\overline{U}$ *is an isolating neighborhood for every* π_λ, $\lambda \in [a, b]$. *Then* $h(I_\lambda; \pi_\lambda)$ *is independent of* $\lambda \in [a, b]$.

Proof: Cf. RYBAKOWSKI [197, Thm. 12.2, p. 65] or WARD [218, Thm. 2] for more general results. □

This continuation property is analogous to the homotopy invariance of the degree of mapping, and it can also be seen as a stability property: $h(\mathrm{inv}(\overline{U}); f)$ does not change, if f is replaced by a nearby g. It is exactly this fact that will allow us in Sect. 5.3 below to define a Conley index for differential inclusions $\dot{y} \in F(y)$ with an upper semicontinuous multi-valued F, since such an F can be approximated well enough by C^1-functions f for which the corresponding Conley index is already defined.

The Conley index as well has an addition property, again (formally) analogous to the addition property of the degree.

Theorem 5.1.2 (Addition Theorem). *If* I_1 *and* I_2 *are disjoint isolated invariant sets, then* $I = I_1 \dot{\cup} I_2$ *is an isolated invariant set, and*

$$h(I_1 \dot{\cup} I_2) = h(I_1) \vee h(I_2).$$

Proof: Cf. SMOLLER [206, Thm. 22.31]. □

Here $I_1 \dot{\cup} I_2$ indicates a disjoint union, and $\vee$ denotes the sum of pointed spaces ; see SMOLLER [206, Def. 12.26]. To be more precise, if (X, x_0) and (Y, y_0) are pointed spaces, then the sum $(X, x_0) \vee (Y, y_0)$ is defined, and this carries over in a well-defined manner to the corresponding equivalence classes by setting $[(X, x_0)] \vee [(Y, y_0)] = [(X, x_0) \vee (Y, y_0)]$, cf. SMOLLER [206, Lemma 22.26].

Next, note that $I = \emptyset$ is an isolated invariant set with isolating block $B = \emptyset$. Thus $h(\emptyset) = [\emptyset/\emptyset] = \bar{0}$. Therefore we also obtain (in accordance to the existence principle for the degree of mapping) the following existence principle.

Theorem 5.1.3 (Existence Principle). *If $I = \mathrm{inv}(\overline{U}) \subset U$ is an isolated invariant set with isolating neighborhood U, and if $h(I) \neq \overline{0}$, then $\mathrm{inv}(\overline{U}) \neq \emptyset$. Hence there exists $y_0 \in U$ with $\mathbb{R} \cdot y_0 \subset U$, i.e., U contains a full orbit.*

After having briefly reviewed some of the main features of classical Conley index theory, we now turn to our main concern, namely multi-valued equations. In the next section, we start by generalizing some of the basic concepts (like isolated invariant sets and isolating neighborhoods) to such multi-valued equations.

5.2 Preliminaries

We consider the differential inclusion

$$\dot{y} \in F(y) \quad \text{a.e.}, \quad y(0) = y_0, \tag{5.2}$$

and we first introduce our basic assumptions on F from (5.2); see Chap. 2 for the terminology.

Basic Assumptions (I).

(i) $F : \mathbb{R}^n \to 2^{\mathbb{R}^n} \setminus \{\emptyset\}$ is ε-δ-usc with closed convex values.
(ii) $\|F(y)\| = \sup\{|z| : z \in F(y)\} \leq \Phi(|y|)$ for some increasing function $\Phi : [0, \infty[\to [0, \infty[$.

Remark 5.2.1. The above basic assumptions (I) imply in particular the following, cf. DEIMLING [65, Cor. 5.2] or FILIPPOV [82, Thm. 1, p. 77] for (a) and (b).

(a) Every initial value problem (5.2) has at least one local solution $y = y(\cdot\,; y_0) :]\omega_y^-, \omega_y^+[\to \mathbb{R}^n$, with a maximal interval of existence $]\omega_y^-, \omega_y^+[\ni 0$.
(b) If $y = y(\cdot\,; y_0) :]\omega_y^-, \omega_y^+[\to \mathbb{R}^n$ is any solution to (5.2) on a maximal interval of existence, and $\sup\{|y(t)| : t \in]\omega_y^-, \omega_y^+[\} < \infty$, then $\omega_y^\pm = \pm\infty$.
(c) For every bounded $D \subset \mathbb{R}^n$,

$$\sup_{y \in D} \|F(y)\| \leq \sup_{y \in D} \Phi(|y|) =: \Phi(D) < \infty$$

is satisfied.   ◇

The Conley index will be defined for isolated invariant sets of (5.2) which we are going to introduce first. For a multi-valued F as above and $N \subset \mathbb{R}^d$, let

$$\mathrm{inv}(N; F) = \Big\{ y_0 \in N : \text{ there exists a solution } y = y(\cdot\,; y_0) :]\omega_y^-, \omega_y^+[\to \mathbb{R}^n$$

$$\text{of (5.2) such that } y(t) \in N,\ t \in]\omega_y^-, \omega_y^+[\Big\}. \tag{5.4}$$

Note that (5.2) may have several solutions, so we have a "weak" concept of a maximal invariant set, i.e., we only require that there is *one* solution through y_0 which remains in N, instead of imposing this condition for *all* solutions; see also MROZEK [151, Sect. 4]. If N is bounded, then $\omega_y^\pm = \pm\infty$ for the solutions appearing in the definition of $\mathrm{inv}(N; F)$, by Rem. 5.2.1(b). Hence $\mathrm{inv}(N; F)$ is "weakly invariant", meaning that for $y_0 \in \mathrm{inv}(N; F)$ and $t_1 \in \mathbb{R}$ there exists a solution $y(\cdot\,; y_0)$ such that $y_1 = y(t_1; y_0) \in \mathrm{inv}(N; F)$, as is a consequence of the following lemma.

Lemma 5.2.1. *Assume* $I = \mathrm{inv}(\overline{U}; F) \subset U$ *for some open bounded* $U \subset \mathbb{R}^n$. *If* $y_0 \in I$ *and* $y = y(\cdot\,; y_0) : \mathbb{R} \to \overline{U}$ *is a solution of (5.2), then* $y(\mathbb{R}) \subset I \subset U$.

Proof: Fix $t_1 \in \mathbb{R}$ and define $y_1 = y(t_1) \in \overline{U}$ as well as $z(t) = y(t_1 + t)$, $t \in \mathbb{R}$. Then $z : \mathbb{R} \to \overline{U}$ is a solution of $\dot{z} \in F(z)$ a.e. with $z(0) = y_1$, hence $y_1 \in \mathrm{inv}(\overline{U}; F) = I$, by the definition of $\mathrm{inv}(\ldots)$, cf. (5.4).   □

We also fix a consequence.

Lemma 5.2.2. *If* $I = \mathrm{inv}(\overline{U}; F) \subset U \cap B_r(0)$ *for* $r > 0$ *and some open bounded* $U \subset \mathbb{R}^n$, *then* $I = \mathrm{inv}(\overline{U \cap B_r(0)}; F) \subset U \cap B_r(0)$.

Proof: On the one hand, $\mathrm{inv}(\overline{U \cap B_r(0)}; F) \subset \mathrm{inv}(\overline{U}; F) = I$. Conversely, if $y_0 \in I$ and $y = y(\cdot\,; y_0) : \mathbb{R} \to \overline{U}$ is a solution of (5.2), then, by Lemma 5.2.1, $y(\mathbb{R}) \subset I \subset U \cap B_r(0)$, and hence $y_0 \in \mathrm{inv}(\overline{U \cap B_r(0)}; F)$.   □

Compactness of N implies compactness of $\mathrm{inv}(N; F)$, as we are going to prove next.

Lemma 5.2.3. *If* $N \subset \mathbb{R}^n$ *is compact, then also* $\mathrm{inv}(N; F)$ *is compact.*

Proof: To show closedness, let $y_0^k \in I = \mathrm{inv}(N; F)$ with $y_0^k \to y_0 \in \mathbb{R}^n$ as $k \to \infty$. By Rem. 5.2.1(b) we find global solutions $y_k = y_k(\cdot\,; y_0^k) : \mathbb{R} \to N$ of $\dot{y} \in F(y)$ a.e. with $y_k(0) = y_0^k$. In particular, by Rem. 5.2.1(c), $|\dot{y}_k(t)| \le \|F(y_k(t))\| \le \Phi(N)$ for $k \in \mathbb{N}$ and a.e. $t \in \mathbb{R}$. Hence the Arzelà–Ascoli theorem and a diagonal argument implies that there exist a subsequence (w.l.o.g. the whole sequence) and a continuous $y : \mathbb{R} \to N$ such that $y_k \to y$ uniformly on every compact subinterval of $\mathbb{R}$. We also may assume that $\dot{y}_k \to \dot{y}$ weakly in $L^2_{\mathrm{loc}}(\mathbb{R}; \mathbb{R}^n)$ as $k \to \infty$, and we have $y(0) = y_0$. Since F is ε-δ-usc and has closed convex values, a standard argument for differential inclusions shows that $\dot{y} \in F(y)$ a.e. in $\mathbb{R}$, i.e., y is a solution to (5.2); see the proof of Thm. 2.2.1. Therefore $y_0 \in I$, by definition. The fact that I is relatively compact is established analogously, since for $(y_0^k) \subset I \subset N$ we may first choose a subsequence converging to some $y_0 \in \mathbb{R}^n$, and then repeat the argument.   □

Now we can define isolated invariant sets and isolating neighborhoods of (5.2).

Definition 5.2.1. *An $I \subset \mathbb{R}^n$ is called an isolated invariant set of (5.2), if $I = \mathrm{inv}(\overline{U}; F) \subset U$ for some open bounded $U \subset \mathbb{R}^n$. In this case, the compact $N = \overline{U}$ is called an isolating neighborhood for F.*

The analogous classical concepts for single-valued equations have been introduced already in Sect. 5.1 above. We repeat that, given an $f \in C^1(\mathbb{R}^n; \mathbb{R}^n)$ which induces a global flow $\pi_f : \mathbb{R} \times \mathbb{R}^n \to \mathbb{R}^n$ via $\dot{y} = f(y)$, one defines for $N \subset \mathbb{R}^n$

$$\mathrm{inv}(N; f) = \{y_0 \in N : \pi_f(\mathbb{R}; y_0) \subset N\}.$$

In addition, we let $\mathrm{inv}(N; \pi) = \{y_0 \in N : \pi(\mathbb{R}; y_0) \subset N\}$, for a flow $\pi : \mathbb{R} \times \mathbb{R}^n \to \mathbb{R}^n$. Note that these sets are invariant w.r. to the corresponding flows, and compact, in case that N is compact.

In a later reduction step we will apply the following simple result, which is in the same spirit as Lemma 5.2.2.

Lemma 5.2.4. *Let $f \in C^1(\mathbb{R}^n; \mathbb{R}^n)$ induce a global flow π_f on $\mathbb{R}^n$, and assume $\mathrm{inv}(\overline{V}; f) \subset U$ for open bounded $V, U \subset \mathbb{R}^n$ such that $V \supset U$. Then*

$$\mathrm{inv}(\overline{V}; f) = \mathrm{inv}(\overline{U}; f) \subset U.$$

Proof: Clearly, $\mathrm{inv}(\overline{V}; f) \supset \mathrm{inv}(\overline{U}; f)$ by definition. Conversely, if $y_0 \in \mathrm{inv}(\overline{V}; f)$, then also $\pi_f(\mathbb{R}; y_0) \subset \mathrm{inv}(\overline{V}; f)$, as $\mathrm{inv}(\dots)$ is invariant. Hence $\pi_f(\mathbb{R}; y_0) \subset U \subset \overline{U}$. $\qquad\qquad\square$

We shall introduce the Conley index for differential inclusions by approximation of the multi-valued mapping through single-valued maps. For this, the following result will be needed

Lemma 5.2.5. *Let $\varepsilon \in {]0, 1]}$ and $r > 0$. Under the basic assumptions (I), there exists a C^∞-function $f : \mathbb{R}^n \to \mathbb{R}^n$ such that $|f|_\infty \leq \Phi(\overline{B}_r(0)) + 1$ and*

$$f(y) \in F(B_\varepsilon(y)) + B_\varepsilon(0), \quad y \in \overline{B}_r(0).$$

Proof: Since F is ε-δ-usc with convex values and $D = \overline{B}_r(0)$ is compact, we find a continuous $g : D \to \mathbb{R}^n$ such that $g(y) \in F(B_\varepsilon(y) \cap D) + B_{\varepsilon/2}(0)$ for $y \in D$; see Lemma 2.1.1. In particular, $|g(y)| \leq \sup_{y \in D} \|F(y)\| + \varepsilon/2 \leq \Phi(D) + 1$ for $y \in D$, with $\Phi(D)$ from Rem. 5.2.1(c). Since D is compact, we may extend g to a continuous function $\tilde{g} : \mathbb{R}^n \to \mathbb{R}^n$ such that $|\tilde{g}(y)| \leq \Phi(D) + 1$ for $y \in \mathbb{R}^n$, see, e.g., DEIMLING [62, Prop. 1.1]. We also have $\tilde{g}(y) \in F(B_\varepsilon(y)) + B_{\varepsilon/2}(0)$ for $y \in D$. This $\tilde{g}$ in turn may be approximated uniformly on the compact D by C^∞-functions $\tilde{g}_j = \tilde{g} * \varphi_j$, φ_j being the standard mollifiers. Thus for $j \in \mathbb{N}$ and $y \in \mathbb{R}^n$

$$|\tilde{g}_j(y)| \leq \int_{\mathbb{R}^n} |\tilde{g}(y)| \, \varphi_j(y-z) \, dz \leq [\Phi(D)+1] \int_{\mathbb{R}^n} \varphi_j(y-z) \, dz$$
$$= \Phi(D) + 1.$$

Consequently, for j large, $f = \tilde{g}_j$ will be as desired.   □

5.3 The Conley index for multi-valued mappings and its basic properties

In this section we will define the Conley index for isolated invariant sets of (5.2), under the basic assumptions (I) on the right-hand side F. First we describe the approximations of F we are going to use for the definition of the index.

Definition 5.3.1. Let $\varepsilon, r > 0$. A (single-valued) function f is called an (ε, r)-approximation of F, if $f \in C^1(\mathbb{R}^n; \mathbb{R}^n)$, $|f|_{\infty, r} = \sup\{|f(y)| : y \in \overline{B}_r(0)\} \leq \Phi(\overline{B}_r(0)) + 1$, f generates a global flow on $\mathbb{R}^n$ via the differential equation $\dot{y} = f(y)$, and if

$$f(y) \in \overline{\mathrm{co}} \, F(B_\varepsilon(y)) + B_\varepsilon(0), \quad y \in \overline{B}_r(0).$$

Remark 5.3.1. For every $\varepsilon \in]0, 1]$ and $r > 0$ there exists an (ε, r)-approximation of F, by Lemma 5.2.5, since in this lemma even $|f|_\infty \leq \Phi(\overline{B}_r(0)) + 1$, hence f generates a global flow.   ◇

Lemma 5.3.1. Let $r > 0$, and assume I is an isolated invariant set of (5.2) with two representations $I = \mathrm{inv}(N_0; F) \subset U_0$ and $I = \mathrm{inv}(N_1; F) \subset U_1$, where $N_i = \overline{U}_i$, $i = 0, 1$, and $U_0, U_1 \subset B_r(0)$ are open. Then there is an $\varepsilon_* > 0$ such that for all $\varepsilon_0, \varepsilon_1 \in]0, \varepsilon_*]$ the following holds. If f_i, $i = 0, 1$, are (ε_i, r)-approximations of F, then

$$\mathrm{inv}(N_i\,; f_\lambda) = \mathrm{inv}(\overline{U}_i\,; f_\lambda) \subset U, \quad i = 0, 1, \quad \lambda \in [0, 1],$$

where $f_\lambda = (1 - \lambda)f_0 + \lambda f_1$ and $U = U_0 \cap U_1$. In particular, if we take $U_0 = U_1 = U$ and $f_0 = f_1 = f$, then $\mathrm{inv}(\overline{U}; f) \subset U$.

Proof: If the first claim were wrong, we would find $\varepsilon_0^k \to 0$ and $\varepsilon_1^k \to 0$ as $k \to \infty$, and sequences (f_i^k) of (ε_i^k, r)-approximations of F, $i = 0, 1$, such that, w.l.o.g. for all $k \in \mathbb{N}$ we have $\mathrm{inv}(N_0; f_k) \not\subset U$, where $f_k = (1 - \lambda_k)f_0^k + \lambda_k f_1^k$ for some sequence $(\lambda_k) \subset [0, 1]$. Hence there is some $y_0^k \in \mathrm{inv}(N_0; f_k) \cap (\mathbb{R}^n \backslash U)$ for every $k \in \mathbb{N}$, and therefore in particular $y_k(\mathbb{R}) \subset N_0 \subset \overline{B}_r(0)$, with $y_k = y_k(\cdot\,; y_0^k)$ being the solution of $\dot{y} = f_k(y)$, $y(0) = y_0^k$. As N_0 is compact, w.l.o.g. $y_0^k \to y_0$ for some $y_0 \in N_0 \cap (\mathbb{R}^n \setminus U)$.

We have $|f_i^k|_{\infty,r} \leq \Phi(\overline{B}_r(0)) + 1 = r_0$ for $i = 0, 1$ and $k \in \mathbb{N}$ by definition of an (ε_i^k, r)-approximation. Therefore we may proceed as in the proof to Lemma 5.2.3: we have

$$|y_k(t) - y_k(s)| = \left| \int_s^t f_k(y_k(\tau)) \, d\tau \right| \leq r_0 |t - s|, \quad k \in \mathbb{N}, \quad t, s \in \mathbb{R},$$

and hence, again as a consequence of the Arzelà–Ascoli theorem and a diagonal sequence argument, there is a function $y : \mathbb{R} \to \mathbb{R}^n$ being Lipschitz with constant r_0 and a subsequence (w.l.o.g. the whole sequence) such that $y_k \to y$ uniformly on every compact subinterval of $\mathbb{R}$; whence $y(0) = y_0$ and $y(\mathbb{R}) \subset N_0$. In addition, we may once more suppose that $\dot{y}_k \to \dot{y}$ weakly in $L^2_{\text{loc}}(\mathbb{R}; \mathbb{R}^n)$. Next, $\varepsilon_k = \varepsilon_0^k \vee \varepsilon_1^k \to 0$ as $k \to \infty$. By definition of an (ε_i^k, r)-approximation we obtain for $i = 0, 1$

$$f_i^k(y) \in \overline{\text{co}}\, F(B_{\varepsilon_k}(y)) + B_{\varepsilon_k}(0), \quad y \in \overline{B}_r(0),$$

and since every $\overline{\text{co}}\, F(B_{\varepsilon_k}(y)) + B_{\varepsilon_k}(0)$ on the right-hand side is convex and $y_k(\mathbb{R}) \subset \overline{B}_r(0)$, we conclude that

$$\dot{y}_k(t) = f_k(y_k(t)) \in \overline{\text{co}}\, F(B_{\varepsilon_k}(y_k(t))) + B_{\varepsilon_k}(0), \quad k \in \mathbb{N}, \ t \in \mathbb{R}.$$

Therefore as $k \to \infty$ it follows that $\dot{y}(t) \in F(y(t))$ for a.e. $t \in \mathbb{R}$, cf. the argument in DEIMLING [65, Lemma 5.1] or the proof of Thm. 2.2.1. Consequently, y is a solution to (5.2) with $y(0) = y_0$, and thus

$$y_0 \in \text{inv}(N_0; F) = I \subset U_0 \cap U_1 = U.$$

On the other hand, $y_0 \in \overline{\mathbb{R}^n \setminus U} = \mathbb{R} \setminus U$, a contradiction. $\qquad \square$

Now we can proceed to define the Conley index for ε-δ-usc multi-valued mappings as above.

Definition 5.3.2. *Let F be a multi-valued mapping satisfying the basic assumptions (I), and let I be an isolated invariant set of (5.2). Hence $I \subset B_r(0)$ for some $r > 0$. First we choose an isolating neighborhood $N = \overline{U}$ for F such that $U \subset B_r(0)$; note that this is possible according to Lemma 5.2.2. Next we choose an (ε, r)-approximation f of F, with $\varepsilon > 0$ being sufficiently small, cf. Rem. 5.3.1 and Lemma 5.3.1. Then we define $H(I) = H(I; F) = h(\text{inv}(N; f); f)$ to be the Conley index of I.*

Since on first sight this definition depends on U and f, we additionally need to verify

Lemma 5.3.2. $H(I)$ *is well-defined.*

Proof: Suppose we have two isolating neighborhoods $N_i = \overline{U_i}$ with $U_i \subset B_r(0)$, $i = 1, 2$. Note that $r > 0$ is determined solely through I. Additionally, let f_i, $i = 1, 2$, be (ε_i, r)-approximations for some $\varepsilon_i \in]0, \varepsilon_*]$, $i = 1, 2$, with $\varepsilon_* > 0$ from Lemma 5.3.1. Hence we conclude that $\mathrm{inv}(N_i\,;f_\lambda) \subset U_0 \cap U_1 = U$ for $\lambda \in [0, 1]$ and $i = 0, 1$, f_λ being the path from f_0 to f_1; whence in particular $\mathrm{inv}(N_i\,;f_\lambda) = \mathrm{inv}(\overline{U}\,;f_\lambda) \subset U$ for $\lambda \in [0, 1]$ and $i = 0, 1$, by Lemma 5.2.4. Therefore we obtain

$$
\begin{aligned}
h(\mathrm{inv}(N_0; f_0); f_0) &= h(\mathrm{inv}(\overline{U}\,; f_0); f_0) = h(\mathrm{inv}(\overline{U}\,; f_1); f_1) \\
&= h(\mathrm{inv}(N_1; f_1); f_1);
\end{aligned}
$$

observe that the middle equality follows from the continuation theorem, cf. Thm. 5.1.1. □

We start with some elementary properties of H.

Remark 5.3.2. The Conley index H is a generalization of the classical Conley index h for single-valued functions $f \in C^1(\mathbb{R}^n; \mathbb{R}^n)$ which generate global flows on $\mathbb{R}^n$.

Proof: If $F(y) = \{f(y)\}$ and $I = \mathrm{inv}(\overline{U}; F) = \mathrm{inv}(\overline{U}; f) \subset U$ is an isolated invariant set, then $H(I; F) = h(I; f)$, since f itself is an admissible (ε, r)-approximation of F for all $\varepsilon, r > 0$. Note that here $\Phi(\rho) = \sup\{|f(y)| : y \in \overline{B}_\rho(0)\}$ in (ii) of the basic assumptions (I). □

In our next result we transfer to H the existence principle from Thm. 5.1.3.

Theorem 5.3.1. *If $H(I) \neq \overline{0}$, then $I \neq \emptyset$, i.e., if $\overline{U}$ is any isolating neighborhood for F which isolates I, then U contains a full orbit.*

Proof: If $I = \emptyset$, then $I = \mathrm{inv}(\overline{U}; F) \subset U$ and also $I = \mathrm{inv}(\overline{U_1}; F) \subset U_1$ with $U_1 = \emptyset$. Since $H(I)$ is independent of the isolating neighborhood, we obtain with some (ε, r)-approximation f of F, for $\varepsilon > 0$ sufficiently small, $H(I) = h(\mathrm{inv}(\overline{U_1}; f)) = h(\emptyset) = \overline{0}$, a contradiction. Hence U contains a full solution, cf. Lemma 5.2.1. □

Again the generalized Conley index H enjoys the addition property, cf. Thm. 5.1.2.

Theorem 5.3.2. *If I_1, I_2 are disjoint isolated invariant sets of (5.2), then $I = I_1 \dot\cup I_2$ is an isolated invariant set of (5.2), and*

$$
H(I_1 \dot\cup I_2) = H(I_1) \vee H(I_2). \tag{5.5}
$$

Proof: Let $I_i = \text{inv}(\overline{U}_i; F) \subset U_i$, $i = 1, 2$. Because I_1 and I_2 are compact and disjoint, we may assume w.l.o.g. that also U_1 and U_2 are disjoint with positive distance, since otherwise we may replace U_i through $U_i \cap (I_i + B_\delta(0))$ for small δ, cf. the argument of Lemma 5.2.2. Then $I = \text{inv}(\overline{U}; F) \subset U$ with $U = U_1 \dot{\cup} U_2$. Indeed, we have $\text{inv}(\overline{U}; F) \supset \text{inv}(\overline{U}_i; F) = I_i$ for $i = 1, 2$, hence $\text{inv}(\overline{U}; F) \supset I$. On the other hand, if $y_0 \in \text{inv}(\overline{U}; F)$ and $y = y(\cdot; y_0) : \mathbb{R} \to \overline{U} = \overline{U}_1 \dot{\cup} \overline{U}_2$ is a solution of (5.2) with $y(0) = y_0$, then either $y(\mathbb{R}) \subset \overline{U}_1$ or $y(\mathbb{R}) \subset \overline{U}_2$, and in both cases we obtain $y_0 \in I_1 \dot{\cup} I_2 = S$.

To prove (5.5), we choose $r > 0$ such that $I \subset B_r(0)$ and $U \subset B_r(0)$. Next we choose an (ε, r)-approximation f of F with $\varepsilon > 0$ small, so that we have

$$H(I) = h(\text{inv}(\overline{U}; f)) \tag{5.6}$$

by definition. Since also $I_i \subset B_r(0)$ and $U_i \subset B_r(0)$, $i = 1, 2$, Definition 5.3.2 shows that additionally

$$H(I_i) = h(\text{inv}(\overline{U}_i; f)), \quad i = 1, 2. \tag{5.7}$$

Moreover, $\text{inv}(\overline{U}; f) = \text{inv}(\overline{U}_1; f) \dot{\cup} \text{inv}(\overline{U}_2; f)$, as follows from $\overline{U}_1 \cap \overline{U}_2 = \emptyset$ and the fact that f generates a global flow, cf. Definition 5.3.2. Because $\text{inv}(\overline{U}; f) \subset U$ by Lemma 5.3.1, we also conclude from $\overline{U}_1 \cap \overline{U}_2 = \emptyset$ that $\text{inv}(\overline{U}_i; f) \subset U_i$, $i = 1, 2$. Therefore (5.6), (5.7), and the addition property of the classical Conley index (Thm. 5.1.2) imply

$$H(I) = h(\text{inv}(\overline{U}; f)) = h\Big(\text{inv}(\overline{U}_1; f) \dot{\cup} \text{inv}(\overline{U}_2; f)\Big)$$
$$= h(\text{inv}(\overline{U}_1; f)) \vee h(\text{inv}(\overline{U}_2; f)) = H(I_1) \vee H(I_2),$$

as was to be shown. $\qquad\square$

This addition property can be used to obtain a generalization of a classical theorem of Conley. For that, we recall that $y_1 \in \mathbb{R}^n$ is an equilibrium of (5.2), if $0 \in F(y_1)$.

Theorem 5.3.3. *Let U be open bounded such $y_1, y_2 \in U$ for exactly two equilibria $y_1 \neq y_2$ of (5.2). Assume that $I = \text{inv}(\overline{U}; F) \subset U$ is an isolated invariant set of (5.2) and that $\{y_i\}$, $i = 1, 2$, are isolated invariant sets of (5.2). If $H(I) \neq H(\{y_1, y_2\})$, then U contains a full nonconstant solution different from y_1 and y_2. In particular, this holds if $H(I) = \bar{0}$, but $H(\{y_i\}) \neq \bar{0}$ for at least one i.*

Proof: By Thm. 5.3.2, $I_* = \{y_1, y_2\}$ is an isolated invariant set of (5.2), and we have $I \supset I_*$. As $H(I) \neq H(I_*)$, $I = I_*$ is impossible. Thus we find $y_0 \in I$ with $y_0 \neq y_1, y_2$, and a solution $y = y(\cdot; y_0) : \mathbb{R} \to \overline{U}$, which exists by definition of $\text{inv}(\ldots)$ and is a full nonconstant solution remaining in U, cf. Lemma 5.2.1. For the second claim we note that

$$H(I_*) = H(\{y_1\}) \vee H(\{y_2\})$$

by Thm. 5.3.2. Thus, if $H(I) = \bar{0}$, then $H(I) \neq H(I_*)$, since otherwise $\bar{0} = H(\{y_1\}) \vee H(\{y_2\})$ would imply $H(\{y_1\}) = \bar{0} = H(\{y_2\})$; see SMOLLER [206, Prop. 12.27(i)]. □

Our next objective is to prove a basic continuation theorem for the generalized Conley index H. For this purpose we first need to introduce parametrized families of differential inclusions

$$\dot{y} \in F(\mu, y) = F_\mu(y) \text{ a.e.}, \quad y(0) = y_0,$$

with $\mu \in [a, b]$. In the following, calligraphic letters like $\mathcal{U}$, etc., will refer to subsets of $[a, b] \times \mathbb{R}^n$, whereas we continue to denote by U, etc., subsets of $\mathbb{R}^n$. In particular, we let

$$\mathcal{B}_r(\mu_0, y_0) = \{(\mu, y) \in [a, b] \times \mathbb{R}^n : |\mu - \mu_0| + |y - y_0| < r\}.$$

Basic Assumptions (II).

(i) $F : [a, b] \times \mathbb{R}^n \to 2^{\mathbb{R}^n} \setminus \{\emptyset\}$ has closed convex values and is such that $F(\mu, \cdot)$ is ε-δ-usc for every $\mu \in [a, b]$, $F(\cdot, y)$ is d_H-continuous for every $y \in \mathbb{R}^n$, and $F(\cdot, y)$ is ε-δ-usc, uniformly for y in bounded subsets of $\mathbb{R}^n$.

(ii) $\|F(\mu, y)\| \leq \Phi(|y|)$ for all $(\mu, y) \in [a, b] \times \mathbb{R}^n$, with some increasing function $\Phi : [0, \infty[\to [0, \infty[$.

Here d_H is the Hausdorff-distance. Continuity of a multi-valued mapping will always be understood as continuity w.r. to d_H. The uniformity condition in (II)(i) means that given $D \subset \mathbb{R}^n$ bounded, $\mu_0 \in [a, b]$, and $\varepsilon > 0$, there is $\delta > 0$ such that $F(\mu, y) \subset F(\mu_0, y) + B_\varepsilon(0)$ for $\mu \in [a, b]$ with $|\mu - \mu_0| < \delta$ and $y \in D$. Thus in particular all assumptions on $F(\cdot, y)$ hold, if $F(\cdot, y)$ is d_H-continuous, uniformly for y in bounded sets.

Lemma 5.3.3. *Assumption (II)(i) in particular implies the following.*

(a) $F : [a, b] \times \mathbb{R}^n \to 2^{\mathbb{R}^n} \setminus \{\emptyset\}$ is jointly ε-δ-usc in (μ, y).

(b) For every $\varepsilon > 0$ and $r > 0$ there exists $\delta > 0$ such that

$$F\Big(([\mu - \delta, \mu + \delta] \cap [a, b]) \times B_\delta(y)\Big) \subset F(\mu, B_\varepsilon(y)) + B_\varepsilon(0) \qquad (5.8)$$

for all $(\mu, y) \in [a, b] \times \overline{B}_r(0)$.

Proof: We only prove (b), since (a) follows from the uniformity condition in (II)(i). If (5.8) were wrong, then we would find $\varepsilon > 0$ and $r > 0$, as well as sequences $\delta_k \to 0^+$, $(\mu_k) \subset [a, b]$ and $(y_k) \subset \overline{B}_r(0)$, and also $\theta_k \in [\mu_k - \delta_k, \mu_k + \delta_k] \cap [a, b]$, $z_k \in B_{\delta_k}(y_k)$, and $w_k \in F(\theta_k, z_k)$ such that $\text{dist}(w_k, F(\mu_k, B_\varepsilon(y_k))) \geq \varepsilon$. W.l.o.g. we may assume $\mu_k \to \mu_0 \in [a, b]$, $y_k \to y_0 \in \overline{B}_r(0)$, $\theta_k \to \mu_0$, and $z_k \to y_0$. By (a), we have $w_k \in F(\theta_k, z_k) \subset F(\mu_0, y_0) + B_{\varepsilon/3}(0)$ for k large, and hence $\text{dist}(\bar{w}_k, F(\mu_k, B_\varepsilon(y_k))) \geq 2\varepsilon/3$ for k large and some sequence $(\bar{w}_k) \subset F(\mu_0, y_0)$. Moreover, $y_0 \in B_\varepsilon(y_k)$

for k large, and consequently $F(\mu_k, B_\varepsilon(y_k))) \supset F(\mu_k, y_0)$, what in turn implies $\text{dist}(\bar{w}_k, F(\mu_k, y_0)) \geq \text{dist}(\bar{w}_k, F(\mu_k, B_\varepsilon(y_k))) \geq 2\varepsilon/3$ for k large. Because $F(\cdot, y_0)$ is d_H-continuous, we have $F(\mu_0, y_0) \subset F(\mu_k, y_0) + B_{\varepsilon/3}(0)$ for k large, and thus $\text{dist}(\bar{w}_k, F(\mu_0, y_0)) \geq \varepsilon/3$ for k large, a contradiction to $\bar{w}_k \in F(\mu_0, y_0)$. $\square$

The following approximation lemma is a parametric version of Lemma 5.2.5, but nevertheless we state it separately to keep things clearer.

Lemma 5.3.4. *Let $\varepsilon \in]0, 1]$, $r > 0$, and suppose that the multi-valued mapping $F = F(\mu, y) : [a, b] \times \mathbb{R}^n \to 2^{\mathbb{R}^n} \setminus \{\emptyset\}$ has convex values and is jointly ε-δ-usc in (μ, y). If (II)(ii) holds, then there exists a continuous function $f : [a, b] \times \mathbb{R}^n \to \mathbb{R}^n$ such that $f(\mu, \cdot) \in C^\infty(\mathbb{R}^n; \mathbb{R}^n)$ and $|f(\mu; \cdot)|_\infty \leq \Phi(\overline{B}_r(0)) + 1$ for $\mu \in [a, b]$, and also*

$$f(\mu, y) \in F\left(([\mu - \varepsilon, \mu + \varepsilon] \cap [a, b]) \times B_\varepsilon(y)\right) + B_\varepsilon(0) \qquad (5.9)$$

for $(\mu, y) \in [a, b] \times \overline{B}_r(0)$. In addition, f can be chosen to satisfy $|f'(\mu, y)| \leq C$, $(\mu, y) \in [a, b] \times \overline{B}_r(0)$, with a constant C depending on f; here f' denotes the Jacobian of f w.r. to y.

Proof: Analogously to the proof to Lemma 5.2.5 we find a continuous function $g : \mathcal{D} \to \mathbb{R}^n$ such that $g(\mu, y) \in F(\mathcal{B}_\varepsilon(\mu, y) \cap \mathcal{D}) + B_{\varepsilon/2}(0)$ for $(\mu, y) \in \mathcal{D} = [a, b] \times \overline{B}_r(0)$. Since $\|F(\mu, y)\| \leq \Phi(\overline{B}_r(0))$ for $(\mu, y) \in \mathcal{D}$, we obtain $|g(\mu, y)| \leq \Phi(\overline{B}_r(0)) + 1$ for $(\mu, y) \in \mathcal{D}$, and again this estimate transfers to a continuous extension $\tilde{g} : [a, b] \times \mathbb{R}^n \to \mathbb{R}^n$ of g, even for all $(\mu, y) \in [a, b] \times \mathbb{R}^n$. We can choose $f(\mu, y) = \int_{\mathbb{R}^n} \tilde{g}(\mu, y) \varphi_j(y - z) \, dz$, $(\mu, y) \in [a, b] \times \mathbb{R}^n$, for some large j, with standard mollifiers φ_j, to obtain a suitable approximating function f which also satisfies (5.9), since $\mathcal{B}_\varepsilon(\mu, y) \cap \mathcal{D} \subset ([\mu - \varepsilon, \mu + \varepsilon] \cap [a, b]) \times B_\varepsilon(y)$ for $(\mu, y) \in \mathcal{D}$. The estimate for $|f'(\mu, y)|$ then is obtained from the defining formula. $\square$

Next we need to introduce parametrized (ε, r)-approximations.

Definition 5.3.3. *Let $\varepsilon, r > 0$. A function $f : [a, b] \times \mathbb{R}^n \to \mathbb{R}^n$ is called an (ε, r)-approximating family of $F = F(\mu, y)$, if*

(a) $f \in C([a, b] \times \mathbb{R}^n; \mathbb{R}^n)$;
(b) for every $\mu \in [a, b]$ we have $f(\mu, \cdot) \in C^1(\mathbb{R}^n; \mathbb{R}^n)$, $|f(\mu; \cdot)|_{\infty, r} \leq \Phi(\overline{B}_r(0)) + 1$, and $f(\mu, \cdot)$ generates a global flow on $\mathbb{R}^n$;
(c) there is a constant $C \geq 0$ such that $|f'(\mu, y)| \leq C$, $(\mu, y) \in [a, b] \times \overline{B}_r(0)$;
(d) we have

$$f(\mu, y) \in \overline{\text{co}}\, F(\mu, B_\varepsilon(y)) + B_\varepsilon(0)\, g, \qquad (\mu, y) \in [a, b] \times \overline{B}_r(0).$$

Note that in this case every $f_\mu = f(\mu, \cdot)$ is an (ε, r)-approximation of $F_\mu = F(\mu, \cdot)$ in the sense of Definition 5.3.1, and the family $(\pi_\mu) = (\pi_{f_\mu})$ of generated global flows is continuous in the sense of Definition 5.1.3.

From now on we suppose the basic assumptions (II) from above to hold for $F = F(\mu, y)$. In particular this implies that every F_μ satisfies the basic assumptions (I).

Remark 5.3.3. For every $\varepsilon \in]0, 1]$ and $r > 0$ there is an (ε, r)-approximating family of F.

Proof: For $\varepsilon, r > 0$ given, we choose the $\delta > 0$ from Lemma 5.3.3(b), and we can assume that $\delta \leq \varepsilon \wedge 1$. Hence, by Lemma 5.3.3(a), application of Lemma 5.3.4 with δ and r is possible, and this yields a function $f : [a, b] \times \mathbb{R}^n \to \mathbb{R}^n$ as is described there. This f is an $(2\varepsilon, r)$-approximating family of F, since (a), (b), and (c) from Definition 5.3.3 hold, and also, by (5.9) and (5.8),

$$f(\mu, y) \in F\Big(([\mu - \delta, \mu + \delta] \cap [a, b]) \times B_\delta(y)\Big) + B_\delta(0)$$
$$\subset F(\mu, B_\varepsilon(y)) + B_\varepsilon(0) + B_\delta(0) \subset F(\mu, B_{2\varepsilon}(y)) + B_{2\varepsilon}(0)$$

for $(\mu, y) \in [a, b] \times \overline{B}_r(0)$. As we can start with $\varepsilon/2$ instead of ε, the claim follows. $\square$

Since ε_* from Lemma 5.3.1 in general depends on I, U_0, and U_1, we also need to prove a "parametrized" version of this result.

Lemma 5.3.5. Let $r > 0$ and assume that $I_\mu = \mathrm{inv}(\overline{U}; F_\mu) \subset U$ for $\mu \in [a, b]$, where $U \subset B_r(0)$ is open. Then there exists an $\varepsilon_* > 0$ such that for every (ε, r)-approximating family f of F we have

$$\mathrm{inv}(\overline{U}; f(\mu; \cdot)) \subset U, \quad \mu \in [a, b].$$

Remark 5.3.4. In fact, this statement holds in the same generality as Lemma 5.3.1, i.e., instead of U one can allow two isolating neighborhoods U_0, U_1, both of which isolate all I_μ, and it is also possible to treat simultaneously two different (ε, r)-approximating families f_0, f_1 of F, defining then $f_\lambda(\mu, y) = (1 - \lambda)f_0(\mu, y) + \lambda f_1(\mu, y)$. We refrained from this generalization in order as to simplify the presentation. $\Diamond$

Proof of Lemma 5.3.5: The method of proof is the same as for Lemma 5.3.1, and we omit the details. It is used that F is jointly ε-δ-usc in (μ, y), cf. Lemma 5.3.3(a), in order to conclude from $\varepsilon_k \to 0^+$, $y_k(0) = y_0^k \to y_0 \in \partial U$,

$$\dot{y}_k(t) = f(\mu_k, y_k(t)) \in \overline{\mathrm{co}} \, F(\mu_k, B_{\varepsilon_k}(y_k(t))) + B_{\varepsilon_k}(0),$$

$y_k \to y$ uniformly on every compact subinterval of $\mathbb{R}$, and $\dot{y}_k \to \dot{y}$ weakly in $L^2_{\mathrm{loc}}(\mathbb{R}; \mathbb{R}^n)$, that y is a solution of (5.2). $\square$

With this preparation we can prove the following basic continuation theorem.

Theorem 5.3.4. *Let the basic assumptions (II) be satisfied for $F = F(\mu, y)$. Assume that there is a bounded open $U \subset \mathbb{R}^n$ such that $I_\mu = \mathrm{inv}(\overline{U}; F_\mu) \subset U$ for every $\mu \in [a, b]$. Then $H(I_\mu; F_\mu)$ is independent of $\mu \in [a, b]$.*

Proof: Choose $r > 0$ such that $U \subset B_r(0)$, hence also $I_\mu \subset B_r(0)$ for $\mu \in [a, b]$. From Definition 5.3.2, Lemma 5.3.5, and Rem. 5.3.4, we conclude that $H(I_\mu; F_\mu) = h(\mathrm{inv}(\overline{U}; f(\mu; \cdot)))$ and $\mathrm{inv}(\overline{U}; f(\mu; \cdot)) \subset U$, $\mu \in [a, b]$, for some (ε, r)-approximating family f of F with $\varepsilon > 0$ sufficiently small, and this is well-defined. Since the family $(\pi_\mu) = (\pi_{f_\mu})$ of generated flows is continuous in the sense of Definition 5.1.3, the claim follows from Thm. 5.1.1. □

In Sect. 5.4 we will give a version of the global bifurcation theorem of WARD [218, Thm. 4] (cf. [219, Thm. 2] for a slightly corrected version) for multi-valued equations. As in WARD [218, 219], this heavily relies on an extension of the continuation theorem, which we shall derive here first. The difference to Thm. 5.3.4 is that it will not be enough to have parametrized isolated invariant sets with a *single* isolating neighbourhood, i.e., $I_\mu = \mathrm{inv}(\overline{U}; F_\mu) \subset U$ with a fixed U, but instead we also have to allow the isolating neighbourhood U to vary with μ. More precisely, the situation will be as follows, cf. WARD [218, Thm. 2]. We are given a relatively open and bounded $\mathcal{U} \subset [a, b] \times \mathbb{R}^n$, and we define for $\mu \in [a, b]$ the sections $\overline{\mathcal{U}}_\mu = \{y \in \mathbb{R}^n : (\mu, y) \in \overline{\mathcal{U}}\}$. To avoid confusion, we want to state explicitly that $\overline{\mathcal{U}}_\mu$ is the section at parameter value μ of the closure of $\mathcal{U}$, but not the closure of the section $\mathcal{U}_\mu = \{y \in \mathbb{R}^n : (\mu, y) \in \mathcal{U}\}$. In general, $\overline{\mathcal{U}_\mu} \subset \overline{\mathcal{U}}_\mu$, but $\overline{\mathcal{U}_\mu} \neq \overline{\mathcal{U}}_\mu$. Let $I_\mu = \mathrm{inv}(\overline{\mathcal{U}}_\mu; F_\mu)$.

Theorem 5.3.5. *In the setting described above, assume the basic assumptions (II) to hold for $F = F(\mu, y)$, and*

$$\mu \in [a, b], \; y \in I_\mu \quad \Longrightarrow \quad (\mu, y) \notin \partial \mathcal{U}, \qquad (5.10)$$

with the boundary being taken w.r. to $[a, b] \times \mathbb{R}^n$. Then $H(I_\mu; F_\mu)$ is defined, $\mu \in [a, b]$, and independent of μ.

Proof: In principle, the proof is analogous to the one of WARD [218, Thm. 2], and we avoid most details. To begin with, (5.10) in particular implies that $\overline{\mathcal{U}}_\mu$ is an isolating neighbourhood for I_μ, and thus every $H(I_\mu; F_\mu)$ is defined, $\mu \in [a, b]$. The idea is now to reduce Thm. 5.3.5 to Thm. 5.3.4 by showing that for every fixed $\mu_0 \in [a, b]$ there is an V_{μ_0} such that V_{μ_0} may serve as an isolating neighbourhood for I_μ with μ in a whole neighbourhood of μ_0. As $H(I_\mu; F_\mu)$ is independent of the special choice of isolating neighbourhood, Thm. 5.3.4 can be applied to yield that $\mu \mapsto H(I_\mu; F_\mu)$ is locally constant close to μ_0.

So let $\mu_0 \in [a, b]$ be fixed and construct the open V_{μ_0} as it was done in the proof of WARD [218, Thm. 2]; whence I_{μ_0} is compactly contained in V_{μ_0}, i.e., there exists $\varepsilon_0 > 0$ with $\text{dist}(y_0, \mathbb{R}^n \setminus V_{\mu_0}) \geq \varepsilon_0$ for all $y_0 \in I_{\mu_0}$. In order to complete the proof it is enough to show $I_\mu \subset V_{\mu_0}$ for μ sufficiently close to μ_0. To see the latter, assume the contrary. Then we would find $\mu_k \to \mu_0$ and $y_0^k \in I_{\mu_k} \cap (\mathbb{R}^n \setminus V_{\mu_0})$, and thus corresponding solutions $y_k : \mathbb{R} \to \overline{U}_{\mu_k}$ of $\dot{y} \in F(\mu_k, y)$ a.e. with $y_k(0) = y_0^k$. Hence by definition $\{\mu_k\} \times y_k(\mathbb{R}) \subset \overline{U}$. Since all $\overline{U}_\mu$ are uniformly bounded, w.l.o.g. we may assume $y_0^k \to y_0$ as $k \to \infty$, and (II)(ii) implies $|\dot{y}_k(t)| \leq \|F(\mu_k, y_k(t))\| \leq \Phi(|y_k(t)|) \leq C$ for some constant $C > 0$, all $k \in \mathbb{N}$, and a.e. $t \in \mathbb{R}$. As in the proof of Lemma 5.3.1, cf. also the proof of Lemma 5.3.5, we thus find a subsequence (relabeled as y_k) and a Lipschitz continuous solution $y : \mathbb{R} \to \mathbb{R}^n$ of $\dot{y} \in F(\mu_0, y)$ a.e., $y(0) = y_0$, such that $y_k \to y$ uniformly on bounded intervals in $\mathbb{R}$. This implies $\{\mu_0\} \times y(\mathbb{R}) \subset \overline{U}$, and therefore $y(\mathbb{R}) \subset \overline{U}_{\mu_0}$, i.e., $y_0 \in I_{\mu_0}$. Consequently, $|y_0 - y_0^k| \geq \varepsilon_0$ for all k, a contradiction. $\square$

5.4 Applications to bifurcation problems

In this section we first state a simple (but useful) bifurcation theorem which is in the same spirit as WARD [218, Thm. 4(c1)], see [219, Thm. 2] for a slightly corrected version. As with degree theory, cf. ZEIDLER [228, Ch. 15.1], one can conclude from a jump of the index (here: the Conley index) in some parameter interval that there has to be a bifurcation point. We remark that we are concerned with bifurcations of non-equilibrium solutions. Afterwards we will apply this bifurcation theorem to a typical non-smooth example problem. Then we prove, along the lines of WARD [218, Thm. 4(c2)] or [219, Thm. 2], a result concerning global bifurcations for non-smooth dynamical systems; in principle, the assertion is analogous to the classical global bifurcation theorem of RABINOWITZ [187].

Our setup is again

$$\dot{y} \in F(\mu, y) = F_\mu(y) \quad \text{a.e.,} \quad \mu \in [a, b], \tag{5.11}$$

with an $F : [a, b] \times \mathbb{R}^n \to 2^{\mathbb{R}^n} \setminus \{\emptyset\}$ satisfying the basic assumptions (II). In addition, we suppose

$$0 \in F(\mu, 0), \quad \mu \in [\mu_1, \mu_2], \tag{E}$$

where $a \leq \mu_1 < \mu_2 \leq b$, i.e., $y = 0$ is an equilibrium solution of (5.11) for $\mu \in [\mu_1, \mu_2]$.

Definition 5.4.1. *The point $(\mu_0, 0) \in [\mu_1, \mu_2] \times \mathbb{R}^n$ is called a bifurcation point of (5.11), if for every $\varepsilon > 0$ there exists $(\mu, y_0) \in [\mu_1, \mu_2] \times (\mathbb{R}^n \setminus \{0\})$ and a global solution $y : \mathbb{R} \to \mathbb{R}^n$ of (5.11) with $y(0) = y_0$ such that $|y(t)| + |\mu - \mu_0| < \varepsilon$ for all $t \in \mathbb{R}$.*

The following theorem implies that a jump of the Conley index causes bifurcation.

Theorem 5.4.1. *Let the basic assumptions (II) and (E) hold for F. If $\{0\}$ is an isolated invariant set for both F_{μ_1} and F_{μ_2}, and if $H(\{0\}; F_{\mu_1}) \neq H(\{0\}; F_{\mu_2})$, then there exists a $\mu_0 \in [\mu_1, \mu_1]$ such that $(\mu_0, 0)$ is a bifurcation point of (5.11).*

Proof: Assume on the contrary that the claim is wrong. Then for every $\mu_0 \in [\mu_1, \mu_2]$ we find $\varepsilon_0 = \varepsilon_0(\mu_0) > 0$ such that for all $(\mu, y_0) \in [\mu_1, \mu_2] \times \mathbb{R}^n$, $y_0 \neq 0$, and global solutions y of (5.11) with initial value y_0 there is a $t \in \mathbb{R}$ such that $|y(t)| + |\mu - \mu_0| \geq 3\varepsilon_0$. This implies that given $\mu_0 \in [\mu_1, \mu_2]$, $H(\{0\}; F_\mu)$ is defined and constant for $\mu \in [\mu_1, \mu_2]$ with $|\mu - \mu_0| \leq \varepsilon_0$. Indeed, the definition of $\mathrm{inv}(\ldots)$ in (5.4), the assumption, and (E) show that $\mathrm{inv}(\overline{B}_{\varepsilon_0}(0); F_\mu) = \{0\}$ for such μ, i.e., $\{0\}$ is an isolated invariant set with uniform isolating neighborhood $\overline{B}_{\varepsilon_0}(0)$. Hence the local constancy of $\mu \mapsto H(\{0\}; F_\mu)$ follows from Thm. 5.3.4. As $[\mu_1, \mu_2]$ is connected, this yields (cf. RYBAKOWSKI [197, p. 66]) that $\mu \mapsto H(\{0\}; F_\mu)$ is constant on $[\mu_1, \mu_2]$, a contradiction. $\quad\square$

Example 5.4.1. For illustration, we return to a special case of (5.1) from p. 141 and consider the differential inclusion

$$\ddot{x} + \mu\dot{x} + x^{2k+1} \in -\varphi(x, \dot{x})\,\mathrm{Sgn}\,\dot{x} \quad \text{a.e.,} \tag{5.12}$$

where $k \in \mathbb{N}_0$. With $y = (x, v) = (x, \dot{x}) \in \mathbb{R}^2$, (5.12) becomes a first-order system in $\mathbb{R}^2$, namely

$$\dot{y} \in F(\mu, y) = \left(v, -\mu v - x^{2k+1} - \varphi(x, v)\,\mathrm{Sgn}\,v\right). \tag{5.13}$$

If $\varphi : \mathbb{R}^2 \to \mathbb{R}$ is continuous, then $F = F(\mu, y) = F(\mu, (x, v))$ may be seen to satisfy the basic assumptions (II), since Sgn is ε-δ-usc; recall Ex. 2.1.2. In addition, (E) holds for arbitrary $\mu_1 < \mu_2$.

The following theorem says that a bifurcation occurs at $\mu_0 = 0$, if φ is sufficiently small near $v = 0$.

Theorem 5.4.2. *Let φ be continuous, and assume there exist $C, \alpha > 0$ such that $|\varphi(x, v)| \leq C|v|^{1+\alpha}$ for (x, v) close to zero in $\mathbb{R}^2$. Then $(\mu_0, (0, 0)) = (0, (0, 0))$ is a bifurcation point of (5.13) resp. (5.12).*

Proof: By Thm. 5.4.1 it is enough to show that $\{(0, 0)\}$ is an isolated invariant set for F_μ with $\mu \neq 0$ and that $H(\{(0, 0)\}; F_\mu) = \Sigma^0$ for $\mu > 0$, but $H(\{(0, 0)\}; F_\mu) = \Sigma^2$ for $\mu < 0$.

We first consider the case $\mu > 0$. Define the Lyapunov function $V(x, v) = \frac{1}{2}v^2 + \frac{1}{2k+2}x^{2k+2}$, $(x, v) \in \mathbb{R}^2$, and $U_r = \{(x, v) \in \mathbb{R}^2 : V(x, v) < r\}$. Then $2C|v|^\alpha \leq \mu$ and $|\varphi(x, v)| \leq C|v|^{1+\alpha}$ for $(x, v) \in \overline{U_r}$, if $r = r(\mu) > 0$ is small enough. We claim that $\text{inv}(\overline{U_r}; F_\mu) = \{(0,0)\}$, i.e., $\{(0,0)\}$ is an isolated invariant set for F_μ with isolating neighborhood $\overline{U_r}$. To see this, assume that there is a solution $y = (x, \dot{x}) : \mathbb{R} \to \overline{U_r}$ of (5.13) with $y(0) = (x_0, v_0) \neq (0,0)$. Then there exists a measurable $w : \mathbb{R} \to [-1,1]$ satisfying $w(t) \in \text{Sgn}\,\dot{x}(t)$ a.e. in $\mathbb{R}$ as well as

$$\ddot{x} + \mu\dot{x} + x^{2k+1} + \varphi(x, \dot{x})\,w(t) = 0 \quad \text{a.e. in } \mathbb{R}. \tag{5.14}$$

By definition of Sgn and choice of r it follows that

$$\frac{d}{dt}V(y(t)) = \dot{x}\,[\ddot{x} + x^{2k+1}] = \dot{x}\,[-\mu\dot{x} - \varphi(x, \dot{x})\,w(t)] = -\mu\,|\dot{x}|^2 - \varphi(x, \dot{x})\,|\dot{x}|$$
$$\leq -\mu\,|\dot{x}|^2 + C\,|\dot{x}|^{2+\alpha} \leq -(\mu/2)\,|\dot{x}|^2. \tag{5.15}$$

This enforces $(x_0, v_0) = (0,0)$, a contradiction; cf. Lemma 5.4.1 below for the technical part. Hence $\{(0,0)\}$ is an isolated invariant set for F_μ when $\mu > 0$. On the other hand, in case that $\mu < 0$, a similar argument may be employed, since then, with analogous notation,

$$\frac{d}{dt}V(y(t)) = -\mu\,|\dot{x}|^2 - \varphi(x, \dot{x})\,|\dot{x}| \geq -\mu\,|\dot{x}|^2 - C\,|\dot{x}|^{2+\alpha} \geq -(\mu/2)\,|\dot{x}|^2.$$

Next we will calculate $H(\{(0,0)\}; F_\mu) = \Sigma^0$ in the case $\mu > 0$. For that, we can use Definition 5.3.2. Let $w_\delta(v) = (2/\pi)\arctan(v/\delta)$, $v \in \mathbb{R}$, be the standard approximations of $\text{Sgn}\,v$. It may be seen that given $\varepsilon > 0$,

$$w_\delta(v) \in \text{Sgn}([v - \varepsilon, v + \varepsilon]) + [-\varepsilon, \varepsilon], \quad v \in \mathbb{R},$$

for δ sufficiently small; whence $f_\delta(x, v) = (v, -\mu v - x^{2k+1} - \varphi(x, v)\,w_\delta(v))$ serves as a suitable (ε, R)-approximation of F_μ, for every $R > 0$ such that $U_r \subset B_R(0)$, with $r = r(\mu)$ from above. Thus $H(\{(0,0)\}; F_\mu) = h(\text{inv}(\overline{U_r}; f_\delta); f_\delta)$ by definition. Since $|w_\delta(v)| \leq 1$, $v \in \mathbb{R}$, V is also a Lyapunov function for the system $\frac{d}{dt}(x, v) = f_\delta(x, v)$. As before we obtain $\text{inv}(\overline{U_r}; f_\delta) = \{(0,0)\}$, and $B = \overline{U_r}$ is a corresponding isolating block with exit set $b^+ = \emptyset$. Hence $H(\{(0,0)\}; F_\mu) = [B/b^+] = \Sigma^0$. In case that $\mu < 0$, we may proceed analogously, because then $b^+ = \partial B$, and therefore $H(\{(0,0)\}; F_\mu) = [B/b^+] = [\overline{B}_1(0)/S^1] = \Sigma^2$. $\qquad\square$

We add some further remarks.

Remark 5.4.1. The method of proof for Thm. 5.4.2 shows that similar results may be derived under various different assumptions on φ, which can be allowed to depend on μ, and also with x^{2k+1} replaced by more general functions. $\qquad\Diamond$

Remark 5.4.2. Note that the isolating neighborhood $N = \overline{U_r}$ from the proof of Thm. 5.4.2 in general will shrink with $\mu > 0$ approaching $\mu = 0$, since we have to ensure $2C|v|^\alpha \leq \mu$ for $(x,v) \in N$. Thus N in fact depends on μ, and therefore the argument of WARD [218, Thm. 5] cannot be applied to yield nontrivial periodic solutions which encircle the origin in $\mathbb{R}^2$, since for this an isolating neighborhood being independent of μ would be needed ($\mu > 0$ in the situation of Thm. 5.4.2 corresponds to $\mu < 0$ in the cited paper and vice versa). Nevertheless, for specific examples it may be verified directly that periodic solutions in fact do bifurcate. Some numerically obtained phase portraits with $\varphi(x, \dot{x}) = \dot{x}^2$ are given in KUNZE/KÜPPER/LI [118].   ◇

The following technical lemma was needed in the proof of Thm. 5.4.2.

Lemma 5.4.1. *In the setting of the proof to Thm. 5.4.2, (5.15) implies* $(x_0, v_0) = (0, 0)$.

Proof: Since $y(\mathbb{R}) \subset \overline{U_r}$ and $t \mapsto V(y(t))$ is non-negative and decreasing, it is enough to show that $V_- = \lim_{t \to -\infty} V(y(t)) = 0$, since then $V(y(t)) = 0$ for $t \in \mathbb{R}$, hence in particular $V(x_0, v_0) = 0$. We have $\dot{x}(t) \to 0$ as $t \to -\infty$. Indeed, suppose $|\dot{x}(t_k)| \geq 2\delta_0$ for some sequence $t_k \downarrow -\infty$ with $t_{k+1} \leq t_k - 1$ and some $\delta_0 > 0$. Because y is globally bounded, (5.14) yields $|\ddot{x}|_{L^\infty(\mathbb{R})} < \infty$, and thus the existence of $\eta \in]0, 1]$ with $|\dot{x}(t)| \geq \delta_0$ for $t \in [t_k, t_k + \eta]$ and all $k \in \mathbb{N}$. Hence we obtain from (5.15)

$$V(y(t_{k-1})) \leq V(y(t_k + \eta)) \leq V(y(t_k)) - (\mu/2)\, \delta_0^2\, \eta, \quad k \in \mathbb{N}.$$

Iteration yields $V(y(t_k)) \geq V(y(t_1)) + (\mu/2)\, \delta_0^2\, \eta\, (k-1)$, a contradiction. Therefore in fact $\dot{x}(t) \to 0$ as $t \to -\infty$, and by definition of V, and since $V(y(t)) \to V_-$, also $x(t) \to x_1$ as $t \to -\infty$ for some $x_1 \in \mathbb{R}$. It follows that $|\int_t^{t+1} x(s)^{2k+1}\, ds - x_1^{2k+1}| \to 0$ as $t \to -\infty$, and thus from integrating (5.14) for sufficiently negative t,

$$|x_1^{2k+1}| \leq \left| \int_t^{t+1} x(s)^{2k+1}\, ds - x_1^{2k+1} \right| + |\dot{x}(t+1) - \dot{x}(t)|$$

$$+ \mu\, |x(t+1) - x(t)| + C \int_t^{t+1} |\dot{x}(s)|^{1+\alpha}\, ds \to 0, \quad t \to -\infty.$$

Hence $x_1 = 0$ yields $V_- = \lim_{t \to -\infty} V(x(t), \dot{x}(t)) = V(0, 0) = 0$ as has been claimed.   □

Now we turn to the investigation of global bifurcations. We introduce

$$\mathcal{S}_0 = \Big\{ (\mu, x_0) \in [a, b] \times (\mathbb{R}^n \setminus \{0\}) : \text{ there exists a global bounded}$$

$$\text{solution } x \text{ of (5.11) with } x(0) = x_0 \Big\}, \quad \text{and}$$

$$\mathcal{S} = \overline{\mathcal{S}_0}^{\,[a,b] \times \mathbb{R}^n}.$$

Note that we can restrict ourselves to $x(0) = x_0$, since (5.11) is an autonomous differential inclusion. As a last result, we obtain the following global bifurcation theorem, cf. WARD [218, Thm. 4] and [219, Thm. 2].

Theorem 5.4.3. *Let the basic assumptions (II) and (E) (with $\mu_1 = a$ and $\mu_2 = b$) hold for F. Assume $\mu_0 \in [a, b]$, $(\mu, 0)$ is not a bifurcation point for $\mu \in [a, b] \setminus \{\mu_0\}$, and $H(\{0\}; F_a) \neq H(\{0\}; F_b)$. Then $(\mu_0, 0)$ is a bifurcation point of (5.11). Moreover, if C denotes the component of S containing $(\mu_0, 0)$, then the following alternative holds: Either $C \subset [a, b] \times \mathbb{R}^n$ is unbounded, or C has nonempty intersection with $\{a, b\} \times \mathbb{R}^n$.*

Proof: We do not give the details here, since we have at hand a continuation theorem that allows a varying isolating neighborhood, cf. Thm. 5.3.5, and thus the proof from WARD [218, Thm. 4] resp. WARD [219, Thm. 2] carries over. Note in particular that $\{0\}$ is an isolated invariant set for every F_μ, $\mu \in [a, b] \setminus \{\mu_0\}$, cf. the proof of Thm. 5.4.1.   □

6. On the application of KAM theory to non-smooth dynamical systems

It is the purpose of this chapter to show how classical KAM (Kolmogorov-Arnold-Moser) theory, which mostly deals with the analysis of smooth systems, can as well be used to prove interesting results for non-smooth dynamical systems. More specifically, we address similar questions as are posed for smooth equations $\ddot{x} + f(x) = p(t)$, i.e., we ask if, for instance, all solutions of the equation are bounded, whether there exist unbounded trajectories, or whether there are "many" periodic and quasiperiodic solutions.

The problem to decide whether all solutions of certain Hamiltonian systems with periodic forcing p have to be bounded (in the $(x, \dot{x})$-phase plane) was posed in LITTLEWOOD [133], cf. LITTLEWOOD [132], and found a first positive answer for the seemingly simple $\ddot{x} + 2x^3 = p(t)$ with a piecewise continuous p in MORRIS [147]; see also DIECKERHOFF/ZEHNDER [72], LAEDERICH/LEVI [122], LEVI [128, 129], ORTEGA [165], and LEVI/YOU [130] for more and refined positive or negative results in this direction. Although the problem seems to be simple, the general method of proof of such boundedness theorems is quite involved, since all these papers rely on the application of suitable variants of Moser's twist (invariant curve) theorem, i.e., on KAM theory. Roughly speaking, this theorem is used to find invariant curves for the Poincaré time-2π return map of the equation, and since solutions cannot leave the flow cylinder comprised by all trajectories up to time 2π starting in the interior of those invariant curves, they have to be bounded.

Following KUNZE/KÜPPER/YOU [119], we are going to illustrate that this approach also works for non-smooth systems, and we consider the example of a forced oscillator, where a nonlinearity is introduced by a jump of the restoring force at $x = 0$, i.e.,

$$\ddot{x} + x + f_a(x) = p(t),$$

with $p \in C^6(\mathbb{R})$ being ω-periodic, and for fixed $a \geq 0$,

$$f_a(x) = \begin{cases} x + a & : & x \geq 0 \\ x - a & : & x < 0 \end{cases}, \tag{6.1}$$

so that the equation is

$$\ddot{x} + x + a\operatorname{sgn} x = p(t) \tag{1.5}$$

if $x(t) \neq 0$; see also p. 3. As was explained in the introduction, for $p = 0$ the set $\{t : x(t) = 0\}$ has measure zero, and thus the value of $\operatorname{sgn}(0)$ does not play a role; cf. Fig. 1.4 for the corresponding phase portrait of (1.5) with $p = 0$.

We recall from Ex. 3.1.2 that all solutions of (1.5) are unbounded in the $(x, \dot{x})$-phase plane in case that $4a < |\int_0^{2\pi} p(t)e^{it} dt|$. This implies that some condition on the size of a (the size of the "gap" of f_a) is needed, and we shall show the boundedness of all solutions of (1.5) as well as the existence of infinitely many periodic and quasiperiodic solutions for all a being sufficiently large. We remark that it follows from a result of LIU [134], relying on a variant of the twist theorem of ORTEGA [166] for maps with small twist, that in fact a does not have to be "very large" to ensure the boundedness of all solutions of (1.5), but that $4a > |\int_0^{2\pi} p(t)e^{it} dt|$ is sufficient.

An application of Moser's invariant curve theorem requires smoothness (see Sect. 6.1), and it is this point which seems to be critical with the above example. To describe how (1.5) nevertheless can be subordinated to an application of a suitable invariant curve theorem, we first remark that this equation is Hamiltonian with

$$H_a(x, y, t) = \frac{1}{2}y^2 + F_a(x) - x\,p(t), \qquad (6.2)$$

where, letting $x_+ = x \vee 0$ and $x_- = (-x) \vee 0$,

$$F_a(x) = \frac{1}{2}\left((x_+ + a)^2 + (x_- + a)^2 - a^2\right) \qquad (6.3)$$

is continuous, but not smooth in x. It is easy to see that moreover $F_a(x) = F_a(-x)$. In the corresponding autonomous equation with $p = 0$, for h being sufficiently large, the level set $\{(x, y) : \frac{1}{2}y^2 + F_a(x) = h\}$ is a closed curve which carries the periodic solution of (1.5) with period denoted by $T(h)$; cf. the phase portrait in Fig. 1.4.

By introducing suitable action-angle variables (I, ϕ) (which will be defined precisely in (6.14), (6.16) below), the Hamiltonian $H_a(x, y, t)$ from (6.2) of the full system will be transformed into

$$H(\phi, I, t) = h_0(I) - x(\phi, I)\,p(t) \qquad (6.4)$$

through some transformation $\Phi_1 : (x, y) \mapsto (\phi, I)$. The new Hamiltonian (6.4) is smooth in I, but only continuous in ϕ. In order to apply KAM theory, the perturbation needs to be sufficiently smooth in the angle variable ϕ. To overcome the difficulty that $x(\phi, I)$ is only continuous with respect to ϕ, we shift the lack of regularity from $x(\phi, I)$ to the forcing p by exchanging the roles of the ϕ and t-variables; see the very illustrative Fig. 1.2 in LEVI [128, p. 47]. This is achieved by the further transformation

$$\Phi_2 : \quad \theta = t, \quad r = h_0(I) - x(\phi, I)\,p(t), \quad \tau = \phi \qquad (6.5)$$

where τ plays the role of new time. We note that this change of variables (and a further transformation) was used in V.I. ARNOLD [12], LEVI [128], or MOSER [150] to get at least the leading term of f [considering $\ddot{x} + f(t,x) = p(t)$] time-independent. Our purpose here is different: since p is assumed to be C^6, we have obtained a new Hamiltonian which is regular enough in (θ, r) to apply Moser's twist theorem to the corresponding Poincaré map.

We will prove the following

Theorem 6.0.4. *For every given ω-periodic function $p \in C^6(\mathbb{R})$ there is a sufficiently large a_* such that for all $a \geq a_*$, every solution of equation (1.5) is bounded, i.e., for every $(t_0, x_0, \dot{x}_0)$ it holds that*

$$\sup_{t \in \mathbb{R}} \left(|x(t; t_0, x_0, \dot{x}_0)| + |\dot{x}(t; t_0, x_0, \dot{x}_0)| \right) < \infty,$$

where $x(t) = x(t; t_0, x_0, \dot{x}_0)$ is the solution with initial values $(x(t_0), \dot{x}(t_0)) = (x_0, \dot{x}_0)$. Moreover, there are infinitely many periodic solutions and quasiperiodic solutions with large amplitude of the form

$$x(t) = f(\lambda t, t/\omega),$$

for some f defined on a 2-torus.

Observe that taking the gap (i.e., a in (6.1)) sufficiently large corresponds to making p small. In fact our system is close to a linear one, since we will show $h_0(I) \sim I$; see Lemma 6.2.2. This once more indicates that we cannot expect to obtain results without some size condition on p, because techniques like DIECKERHOFF/ZEHNDER [72, Prop. 1] cannot be applied to improve the perturbation step by step. In this respect, we are in a position comparable to ORTEGA [165], where similar results were obtained for $\ddot{x} + f(x) = 1 + p(t)$, with the continuous

$$f(x) = \begin{cases} ax & : & x \geq 0 \\ bx & : & x < 0 \end{cases}, \quad a \neq b.$$

Note also that, since we want to show the boundedness of *all* solutions, we cannot use KAM theory in a fixed compact domain, but we have to derive estimates on the perturbation. It should moreover be remarked that it is enough to assume $p \in C^4(\mathbb{R})$ in Thm. 6.0.4, since the version of Moser's twist theorem which will be applied below (cf. Thm. 6.2.1) holds if the Poincaré map is only C^3-regular; see HERMAN [99].

This chapter is organized as follows. We first explain, on a very basic level, some aspects of KAM theory (to be more precise, of Moser's twist theorem) in Sect. 6.1. Then in Sect. 6.2 we describe the necessary coordinate changes to transform the system into a smooth and nearly integrable one. We also state the corresponding estimates for the application of the twist theorem, postponing their proofs to an appendix, Sect. 6.3. These estimates are somewhat lengthy, but crucial in the procedure to fit the problem into the framework of KAM theory.

6.1 A very brief introduction to Moser's twist theorem

Originally KAM theory has been developed from questions concerning the stability of Hamiltonian systems w.r. to small perturbations. A primary motivation for studying such problems was (and still is) the classical three body problem of celestial mechanics, where the system consisting of sun and earth is perturbed by an additional moon which is small compared to sun and earth. Starting with the work of POINCARÉ [180] and BIRKHOFF [25], see also SIEGEL/MOSER [203], this question of the stability of the solar system had an enormous impact on the development of the whole theory of dynamical systems.

The following few remarks are meant to provide a short introduction to the twist theorem from MOSER [148], since this theorem (in one of its variants) will play the central role in our analysis of (1.5).

We consider the example

$$\ddot{x} + x^3 = 0 \tag{6.6}$$

which is a Hamiltonian system, because for a solution to (6.6) the Hamiltonian $H_0(x,y) = \frac{1}{2}y^2 + \frac{1}{4}x^4$ is constant along solutions, i.e., if $t \mapsto x(t)$ is a solution of (6.6) on some time interval, then $\frac{d}{dt}H_0(x(t), \dot{x}(t)) = 0$ on this time interval. This fact already has an important consequence, since all solutions have to lie on some level set $\{(x,y) : H_0(x,y) = h\}$ in the $(x,\dot{x})$-phase plane, and the appropriate energy level $h \geq 0$ is determined solely through the initial state $(x(0), \dot{x}(0))$. A sketch of the phase portrait of (6.6) is shown in Fig. 6.1.

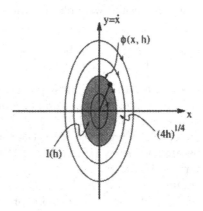

Fig. 6.1. Phase portrait of (6.6)

It can be seen that the phase plane is filled with periodic solutions of (6.6), and the period $T(h)$ on the energy level h is, cf. HALE/KOÇAK [97, Exerc. 7.15],

$$T(h) = 4 \int_0^{(4h)^{1/4}} \frac{dx}{\sqrt{2(h - \frac{1}{4}x^4)}} = 4h^{-1/4} \int_0^1 \frac{dy}{\sqrt{1 - y^4}} = B\left(\frac{1}{4}, \frac{1}{2}\right) h^{-1/4},$$

with $B(p, q) = \int_0^1 t^{p-1}(1 - t)^{q-1} dt$ being Euler's beta-function.

From Fig. 6.1 it is clear that solutions may equivalently be described through so-called action-angle variables (ϕ, I); action I is the area encircled by the solution orbit, whereas $\phi \in [0, 2\pi[$ is the clockwise angle with the y-axis $\{x = 0\}$. Hence

$$I(h) = 4 \int_0^{(4h)^{1/4}} \sqrt{2(h - \frac{1}{4}x^4)} \, dx = 8h^{3/4} \int_0^1 \sqrt{1 - y^4} \, dy$$

$$= 2B\left(\frac{1}{4}, \frac{3}{2}\right) h^{3/4},$$

and for $x, y \geq 0$ and $h = \frac{1}{2}y^2 + \frac{1}{4}x^4$

$$\phi(x, h) = \frac{2\pi}{T(h)} \int_0^x \frac{d\xi}{\sqrt{2(h - \frac{1}{4}\xi^4)}} = 2\pi B\left(\frac{1}{4}, \frac{1}{2}\right)^{-1} \left(\int_0^{x/(4h)^{1/4}} \frac{dy}{\sqrt{1 - y^4}}\right).$$

Applying the transformation $\Phi : (x, y) = (x, \dot{x}) \mapsto (\phi, I)$ to the Hamiltonian, H_0 in new coordinates becomes

$$H_0(\phi, I) \cong H_0(x, y) \cong h \cong h(I) = \left(2B\left(\frac{1}{4}, \frac{3}{2}\right)\right)^{-4/3} I^{4/3},$$

since $I(h) = 2B(\frac{1}{4}, \frac{3}{2})h^{3/4}$. Therefore the corresponding equations of motions are

$$\dot{\phi} = \frac{\partial}{\partial I} H_0(\phi, I) = \frac{2^{2/3}}{3} B\left(\frac{1}{4}, \frac{3}{2}\right)^{-4/3} I^{1/3} = h_0(I),$$

$$\dot{I} = -\frac{\partial}{\partial \phi} H_0(\phi, I) = 0, \tag{6.7}$$

in accordance with the fact that by definition I does not change on a solution orbit. Let $P_0 : [0, 2\pi] \times [0, \infty[\to [0, 2\pi] \times [0, \infty[$, $P_0 : (\phi, I) \mapsto (\phi_1, I_1)$, be the Poincaré map of (6.7), i.e., (ϕ_1, I_1) is the point where a trajectory starting in (ϕ, I) arrives at time $t = 2\pi$; observe that here $[0, 2\pi] \times [0, \infty[$ in fact stands for the infinite (periodic) cylinder $S^1 \times [0, \infty[$, since in particular the equations are 2π-periodic in ϕ. Hence we have

$$P_0(\phi, I) = (\phi_1(\phi, I), I_1(\phi, I)) = (\phi + 2\pi h_0(I), I), \tag{6.8}$$

and thus

$$\det DP_0(\phi, I) = \det \begin{pmatrix} 1 & 2\pi h_0'(I) \\ 0 & 1 \end{pmatrix} = 1,$$

i.e., P_0 is an area-preserving map.

From (6.8) it is obvious why such maps are called twist maps: an application of P_0 corresponds to a clockwise twist of angle $2\pi h_0(I)$. Note also that trivially $P_0(\{(\phi, I) : I = I_0\}) \subset \{(\phi, I) : I = I_0\}$, and consequently for every $I_0 > 0$, $\{(\phi, I) : I = I_0\}$ is an invariant circle of P_0. On this circle the rotation number is, by definition,

$$\alpha(I_0) = \lim_{n \to \infty} \frac{1}{n} [\phi_1^n(\phi, I_0) - \phi] = \lim_{n \to \infty} \frac{1}{n} [2n\pi h_0(I_0)] = 2\pi h_0(I_0)$$

$$= \frac{2^{5/3}}{3} \pi B\left(\frac{1}{4}, \frac{3}{2}\right)^{-4/3} I_0^{1/3},$$

with $f^2 = f \circ f$ and $f^n = f^{n-1} \circ f$. Hence for a.e. (w.r. to Lebesgue measure) I_0, $\alpha(I_0)/2\pi = h_0(I_0)$ is an irrational number, and there are countably many I_0 such that $\alpha(I_0)/2\pi \in \mathbb{Q}$.

Now the question that Moser's twist theorem addresses is whether such invariant circles do persist if a small perturbation is added to the Poincaré map, or equivalently, if the Hamiltonian $H_0(x, y)$ is perturbed to $H(x, y, t, \varepsilon) = H_0(x, y) + H_1(x, y, t, \varepsilon)$, with a sufficiently small perturbation H_1 for ε small. Letting $\alpha(I) = 2\pi h_0(I)$, we now consider the perturbed

$$P_\varepsilon : \quad (\phi, I) \mapsto (\phi_1, I_1) = (\phi + \alpha(I) + \varepsilon f(\phi, I), I + \varepsilon g(\phi, I)) \tag{6.9}$$

with $f, g : [0, 2\pi] \times [a, b] \to \mathbb{R}$ being 2π-periodic in ϕ and sufficiently often differentiable. Note that here we restricted ourselves to a finite cylinder $S^1 \times [a, b]$ with $0 < a < b$, i.e., P_ε is defined on an annulus.

The twist theorem states that all invariant curves $\{(\phi, I) : I = I_0\}$ survive the perturbation (in the sense that P_ε will have an invariant curve in $[0, 2\pi] \times [a, b]$), provided that $\alpha(I_0)/2\pi$ is "sufficiently irrational". Thus "almost all" invariant curves are going to survive the perturbation. The precise statement, of course not restricted to our example, is as follows.

Theorem 6.1.1 (Moser's Twist Theorem). *Assume $\alpha'(I) \neq 0$, $I \in [a, b]$. Then there exist $C_* > 0$ and $\varepsilon_* > 0$ such that the following holds. Let P_ε be a map of the form (6.9) with $|\varepsilon| \leq \varepsilon_*$ and $f, g \in C^N$,*

$$|f|_{C^N} + |g|_{C^N} \leq C_*,$$

and, moreover, suppose that P_ε has the intersection property, i.e., $P_\varepsilon(C) \cap C \neq \emptyset$ for every embedded circle C in $[0, 2\pi] \times [a, b]$ which is homotopic to a circle $I = $ const. in $[0, 2\pi] \times [a, b]$. Let ω between $\alpha(a)$ and $\alpha(b)$ be incommensurable with 2π such that

$$\exists c_\omega > 0 \ \forall p, q \in \mathbb{N} : \quad \left| \frac{\omega}{2\pi} - \frac{p}{q} \right| \geq c_\omega |q|^{-5/2}. \tag{6.10}$$

Then there exists a differentiable closed curve

$$\phi = \xi + F(\xi, \varepsilon), \quad I = G(\xi, \varepsilon),$$

(with F, G of period 2π in the curve parameter ξ) which is invariant for P_ε, and P_ε restricted to this curve has rotation number ω.

Note that Thm. 6.1.1 in particular applies to our above example, since in that case

$$\alpha'(I) = \frac{2^{5/3}}{9} \pi B\left(\frac{1}{4}, \frac{3}{2}\right)^{-4/3} I^{-2/3} \neq 0, \quad I \in [a, b],$$

for every $0 < a < b$.

Theorem 6.1.1 originally appeared in MOSER [148] with $N \geq 333$, the number "333" being more or less a joke, as e.g. $N \geq 17$ would have been enough; cf. also V.I. ARNOLD [13, 14] and ARROWSMITH/PLACE [16, p. 330 ff.]. If the diophantine condition (6.10) is modified to

$$\exists \tau > 0 \; \exists c_\omega > 0 \; \forall p, q \in \mathbb{N} : \quad \left| \frac{\omega}{2\pi} - \frac{p}{q} \right| \geq c_\omega |q|^{-\tau},$$

then $N \geq 5$ is sufficient; see MOSER [149] and HERMAN [99]. I fact there have been developed a lot of versions of the twist theorem (also called invariant curve theorem), and the one appropriate for our purpose will be given in Thm. 6.2.1 below; it is from ORTEGA [165] and follows from results of HERMAN [99].

To end this short introductory section, we remark that Thm. 6.1.1 often can be used to prove the boundedness of solutions in the $(x, \dot{x})$-phase plane, since, roughly speaking, an application to every cylinder $[0, 2\pi] \times [K, K+1]$, $K \geq 1$, will produce a closed invariant curve $\mathcal{I}_K$ for the Poincaré map P_ε. Transformed back to $(x, y) = (x, \dot{x})$-geometry, $\mathcal{I}_K$ is a curve encircling the origin with the property that whenever a solution hits this curve, it will return to the curve after time 2π. Hence, if $\dot{I} = -\frac{\partial}{\partial \phi} H(\phi, I)$ (if the perturbed Hamiltonian $H = H_0 + H_1$ is written in action-angle variables) is bounded on bounded I-intervals, the solution cannot get arbitrarily far in bounded time-intervals, in particular, in time 2π, and this means that solutions starting in $\mathcal{I}_K$ have to be bounded. As $\mathcal{I}_K$ will become arbitrarily large for $K \to \infty$, this forces all solutions to be bounded. For equations of type $\ddot{x} + V(t, x) = p(t)$ with a potential V being "superquadratic" in x, one can prove this boundedness, under certain conditions, even without the assumption that p be suitably small. The main idea is to introduce a sequence of further coordinate changes, which somehow decrease the "order of the errors" in each step.

6.2 Coordinate transformations, estimates, and the proof of Theorem 6.0.4

In this section we give some preliminary results and introduce in detail suitable transformations which will allow us to apply Moser's twist theorem. To simplify notation, we do not consider (1.5) directly, but the equivalent

$$\ddot{x} + x + \operatorname{sgn} x = \varepsilon p(t) \tag{6.11}$$

where $\varepsilon = \frac{1}{a}$. In the autonomous case $p = 0$, for energies $h > 1/2$, the closed curve $H_1^0(x, y) = \frac{1}{2} y^2 + F_1(x) = h$ with F_1 from (6.3) carries the periodic solution of (1.5) with period $T(h)$. If we let $\alpha(h) = \sqrt{2h} - 1 > 0$, then the intersections of this level set with the x-axis in the phase plane are $(\pm \alpha(h), 0)$; see Fig. 6.1. Hence symmetry of F_1 yields

$$T(h) = 4 \int_0^{\alpha(h)} \frac{dx}{\sqrt{2(h - F_1(x))}} = 2\pi - 4 \arcsin (2h)^{-1/2}$$

$$= 4 \arccos (2h)^{-1/2} \quad (h > 1/2). \tag{6.12}$$

Some properties of $T(h)$ are collected in

Lemma 6.2.1. $T(\cdot)$ *is smooth in h, $0 \leq T(h) \leq 2\pi$, $T'(h) > 0$, and*

$$T^{-1}(\rho) = \frac{1}{2 \cos^2(\rho/4)}, \quad 0 < \rho < 2\pi.$$

In addition, the following estimates hold:

$$c_i h^{-\frac{1}{2}-i} \leq |D^i T(h)| \leq C_i h^{-\frac{1}{2}-i} \quad (i \geq 1) \tag{6.13a}$$

$$c_0 |2\pi - \rho|^{-2} \leq T^{-1}(\rho) \leq C_0 |2\pi - \rho|^{-2},$$

$$c_i |2\pi - \rho|^{-(i+2)} \leq |D^i T^{-1}(\rho)| \leq C_i |2\pi - \rho|^{-(i+2)} \quad (i \geq 1) \tag{6.13b}$$

for some constants $C_i, c_i > 0$ and h sufficiently large resp. $2\pi - \rho$ sufficiently small; here D^i denotes the i'th derivative. Moreover, let $\delta_n = 1/n$. Then for sufficiently large $n \in \mathbb{N}$ the intervals $[b_n^-, b_n^+] = T^{-1}([2\pi - 2\delta_n, 2\pi - \delta_n]) \subset (1/2, \infty)$ are disjoint, and $b_n^-, b_n^+, b_n^+ - b_n^- \to \infty$ as $n \to \infty$.

Proof: From (6.12) we have

$$T'(h) = \frac{2}{h \sqrt{2h - 1}}.$$

Thus it follows by induction that $D^i T(h) = P_{i-1}(h) h^{-i} (2h - 1)^{1/2-i}$ for $i \geq 1$, with P_{i-1} being a polynomial of degree i, and hence we obtain the claimed estimates for $D^i T$. The estimates for $D^i T^{-1}$ are obtained inductively by differentiation of $T^{-1}(T(h)) \equiv h$ and by (6.13a), cf. LEVI [128, (A1.5), p. 73] for the corresponding formula. The claim concerning the intervals $[b_n^-, b_n^+]$ can also be derived from the explicitly known T^{-1}.   □

To construct the action-angle variables (cf. V.I. ARNOLD [15] or LEVI [128, Sect. 2] for more information) and to transform the Hamiltonian into a nearly integrable one, let

$$\Phi_1 : \quad (x, y) \mapsto (\phi, I)$$

be defined implicitly by the following two equations: For $h > 1/2$ and $|x| \leq \alpha(h)$ set

$$\phi(x, h) = \begin{cases} \phi_1(x, h) & : \ x, y \geq 0, \\ \pi - \phi_1(x, h) & : \ x \geq 0, y \leq 0, \\ \pi + \phi_1(-x, h) & : \ x, y \leq 0, \\ 2\pi - \phi_1(-x, h) & : \ x \leq 0, y \geq 0, \end{cases} \tag{6.14}$$

where for $0 \leq x \leq \alpha(h)$

$$\begin{aligned} \phi_1(x, h) &= \frac{2\pi}{T(h)} \int_0^x \frac{d\xi}{\sqrt{2(h - F_1(\xi))}} \\ &= \frac{2\pi}{T(h)} \left(\arcsin \frac{1 + x}{\sqrt{2h}} - \arcsin \frac{1}{\sqrt{2h}} \right), \end{aligned} \tag{6.15}$$

and

$$\begin{aligned} I(h) &= 4 \int_0^{\alpha(h)} \sqrt{2(h - F_1(x))}\, dx = 4 \int_1^{1+\alpha(h)} \sqrt{2h - x^2}\, dx \\ &= 2h\pi - 2\sqrt{2h - 1} - 4h \arcsin (2h)^{-1/2} = h\, T(h) - 2\sqrt{2h - 1}. \end{aligned} \tag{6.16}$$

So by (6.14), (6.15), and (6.16) we have obtained concrete formulae for the action-angle transformation Φ_1. Geometrically, $I(h)$ is the area surrounded by $\{(x, y) : \frac{1}{2}y^2 + F_1(x) = h\}$.

Before turning our attention to Φ_1, we first collect some properties of $I(h)$. Here and in the sequel, positive constants not depending on important quantities are denoted by the same symbols $c, c_i, C_i \dots$ etc.

Lemma 6.2.2. $I(\cdot)$ *as well as its inverse* $I \mapsto h(I) = h_0(I)$ *are smooth, and* $I'(h) = T(h) > 0$. *Moreover, for suitable positive constants and for sufficiently large* h *resp.* I

$$\begin{aligned} & c_0\, h \leq I(h) \leq C_0\, h, \quad c_1 \leq I'(h) \leq C_1, \\ & c_i\, h^{\frac{1}{2} - i} \leq D^i I(h) \leq C_i\, h^{\frac{1}{2} - i} \quad (i \geq 2) \end{aligned} \tag{6.17a}$$

$$\begin{aligned} & c_0\, I \leq h_0(I) \leq C_0\, I, \quad c_1 \leq h_0'(I) \leq C_1, \\ & |D^i h_0(I)| \leq C_i\, I^{\frac{1}{2} - i} \quad (i \geq 2). \end{aligned} \tag{6.17b}$$

Proof: First, (6.16) gives $I' = T$, and the estimates for I follow from (6.16), (6.12), and (6.13a). Second, the desired estimates for h_0 are again obtained by successive differentiation of $h_0(I(h)) \equiv h$ using (6.17a).   □

Hence $h_0(\cdot)$ behaves like $I(\cdot)$ itself. Now we can investigate Φ_1 in greater detail.

Lemma 6.2.3. Φ_1 *is a homeomorphism from* $\mathbb{R}^2 \backslash \{0\}$ *to the cylinder* $[0, 2\pi] \times (1/2, \infty)$. *Moreover, if* $(x, y)(\phi, I) = \Phi_1^{-1}(\phi, I)$, *then* $x(\phi, I)$ *is smooth in* I *for every fixed* ϕ, *and*

$$|\partial_I^i x(\phi, I)| \leq C_i I^{\frac{1}{2} - i} \quad (i \geq 0). \qquad (6.18)$$

Proof: By construction of (ϕ, I) it is clear that Φ_1 is continuous, onto and one-to-one. In addition, we note that $x(\phi, I) = \tilde{x}(\phi, h_0(I))$, where we have e.g. in the first quadrant $\phi = \phi_1(\tilde{x}, h)$ with $h = h_0(I)$ and ϕ_1 from (6.15). Therefore for $\phi \in (0, \pi/2)$, by solving $\phi = \phi_1(\tilde{x}, h)$ w.r. to $\tilde{x}$,

$$\tilde{x}(\phi, h) = -1 + \sqrt{2h} \, \sin\left(\frac{T(h)}{2\pi} \phi + \arcsin(2h)^{-1/2}\right)$$

$$= -1 + \sqrt{2h - 1} \, \sin\left(\frac{T(h)}{2\pi} \phi\right) + \cos\left(\frac{T(h)}{2\pi} \phi\right), \qquad (6.19)$$

where we used the formula $\sin(v + \arcsin(u)) = \sqrt{1 - u^2} \, \sin(v) + u \cos(v)$ for $u \in [0, 1]$, $v \in \mathbb{R}$. In the same way we obtain for $\phi \in (3\pi/2, 2\pi)$

$$\tilde{x}(\phi, h) = 1 - \sqrt{2h - 1} \, \sin\left(\frac{T(h)}{2\pi} [2\pi - \phi]\right) - \cos\left(\frac{T(h)}{2\pi} [2\pi - \phi]\right),$$

and this proves the continuity of $\tilde{x}(\phi, h)$ in ϕ at $\phi = 0 \cong 2\pi$. Hence it follows that $\tilde{x}(\phi, h)$ is continuous in (ϕ, h), and this carries over to x because of Lemma 6.2.2. Moreover, (6.19) shows that $\tilde{x}(\phi, h)$ is smooth in h for fixed $\phi \in [0, \pi/2]$, and thus $x(\phi, I)$ is smooth in I in view of Lemma 6.2.2.

To verify the desired estimates, we first claim that

$$|\partial_h^i \tilde{x}(\phi, h)| \leq C_i h^{\frac{1}{2} - i} \quad (i \geq 0). \qquad (6.20)$$

For $\phi \in [0, \pi/2]$ this can be derived from (6.19) and (6.13a). Here we omit the details, since the required technique will be applied once more below; see the appendix, Sect. 6.3, or DIECKERHOFF/ZEHNDER [72], LEVI [128], MORRIS [147], ORTEGA [165], and related papers. Usually, these estimates can be derived by induction, using Leibniz' rule and the formulas from Lemma 6.3.1. Since $x(\phi, I) = \tilde{x}(\phi, h_0(I))$, the estimate (6.20) also carries over to x, by Lemma 6.3.1 and the estimates for $D^i h_0$ from (6.17b).   □

Applying the action-angle transformation to the full system, the Hamiltonian $H_1(x, y, t) = \frac{1}{2}y^2 + F_1(x) - \varepsilon x p(t)$ which corresponds to (6.11) is transformed into a new function in (ϕ, I), that is

$$H(\phi, I, t) = h_0(I) - \varepsilon x(\phi, I)p(t). \tag{6.21}$$

From Lemma 6.2.3 it follows that $H(\phi, I, t)$ is smooth in I and continuous in ϕ. Moreover, $|\partial_I x(\phi, I)| \leq C I^{-1/2}$ by (6.18). Therefore (6.17b) implies that $H(\phi, \cdot, t)$ is invertible for sufficiently large I.

Next we exchange the roles of ϕ and t by means of

$$\Phi_2: \quad (\phi, I, t) \mapsto (\theta, r, \tau) = (t, H(\phi, I, t), \phi),$$

cf. LEVI [128, Sect. 3]. This transformation again leads to a Hamiltonian system, where the new Hamiltonian is

$$\mathcal{H}(\theta, r, \tau) = [H(\tau, \cdot, \theta)]^{-1}(r). \tag{6.22}$$

Thus $\mathcal{H}(\theta, r, \tau)$ is ω-periodic in θ, 2π-periodic in τ, and smooth in (θ, r). We write

$$\mathcal{H}(\theta, r, \tau) = I(r) + \varepsilon \mathcal{H}_1(\theta, r, \tau), \tag{6.23}$$

i.e., $\mathcal{H}_1$ is defined by this formula. Some estimates of $\mathcal{H}_1$ are given in

Lemma 6.2.4. *For r sufficiently large,*

$$|\partial_\theta^i \partial_r^j \mathcal{H}_1(\theta, r, \tau)| \leq C_{i,j} \, r^{\frac{1}{2} - j} \quad (0 \leq i + j \leq 6), \tag{6.24}$$

where $C_{i,j}$ depends on $|p|_{C^i([0,\omega])}$.

Proof: Cf. p. 178 in the appendix, Sect. 6.3. □

Since $I' = T$, the equations of motion corresponding to the Hamiltonian $\mathcal{H}$ from (6.23) are

$$\frac{d\theta}{d\tau} = T(r) + \varepsilon \, \partial_r \mathcal{H}_1(\theta, r, \tau), \qquad \frac{dr}{d\tau} = -\varepsilon \, \partial_\theta \mathcal{H}_1(\theta, r, \tau), \tag{6.25}$$

with τ serving as time in the equation and $\mathcal{H}_1(\theta, r, \tau)$ being ω-periodic in θ and 2π-periodic in τ. Moreover, since $\partial_r \mathcal{H}_1$ resp. $\partial_\theta \mathcal{H}_1$ are C^6 resp. C^5 in θ, smooth in r, and continuous in τ, we note that in particular solutions $\tau \mapsto (\theta(\tau), r(\tau))$ of (6.25) exist. In addition, by Lemma 6.2.4 and the equation for $dr/d\tau$, there are (a large) $r_* > 0$ and a constant C_* such that all solutions of (6.25) with $\varepsilon \leq 1$ and initial values $r_0 \geq r_*$ are defined on the whole of $[0, 2\pi]$ and satisfy there $|r(\tau) - r_0| \leq C_* r_0$, hence $[C_* - 1]r_0 \leq r(\tau) \leq [C_* + 1]r_0$. Thus, letting $\delta_n = 1/n$ and retaining the notation of Lemma 6.2.1, we obtain for initial values

$$r_0 \in J_n = [b_n^-/(C_* - 1), b_n^+/(C_* + 1)] \tag{6.26}$$

that

$$T(r(\tau)) \in [2\pi - 2\delta_n, 2\pi - \delta_n], \quad \tau \in [0, 2\pi]. \tag{6.27}$$

Also note that $b_n^-/(C_* - 1) \to \infty$, $b_n^+/(C_* + 1) \to \infty$, and $b_n^+/(C_* + 1) - b_n^-/(C_* - 1) \to \infty$ as $n \to \infty$. In particular we may assume that the initial values r_0 are large enough that all estimates derived so far hold. Suppressing the index n, we rescale, cf. LEVI [128, Sect. 4.2],

$$\bar{\theta}(\tau) = \omega^{-1}\,\theta(\tau) \quad \text{and} \quad \bar{\rho}(\tau) = \delta^{-1}\left[T(r(\tau)) - 2\pi\right]. \tag{6.28}$$

Thus

$$\bar{\rho} \in [-2, -1] \tag{6.29}$$

by (6.27), and differentiation yields

$$\frac{d\bar{\theta}}{d\tau} = \frac{\delta}{\omega}\,\bar{\rho} - \frac{2\pi}{\omega} + \varepsilon\,f_1(\bar{\theta}, \bar{\rho}, \tau) \quad \text{and} \quad \frac{d\bar{\rho}}{d\tau} = \varepsilon\,f_2(\bar{\theta}, \bar{\rho}, \tau) \tag{6.30}$$

with

$$f_1(\bar{\theta}, \bar{\rho}, \tau) = \omega^{-1}\,\partial_r \mathcal{H}_1(\omega\bar{\theta}, r(\bar{\rho}), \tau), \quad \text{and}$$
$$f_2(\bar{\theta}, \bar{\rho}, \tau) = -\delta^{-1}\,T'(r(\bar{\rho}))\,\partial_\theta \mathcal{H}_1(\omega\bar{\theta}, r(\bar{\rho}), \tau),$$

where we let $r(\bar{\rho}) = T^{-1}(\delta\bar{\rho} + 2\pi)$. Then f_1 and f_2 are 1-periodic in $\bar{\theta}$ and 2π-periodic in τ. Estimates on f_1 and f_2 are given in

Lemma 6.2.5. *There is a C depending on $|p|_{C^6([0,\omega])}$ (but not on δ) such that*

$$|\partial_{\bar{\theta}}^i \partial_{\bar{\rho}}^j f_1(\bar{\theta}, \bar{\rho}, \tau)| + |\partial_{\bar{\theta}}^i \partial_{\bar{\rho}}^j f_2(\bar{\theta}, \bar{\rho}, \tau)| \leq C\delta \quad (0 \leq i + j \leq 5). \tag{6.31}$$

Proof: Cf. p. 181 in the appendix, Sect. 6.3.   □

Let $\bar{P} : \mathbb{R} \times [-2, -1] \to \mathbb{R} \times \mathbb{R}$ denote the time-2π-map of (6.30). Then this Poincaré map is of class C^5 in $(\bar{\theta}, \bar{\rho})$, 1-periodic in $\bar{\theta}$, and it can be written in the following form

$$\bar{P} : \quad \begin{array}{rcl} \bar{\theta}_1 & = & \bar{\theta} + \alpha + \bar{\delta}\bar{\rho} + \bar{\delta}\varepsilon\,\bar{P}_1(\bar{\theta}, \bar{\rho}) \\ \bar{\rho}_1 & = & \bar{\rho} + \bar{\delta}\varepsilon\,\bar{P}_2(\bar{\theta}, \bar{\rho}) \end{array}, \tag{6.32}$$

i.e., we define $\bar{P}_1$ and $\bar{P}_2$ through this relations. Here $\alpha = -4\pi^2/\omega$ and $\bar{\delta} = 2\pi\delta/\omega$. Recall that ω is the period of the forcing p, and also notice that $\bar{P}$, $\bar{P}_1$, $\bar{P}_2$, and $\bar{\delta}$ in fact depend on n. For instance, $\bar{\delta}_n = 2\pi\delta_n/\omega = 2\pi/\omega n$. The next result gives estimates on $\bar{P}_1$ and $\bar{P}_2$.

Lemma 6.2.6. *For sufficiently large* $n \in \mathbb{N}$, *i.e., sufficiently small* $\bar{\delta} = \bar{\delta}_n$,

$$|\partial_{\bar{\theta}}^i \partial_{\bar{\rho}}^j \bar{P}_1(\bar{\theta}, \bar{\rho})| + |\partial_{\bar{\theta}}^i \partial_{\bar{\rho}}^j \bar{P}_2(\bar{\theta}, \bar{\rho})| \leq C \quad (0 \leq i+j \leq 5), \tag{6.33}$$

with C *depending on* $|p|_{C^6([0,\omega])}$, *but not on* n.

Proof: Cf. p. 182 in the appendix, Sect. 6.3. $\qquad\square$

Up to one further transformation, we are now in the position to apply Moser's twist theorem, first stated in MOSER [148]. Below we are going to use this theorem in a form analogously to the one in ORTEGA [165, Thm. 4.5], which in turn was derived from the results of HERMAN [99, Sect. 5]. We first need to introduce irrational numbers of constant type.

Definition 6.2.1. *An* $\alpha \in \mathbb{R} \setminus \mathbb{Q}$ *is of constant type, if*

$$\gamma = \inf \{q^2|\alpha - p/q| : p \in \mathbb{Z}, q \in \mathbb{N}\} > 0.$$

In this case, γ *is called the Markoff constant of* α, *and* α *is called of constant-* γ-*type.*

Then $|\alpha - p/q| \geq \gamma q^{-2}$ for all $p \in \mathbb{Z}$ and $q \in \mathbb{N}$, i.e., γ gives some bound on how α can be approximated by rational numbers. We moreover remark that if $\alpha \in \mathbb{R} \setminus \mathbb{Q}$ has the continued fraction expansion

$$\alpha = [a_0; a_1, a_2, \ldots] = a_0 + \cfrac{1}{a_1 + \cfrac{1}{a_2 + \cdots}},$$

then α is of constant type if and only if $C = \sup\{a_i : i \geq 1\} < \infty$; in that case $1/(C+2) \leq \gamma \leq 1/C$; see HERMAN [99, p. 160]. An example of an irrational of constant type is $\frac{1}{2}(\sqrt{5} - 1)$, since $\frac{1}{2}(\sqrt{5} - 1) = [0; 1, 1, 1, 1, 1, \ldots]$.
In fact, a quite precise approximation of real numbers through irrationals of constant type is possible.

Lemma 6.2.7. *For every interval* $[a, b] \subset [0, 1]$ *with* $b - a \geq \varepsilon > 0$ *there exists* $\alpha_* \in [a, b] \setminus \mathbb{Q}$ *of constant type such that the corresponding Markoff constant* γ_* *satisfies* $\varepsilon/16 \leq \gamma_* \leq \varepsilon/4$.

Proof: This is ORTEGA [165, Lemma 4.4]. $\qquad\square$

The following is a version of Moser's twist theorem. The term "intersection property" is made precise in Lemma 6.2.8 below; cf. also Thm. 6.1.1 in Sect. 6.1.

Theorem 6.2.1 (Invariant Curve Theorem). *Let* $P : \mathbb{R} \times [-6, -3] \to \mathbb{R}^2$, $(u, v) \mapsto (u_1, v_1)$, *be of class* C^5, *one-to-one, and 1-periodic in* u. *In addition, assume that* P *has the intersection property, and that* P *may be written in the form*

$$u_1 = u + \beta + \bar{\delta} v + \bar{\delta} F_1(u, v), \quad v_1 = v + \bar{\delta} F_2(u, v)$$

where $\bar{\delta} \in (0, 2)$, *and* β *is an irrational of constant type with Markoff constant* γ *satisfying*

$$\gamma \leq \bar{\delta} \leq M\gamma \tag{6.34}$$

for some fixed constant M. *Then there is a positive constant* M_*, *depending only on* M, *such that if*

$$|F_1|_{C^5} + |F_2|_{C^5} \leq M_*,$$

one finds $\mu \in C^3(\mathbb{R}/\mathbb{Z})$ *such that the curve* $\mathcal{I}_\mu = \{(u, \mu(u)) : u \in \mathbb{R}\}$ *is invariant under* P, *and* $P|_{\mathcal{I}_\mu}$ *has rotation number* β.

Now we can carry out the

Proof of Theorem 6.0.4:

The proof of the boundedness of all solutions is similar to the one of ORTEGA [165, Thm. 4.1]. We will no longer suppress the index n, to emphasize the dependence of $\bar{P}$, $\bar{P}_1$, $\bar{P}_2$, and $\bar{\delta}$ on n. In order to apply Thm. 6.2.1, we need to approximate $\alpha = -4\pi^2/\omega$ by irrationals of constant type with suitable Markoff constants. For every $n \in \mathbb{N}$ we can use Lemma 6.2.7 with $\varepsilon = 2\bar{\delta}_n = 4\pi/\omega n$, $b = \alpha + 2\varepsilon$, and $a = \alpha + \varepsilon$ to find irrationals α_n of constant type with Markoff constants γ_n such that

$$2\bar{\delta}_n = \frac{4\pi}{\omega n} \leq \alpha_n - \alpha \leq \frac{8\pi}{\omega n} = 4\bar{\delta}_n \quad \text{and} \quad \frac{\pi}{4\omega n} \leq \gamma_n \leq \frac{\pi}{\omega n}, \quad \text{hence}$$

$$\bar{p} + \frac{\alpha - \alpha_n}{\bar{\delta}_n} \in [-6, -3] \tag{6.35}$$

by (6.29). We obtain from (6.32) that

$$\bar{P}_{(n)} : \quad \begin{aligned} \bar{\theta}_1 &= \bar{\theta} + \alpha_n + \bar{\delta}_n \left(\bar{p} + \frac{\alpha - \alpha_n}{\bar{\delta}_n} \right) + \bar{\delta}_n \, \varepsilon \bar{P}_{1,(n)}(\bar{\theta}, \bar{p}) \\ \bar{p}_1 &= \bar{p} + \bar{\delta}_n \varepsilon \, \bar{P}_{2,(n)}(\bar{\theta}, \bar{p}) \end{aligned},$$

and this implies that we finally can transform

$$u = \bar{\theta}, \quad v = \bar{p} + \frac{\alpha - \alpha_n}{\bar{\delta}_n} \tag{6.36}$$

to obtain in new coordinates (u, v) the Poincaré maps $P_{(n)} : \mathbb{R} \times [-6, -3] \to \mathbb{R}^2$, $(u, v) \mapsto (u_1, v_1)$, where

$$P_{(n)} : \quad u_1 = u + \alpha_n + \bar{\delta}_n v + \bar{\delta}_n \varepsilon P_{1,(n)}(u, v), \quad v_1 = v + \bar{\delta}_n \varepsilon P_{2,(n)}(u, v),$$

with $P_{j,(n)}(u, v) = \bar{P}_{j,(n)}(u, v - (\alpha - \alpha_n)/\bar{\delta}_n)$ for $j = 1, 2$. We also have

Lemma 6.2.8. *Every* $P = P_{(n)}$ *has the intersection property, i.e., if an embedded circle* C *in* $\mathbb{R} \times [-6, -3]$ *is homotopic to a circle* $v = $ const, *then* $P(C) \cap C \neq \emptyset$.

Proof: Let $P_{(6.25)}$ be the time-2π-map of (6.25). Since (6.25) comes from a Hamiltonian system, $P_{(6.25)}$ has the intersection property, cf. DIECKERHOFF/ZEHNDER [72, Lemma 5]. Let $\Phi_3 : (\theta, r) \mapsto (\bar{\theta}, \bar{\rho}) \mapsto (u, v)$ denote the transformations from (6.28) and (6.36). Then $P = \Phi_3 \circ P_{(6.25)} \circ \Phi_3^{-1}$, i.e., P and $P_{(6.25)}$ are conjugated, and therefore P as well has the intersection property. $\qquad\square$

Next we note that by (6.35)

$$\gamma_n \leq \frac{2\pi}{\omega n} = \bar{\delta}_n = M \frac{\pi}{4\omega n} \leq M \gamma_n \quad \text{with} \quad M = 8.$$

Hence we may choose M_*, independent of n, such that the claim of the invariant curve theorem holds (where $\beta = \alpha_n$), since for large n we also have $\bar{\delta}_n \in (0, 2)$. Thus by Lemma 6.2.6, for sufficiently small $\varepsilon > 0$ (corresponding to sufficiently large a in (1.5)) and all large n, every $P_{(n)}$ has an invariant curve. Transformed back to the original system this means that we have found arbitrary large invariant tori in $(x, \dot{x}, t \bmod \omega)$ space, and this implies the boundedness of all solutions; see DIECKERHOFF/ZEHNDER [72], LEVI [128], or ORTEGA [165].

Finally we remark that the existence of infinitely many periodic solutions is obtained via the Poincaré–Birkhoff theorem. This theorem may be applied to $P_{(6.25)}$, the area-preserving time-2π-map of the Hamiltonian system (6.25), just like in DIECKERHOFF/ZEHNDER [72, pp. 92/93], cf. also LEVI [128, Thm. 1]. Moreover, the existence of quasiperiodic solutions may also be shown analogously to these papers, and consequently the proof of Thm. 6.0.4 is complete. $\qquad\square$

6.3 Appendix: Some technicalities

In this appendix we give the proofs of the estimates in Lemma 6.2.4, 6.2.5, and 6.2.6 stated in Sect. 6.2. Before doing this, we first include, for convenience of the reader, a result on the differentiation of chain functions.

Lemma 6.3.1. *Let* $F : \mathbb{R}^2 \to \mathbb{R}$ *and* $f, g : \mathbb{R}^2 \to \mathbb{R}$ *be sufficiently smooth. Then for* $i + j \geq 1$

$$\partial_x^i \partial_y^j [F \circ (f, g)] =$$

$$\sum_{\substack{(k,p) \in \mathbb{N}_0^2 : 1 \leq k+p \leq i+j, \\ i = (i_1, \dots, i_{k+p}), |i| = i, \\ j = (j_1, \dots, j_{k+p}), |j| = j}} c_{k,p,\mathbf{i},\mathbf{j}} \left(\partial_1^k \partial_2^p F(f, g) \right) \left(\partial_x^{i_1} \partial_y^{j_1} f \right) \dots \left(\partial_x^{i_k} \partial_y^{j_k} f \right)$$
$$\times \left(\partial_x^{i_{k+1}} \partial_y^{j_{k+1}} g \right) \dots \left(\partial_x^{i_{k+p}} \partial_y^{j_{k+p}} g \right),$$

with integer coefficients $c_{k,p,i,j}$ satisfying $c_{k,p,i,j} = 0$ if $i_l = j_l = 0$ for some $1 \leq l \leq k+p$, i.e., no terms with $i_l = j_l = 0$ will appear. In particular, the above formula gives for $F: \mathbb{R} \to \mathbb{R}$ and $f: \mathbb{R}^2 \to \mathbb{R}$

$$\partial_x^i \partial_y^j [F \circ f] = \sum_{\substack{1 \leq k \leq i+j, \\ i=(i_1,\dots,i_k),\, |i|=i, \\ j=(j_1,\dots,j_k),\, |j|=j}} c_{k,i,j} \left(D^k F(f)\right) \left(\partial_x^{i_1} \partial_y^{j_1} f\right) \dots \left(\partial_x^{i_k} \partial_y^{j_k} f\right).$$

Proof: The proof is omitted, cf. ABRAHAM/ROBBIN [1, p. 3], DIECKER-HOFF/ZEHNDER [72, p. 88], LEVI [128] or ORTEGA [165, Lemma 3.8] for similar results. The claim follows inductively using Leibniz' rule for the differentiation of products of $n \geq 2$ functions. □

Proof of Lemma 6.2.4:

We obtain from (6.22) and (6.21) with $\mathcal{H}_1 = \mathcal{H}_1(\theta, r, \tau)$

$$r = H(\tau, I(r) + \varepsilon \mathcal{H}_1, \theta) = h_0(I(r) + \varepsilon \mathcal{H}_1) - \varepsilon x(\tau, I(r) + \varepsilon \mathcal{H}_1)p(\theta),$$

and thus, since $I(\cdot) = h_0^{-1}$,

$$\varepsilon \mathcal{H}_1 = I\big(r + \varepsilon x(\tau, I(r) + \varepsilon \mathcal{H}_1)p(\theta)\big) - I(r).$$

Consequently, differentiation of $f(\rho) = I\big(r + \rho \varepsilon x(\dots)p(\theta)\big)$ and $I' = T$ yield

$$\mathcal{H}_1 = x\big(\tau, I(r) + \varepsilon \mathcal{H}_1\big) p(\theta) \int_0^1 T\Big(r + \rho \varepsilon x\big(\tau, I(r) + \varepsilon \mathcal{H}_1\big) p(\theta)\Big) d\rho. \quad (6.37)$$

From this equation the claimed estimates will be derived inductively, always assuming that r is large enough and that $\varepsilon \leq 1$.

For that, we consider first the case $i = j = 0$. Then (6.37), (6.18) and the boundedness of T imply $|\mathcal{H}_1| \leq C |x(\tau, \mathcal{H})| \leq C |\mathcal{H}|^{1/2}$. Consequently, by (6.23), and since $|I(r)| \leq C r$, we find $|\mathcal{H}_1| \leq C r^{1/2}$ as desired. In particular, (6.17a) and (6.23) yield

$$cr \leq |\mathcal{H}| \leq Cr, \quad (6.38)$$

and thus

$$|\partial_I^k x(\tau, \mathcal{H})| \leq C r^{\frac{1}{2} - k} \quad (k \geq 0) \quad (6.39)$$

because of (6.18). For the induction step we assume that (6.24) already holds for all $0 \leq i + j \leq N$ and is to be shown for some fixed $i^* + j^* = N + 1$. We start to estimate the derivatives of the ingredients of the right-hand side of (6.37). First we will show

$$|\partial_\theta^i \partial_r^j [x(\tau, \mathcal{H})]| \leq C \left(r^{-\frac{1}{2}} |\partial_\theta^i \partial_r^j \mathcal{H}| + r^{\frac{1}{2} - j}\right) \quad (0 \leq i + j \leq N + 1). \quad (6.40)$$

By induction assumption this will yield in particular

$$|\partial_\theta^i \partial_r^j [x(\tau, \mathcal{H})]| \leq C \left(r^{-\frac{1}{2}} r^{\frac{1}{2} - j} + r^{\frac{1}{2} - j}\right) \leq C r^{\frac{1}{2} - j} \quad (0 \leq i + j \leq N). \quad (6.41)$$

To prove (6.40), we write, using the second formula from Lemma 6.3.1,

$$\partial_\theta^i \, \partial_r^j \, [x(\tau, \mathcal{H})] =$$
$$C \left(\partial_I \, x(\tau, \mathcal{H}) \right) \left(\partial_\theta^i \, \partial_r^j \, \mathcal{H} \right)$$
$$+ \sum_{\substack{2 \le k \le i+j, \\ \mathbf{i}=(i_1,\dots,i_k),\,|\mathbf{i}|=i, \\ \mathbf{j}=(j_1,\dots,j_k),\,|\mathbf{j}|=j}} c_{k,\mathbf{i},\mathbf{j}} \left(\partial_I^k \, x(\tau, \mathcal{H}) \right) \left(\partial_\theta^{i_1} \, \partial_r^{j_1} \, \mathcal{H} \right) \cdots \left(\partial_\theta^{i_k} \, \partial_r^{j_k} \, \mathcal{H} \right). \quad (6.42)$$

Since due to (6.39) the first term is dominated by $C \, r^{-\frac{1}{2}} |\partial_\theta^i \, \partial_r^j \, \mathcal{H}|$, we only have to show that every term in the $\sum$ is $\le C \, r^{\frac{1}{2}-j}$. To do this, we note that because of Lemma 6.3.1, the non-zero terms in this sum have $i_l + j_l \ge 1$ for every $1 \le l \le k$, and thus $i_l + j_l \le N$, since in the opposite case one would obtain $i + j = i_l + j_l = N + 1$, and thus $\mathbf{i} = i \, e_l \in \mathbb{R}^k$ and $\mathbf{j} = j \, e_l \in \mathbb{R}^k$, with e_l being the l'th unit vector. But since $k \ge 2$, this would imply that at least either $i_{l+1} = j_{l+1} = 0$ or $i_{l-1} = j_{l-1} = 0$, a contradiction. Therefore $i_l + j_l \le N$ in the non-vanishing terms, and hence we can apply the induction hypotheses to estimate $|\partial_\theta^{i_l} \, \partial_r^{j_l} \, \mathcal{H}_1| \le C \, r^{\frac{1}{2}-j_l}$ for those indices. Thus if $i_l \ge 1$ we obtain from (6.23) that $|\partial_\theta^{i_l} \, \partial_r^{j_l} \, \mathcal{H}| = \varepsilon |\partial_\theta^{i_l} \, \partial_r^{j_l} \, \mathcal{H}_1| \le C \, r^{\frac{1}{2}-j_l}$. On the other hand, if $i_l = 0$, then

$$|\partial_r^{j_l} \, \mathcal{H}| = |D^{j_l} I + \varepsilon \partial_r^{j_l} \, \mathcal{H}_1| \le \begin{cases} C + C \, r^{-1/2} & : \quad j_l = 1 \\ C \, r^{1/2-j_l} + C \, r^{1/2-p} & : \quad j_l \ge 2 \end{cases}$$
$$\le \begin{cases} C & : \quad j_l = 1 \\ C \, r^{1/2-j_l} & : \quad j_l \ge 2 \end{cases},$$

where we have used (6.17a). To sum up, we have shown that for all non-vanishing terms in the sum in (6.42) one has

$$|\partial_\theta^{i_l} \, \partial_r^{j_l} \, \mathcal{H}| \le \begin{cases} C & : \quad (i_l, j_l) = (0, 1) \\ C \, r^{1/2-j_l} & : \quad (i_l, j_l) \ne (0, 1) \end{cases}, \quad 1 \le l \le k. \quad (6.43)$$

This information can be used as follows. If in one of the non-zero terms there are no index-pairs $(i_l, j_l) = (0, 1)$, then by (6.39) and (6.43)

$$\left| \left(\partial_I^k \, x(\tau, \mathcal{H}) \right) \left(\partial_\theta^{i_1} \, \partial_r^{j_1} \, \mathcal{H} \right) \cdots \left(\partial_\theta^{i_k} \, \partial_r^{j_k} \, \mathcal{H} \right) \right| \le C \, r^{\frac{1}{2}-k} \, r^{1/2-j_1} \cdots r^{1/2-j_k}$$
$$= C \, r^{\frac{1}{2}(1-k)-|\mathbf{j}|} = C \, r^{\frac{1}{2}(1-k)-j},$$

but for every appearing pair $(i_l, j_l) = (0, 1)$ the corresponding term $r^{1/2-j_l} = r^{-1/2}$ in this estimate has to be replaced by a constant. Thus, if in a general non-zero term in the sum in (6.42) there are $1 \le M \le k$ such pairs, then we obtain the estimate

$$|\dots| \le C \, r^{\frac{1}{2}(1-k)-j} \, r^{\frac{M}{2}} = C \, r^{\frac{1}{2}(1-k+M)-j}.$$

Since $1 - k + M \le 1$, we finally conclude that $|\dots| \le C \, r^{\frac{1}{2}-j}$, as was desired to finish the proof of (6.40) and (6.41).

Now we intend to show that with $U(\theta, r, \tau) = x(\tau, \mathcal{H}(\theta, r, \tau)) \, p(\theta)$ we also have

$$|\partial_\theta^i \partial_r^j \, U| \leq C \left(r^{-\frac{1}{2}} |\partial_\theta^i \partial_r^j \, \mathcal{H}| + r^{\frac{1}{2}-j} \right) \quad (0 \leq i + j \leq N + 1), \qquad (6.44)$$

and therefore again by induction assumption

$$|\partial_\theta^i \partial_r^j \, U| \leq C \, r^{\frac{1}{2}-j} \quad (0 \leq i + j \leq N). \qquad (6.45)$$

Estimate (6.44) is in fact a direct consequence of Leibniz' rule, since by (6.41) and (6.40)

$$|\partial_\theta^i \partial_r^j \, U| = |\partial_\theta^i (\partial_r^j \, [x(\tau, \mathcal{H})] \, p)|$$

$$\leq C \left(\sum_{k=0}^{i-1} |\partial_\theta^k \partial_r^j \, [x(\tau, \mathcal{H})]| + |\partial_\theta^i \partial_r^j \, [x(\tau, \mathcal{H})]| \right)$$

$$\leq C \left(\sum_{k=0}^{i-1} r^{\frac{1}{2}-j} + r^{-\frac{1}{2}} |\partial_\theta^i \partial_r^j \, \mathcal{H}| + r^{\frac{1}{2}-j} \right)$$

$$\leq C \left(r^{-\frac{1}{2}} |\partial_\theta^i \partial_r^j \, \mathcal{H}| + r^{\frac{1}{2}-j} \right).$$

Our next step towards the estimation of $|\partial_\theta^{i^*} \partial_r^{j^*} \, \mathcal{H}_1|$ by means of (6.37) will be to prove that for fixed $\rho \in [0, 1]$

$$|\partial_\theta^i \partial_r^j \, [T(r + \rho \varepsilon U)]| \leq C \left(r^{-2} |\partial_\theta^i \partial_r^j \, \mathcal{H}| + r^{-1/2-j} \right) \quad (0 \leq i + j \leq N + 1). \qquad (6.46)$$

Again this implies by induction assumption

$$|\partial_\theta^i \partial_r^j \, [T(r + \rho \varepsilon U)]| \leq C \, r^{-1/2-j} \quad (0 \leq i + j \leq N). \qquad (6.47)$$

To see (6.46), we first remark that $|U| \leq C \left(r^{-1/2} |\mathcal{H}| + r^{1/2} \right) \leq C \, r^{1/2}$ by (6.44) and (6.38). Therefore $|r + \rho \varepsilon U| \geq cr$, and this in turn yields by means of (6.13a)

$$|D^k T(r + \rho \varepsilon U)| \leq C \, |r + \rho \varepsilon U|^{-\frac{1}{2}-k} \leq C \, r^{-\frac{1}{2}-k} \quad (k \geq 1). \qquad (6.48)$$

Moreover,

$$\partial_\theta^i \partial_r^j \, [r + \rho \varepsilon U] \leq \begin{cases} 1 + \rho \varepsilon (\partial_r \, U) & : \quad i = 0, j = 1 \\ \rho \varepsilon (\partial_\theta^i \partial_r^j \, U) & : \quad \text{otherwise for } i + j \geq 1 \end{cases}$$

First we show (6.46) for $i = 0$, $j = 1$. In this case by (6.48) and (6.44)

$$|\partial_r \, [T(r + \rho \varepsilon U)]| \leq C \, |T'(r + \rho \varepsilon U)| \, (1 + |\partial_r U|)$$

$$\leq C \, r^{-3/2} \left(1 + r^{-1/2} |\partial_r \mathcal{H}| + r^{-1/2} \right)$$

$$\leq C \left(r^{-2} |\partial_r \mathcal{H}| + r^{-3/2} \right).$$

So we can turn to the general case $i + j \geq 1$ and $(i, j) \neq (0, 1)$ where we have $|\partial_\theta^i \partial_r^j [r + \rho \varepsilon U]| \leq C |\partial_\theta^i \partial_r^j U|$. By Lemma 6.3.1 we obtain, analogously to (6.42)

$$\partial_\theta^i \partial_r^j [T(r + \rho \varepsilon U)] =$$
$$C T'(r + \rho \varepsilon U) (\partial_\theta^i \partial_r^j [r + \rho \varepsilon U])$$
$$+ \sum_{\substack{2 \leq k \leq i+j, \\ i=(i_1,\ldots,i_k),\, |i|=i, \\ j=(j_1,\ldots,j_k),\, |j|=j}} c_{k,i,j} D^k T(r + \rho \varepsilon U) (\partial_\theta^{i_1} \partial_r^{j_1} [r + \rho \varepsilon U]) \ldots (\partial_\theta^{i_k} \partial_r^{j_k} [r + \rho \varepsilon U]).$$

As a consequence of (6.48) and (6.44) the first term may be estimated by

$$C r^{-\frac{3}{2}} \left(r^{-\frac{1}{2}} |\partial_\theta^i \partial_r^j \mathcal{H}| + r^{\frac{1}{2}-j} \right) \leq C \left(r^{-2} |\partial_\theta^i \partial_r^j \mathcal{H}| + r^{-1-j} \right).$$

Concerning the sum, using (6.48) and (6.45) we may argue completely analogous to the estimation of (6.42) to obtain the bound $C r^{-1/2-j}$ for every non-zero term, and from this (6.46) follows.

Finally we turn to prove the claim of the induction step by means of (6.37). For that, we let $\tilde{T}(\rho) = T(r + \rho \varepsilon U)$. In this notation, (6.37) reads as $\mathcal{H}_1 = U \int_0^1 \tilde{T}(\rho) \, d\rho$. Therefore, by Leibniz' rule, (6.44), (6.46), (6.45) and (6.47),

$$|\partial_\theta^{i^*} \partial_r^{j^*} \mathcal{H}_1|$$
$$\leq |\partial_\theta^{i^*} \partial_r^{j^*} U| + \int_0^1 |\partial_\theta^{i^*} \partial_r^{j^*} [T(r + \rho \varepsilon U)]| \, d\rho$$
$$+ C \sum_{\substack{p=0,\ldots,i^* \\ k=0,\ldots,j^* \\ (p,k) \neq (0,0),\, (p,k) \neq (i^*,j^*)}} |\partial_\theta^p \partial_r^k U| \left(\int_0^1 |\partial_\theta^{i^*-p} \partial_r^{j^*-k} [T(r + \rho \varepsilon U)]| \, d\rho \right)$$
$$\leq C \left(r^{-1/2} |\partial_\theta^{i^*} \partial_r^{j^*} \mathcal{H}| + r^{1/2-j^*} + r^{-2} |\partial_\theta^{i^*} \partial_r^{j^*} \mathcal{H}| + r^{-1/2-j^*} \right.$$
$$\left. + \sum_{(p,k)} r^{1/2-k} r^{-1/2-[j^*-k]} \right)$$
$$\leq C \left(r^{-1/2} |\partial_\theta^{i^*} \partial_r^{j^*} \mathcal{H}| + r^{1/2-j^*} \right).$$

Therefore we can choose r as large as is necessary to ensure $C r^{-1/2} \leq 1/2$, and thus to obtain the claim of (6.24).   □

Proof of Lemma 6.2.5:

We first remark that by (6.13b), (6.29) and the definition of $r(\bar{\rho})$

$$c \delta^{-2} \leq c_0 \delta^{-2} |\bar{\rho}|^{-2} \leq r(\bar{\rho}) \leq C_0 \delta^{-2} |\bar{\rho}|^{-2} \leq C \delta^{-2}, \tag{6.49}$$
$$|D^i [r(\bar{\rho})]| = \delta^i |D^i T^{-1}(\delta \bar{\rho} + 2\pi)| \leq C \delta^i (\delta |\bar{\rho}|)^{-(2+i)} \leq C \delta^{-2} \quad (i \geq 1).$$

Consequently (6.13a) implies

$$|D^i T(\bar{\rho})| \le C\,|\bar{\rho}|^{-\frac{1}{2}-i} \le C\,\delta^{1+2i} \quad (i \ge 1)\,.$$

Therefore in particular $|T'(\bar{\rho})| \le C\,\delta^3$, and the second formula in Lemma 6.3.1 yields for $i \ge 1$

$$|D^i[T'(r(\bar{\rho}))]| = \left| \sum_{\substack{1 \le k \le i, \\ \mathbf{i}=(i_1,\dots,i_k),\,|\mathbf{i}|=i, \\ \forall 1 \le l \le k:\, i_l \ge 1}} c_{k,\mathbf{i}}\, D^{k+1}T(r(\bar{\rho}))\,(D^{i_1}[r(\bar{\rho})])\,\dots\,(D^{i_k}[r(\bar{\rho})]) \right|$$

$$\le C \sum_k \delta^{3+2k}\,\delta^{-2}\dots\delta^{-2} \le C\,\delta^3,$$

i.e., $|D^i[T'(r(\bar{\rho}))]| \le C\,\delta^3$ for $i \ge 0$. Next, it follows from the second formula in Lemma 6.3.1, Lemma 6.2.4, and (6.49), that

$$\left| \partial_{\bar{\theta}}^i\, \partial_{\bar{\rho}}^j\, [\partial_\theta\, \mathcal{H}_1(\omega\bar{\theta}, r(\bar{\rho}), \tau)] \right|$$

$$= \omega^i\, \left| \partial_{\bar{\rho}}^j\, [\partial_\theta^{i+1}\, \mathcal{H}_1(\omega\bar{\theta}, r(\bar{\rho}), \tau)] \right|$$

$$= \omega^i\, \left| \sum_{\substack{1 \le k \le j, \\ \mathbf{j}=(j_1,\dots,j_k),\,|\mathbf{j}|=j, \\ \forall 1 \le l \le k:\, j_l \ge 1}} c_{k,\mathbf{j}}\, (\partial_r^k\, \partial_\theta^{i+1}\, \mathcal{H}_1(\omega\bar{\theta}, r(\bar{\rho}), \tau))\,(D^{j_1}[r(\bar{\rho})])\,\dots\,(D^{j_k}[r(\bar{\rho})]) \right|$$

$$\le C \sum_k r(\bar{\rho})^{\frac{1}{2}-k}\,\delta^{-2}\dots\delta^{-2} \le C\,\delta^{-1}.$$

Since the same reasoning applies to give $\left| \partial_{\bar{\theta}}^i\, \partial_{\bar{\rho}}^j\, [\partial_r\, \mathcal{H}_1(\omega\bar{\theta}, r(\bar{\rho}), \tau)] \right| \le C\,\delta$, only the estimate for $|\partial_{\bar{\theta}}^i\, \partial_{\bar{\rho}}^j\, f_2|$ is still to be verified. For that, we note that due to the above estimates and Leibniz' rule

$$\left| \partial_{\bar{\theta}}^i\, \partial_{\bar{\rho}}^j\, [T'(r(\bar{\rho}))\, \partial_\theta \mathcal{H}_1(\omega\bar{\theta}, r(\bar{\rho}), \tau)] \right|$$

$$\le C \sum_{k=0}^j |D^{j-k}[T'(r(\bar{\rho}))]|\, \left| \partial_{\bar{\theta}}^i\, \partial_{\bar{\rho}}^k\, [\partial_\theta\, \mathcal{H}_1(\omega\bar{\theta}, r(\bar{\rho}), \tau)] \right|$$

$$\le C \sum_k \delta^3\,\delta^{-1} \le C\,\delta^2.$$

Taking into consideration that $|\partial_{\bar{\theta}}^i\, \partial_{\bar{\rho}}^j\, f_2|$ has an extra δ^{-1} in comparison with the estimated expression, we obtain the claim. $\qquad\square$

Proof of Lemma 6.2.6:

We make the ansatz, cf. DIECKERHOFF/ZEHNDER [72, Lemma 4],

$$\bar{\theta}(\tau) = \bar{\theta} + \frac{\alpha}{2\pi}\tau + \frac{\bar{\delta}}{2\pi}\bar{\rho}\tau + \bar{\delta}\varepsilon\, A(\bar{\theta}, \bar{\rho}, \tau)\,, \quad \bar{\rho}(\tau) = \bar{\rho} + \bar{\delta}\varepsilon\, B(\bar{\theta}, \bar{\rho}, \tau) \quad (6.50)$$

for the solution $\tau \mapsto (\bar{\theta}(\tau), \bar{\rho}(\tau))$ of (6.30) with initial values $(\bar{\theta}, \bar{\rho})$. Here $A(\bar{\theta}, \bar{\rho}, \tau)$ and $B(\bar{\theta}, \bar{\rho}, \tau)$ are suitable functions which are defined through (6.50). Since $\bar{P}_1(\bar{\theta}, \bar{\rho}) = A(\bar{\theta}, \bar{\rho}, 2\pi)$ as well as $\bar{P}_2(\bar{\theta}, \bar{\rho}) = B(\bar{\theta}, \bar{\rho}, 2\pi)$, we need to estimate the derivatives of A and B. For that, we first remark that due to (6.30)

$$\bar{\delta} A(\bar{\theta}, \bar{\rho}, \tau) = \frac{\bar{\delta}^2}{2\pi} \int_0^\tau B(\bar{\theta}, \bar{\rho}, s) \, ds + \int_0^\tau f_1(\bar{\theta}(s), \bar{\rho}(s), s) \, ds, \quad (6.51a)$$

$$\bar{\delta} B(\bar{\theta}, \bar{\rho}, \tau) = \int_0^\tau f_2(\bar{\theta}(s), \bar{\rho}(s), s) \, ds. \quad (6.51b)$$

Inserting (6.50) for $\bar{\theta}(s)$ and $\bar{\rho}(s)$, we have obtained a system of two integral equations for A and B, from which the desired estimates inductively can be derived as follows. If we let $\|B\|_N = \sup_{0 \le i+j \le N} |\partial_{\bar{\theta}}^i \partial_{\bar{\rho}}^j B|_\infty$, and analogously for A, then (6.51b), (6.31) and $c\bar{\delta} \le \delta \le C\bar{\delta}$ imply $\|B\|_0 \le C$, hence (6.51a) and (6.31) yield $\bar{\delta}\|A\|_0 \le C\bar{\delta}^2 + C\bar{\delta}$, and therefore also $\|A\|_0 \le C$. Now we assume that we already have shown $\|A\|_N + \|B\|_N \le C$ for some N, and we fix indices i^*, j^* with $1 \le i^* + j^* = N + 1$. We have by (6.50)

$$\partial_{\bar{\theta}}[\bar{\theta}(\tau; \bar{\theta}, \bar{\rho})] = 1 + \bar{\delta}\varepsilon (\partial_{\bar{\theta}} A), \quad \partial_{\bar{\theta}}[\bar{\rho}(\tau; \bar{\theta}, \bar{\rho})] = \bar{\delta}\varepsilon (\partial_{\bar{\theta}} B), \quad (6.52a)$$

$$\partial_{\bar{\rho}}[\bar{\theta}(\tau; \bar{\theta}, \bar{\rho})] = \frac{\bar{\delta}}{2\pi} \tau + \bar{\delta}\varepsilon (\partial_{\bar{\rho}} A), \quad \partial_{\bar{\rho}}[\bar{\rho}(\tau; \bar{\theta}, \bar{\rho})] = 1 + \bar{\delta}\varepsilon (\partial_{\bar{\rho}} B), \quad (6.52b)$$

and for $i + j \ge 2$

$$\partial_{\bar{\theta}}^i \partial_{\bar{\rho}}^j [\bar{\theta}(\tau; \bar{\theta}, \bar{\rho})] = \bar{\delta}\varepsilon (\partial_{\bar{\theta}}^i \partial_{\bar{\rho}}^j A), \quad \partial_{\bar{\theta}}^i \partial_{\bar{\rho}}^j [\bar{\rho}(\tau; \bar{\theta}, \bar{\rho})] = \bar{\delta}\varepsilon (\partial_{\bar{\theta}}^i \partial_{\bar{\rho}}^j B), \quad (6.53)$$

hence in particular for $i + j \ge 1$

$$|\partial_{\bar{\theta}}^i \partial_{\bar{\rho}}^j [\bar{\theta}(\tau; \bar{\theta}, \bar{\rho})]| + |\partial_{\bar{\theta}}^i \partial_{\bar{\rho}}^j [\bar{\rho}(\tau; \bar{\theta}, \bar{\rho})]| \le C(1 + \|A\|_{i+j} + \|B\|_{i+j}). \quad (6.54)$$

Let F denote f_1 or f_2. Then by the first formula in Lemma 6.3.1

$$\left| \partial_{\bar{\theta}}^{i^*} \partial_{\bar{\rho}}^{j^*} [F(\bar{\theta}(\tau), \bar{\rho}(\tau), \tau)] \right|$$

$$\le \left| c_1 \, \partial_{\bar{\theta}} F(\bar{\theta}(\tau), \bar{\rho}(\tau), \tau) \, \partial_{\bar{\theta}}^{i^*} \partial_{\bar{\rho}}^{j^*} [\bar{\theta}(\tau)] \right|$$

$$+ \left| c_2 \, \partial_{\bar{\rho}} F(\bar{\theta}(\tau), \bar{\rho}(\tau), \tau) \, \partial_{\bar{\theta}}^{i^*} \partial_{\bar{\rho}}^{j^*} [\bar{\rho}(\tau)] \right|$$

$$+ \left| \sum_{\substack{(k,p) \in \mathbb{N}_0^2 : 2 \le k+p \le i^*+j^*, \\ \mathbf{i}=(i_1,\ldots,i_{k+p}), |\mathbf{i}|=i^*, \\ \mathbf{j}=(j_1,\ldots,j_{k+p}), |\mathbf{j}|=j^*}} \left(c_{k,p,\mathbf{i},\mathbf{j}} \, \partial_{\bar{\theta}}^k \partial_{\bar{\rho}}^p F(\bar{\theta}(\tau), \bar{\rho}(\tau), \tau) \right.$$

$$\times (\partial_{\bar{\theta}}^{i_1} \partial_{\bar{\rho}}^{j_1} [\bar{\theta}(\tau)]) \cdots (\partial_{\bar{\theta}}^{i_k} \partial_{\bar{\rho}}^{j_k} [\bar{\theta}(\tau)])$$

$$\left. \times (\partial_{\bar{\theta}}^{i_{k+1}} \partial_{\bar{\rho}}^{j_{k+1}} [\bar{\rho}(\tau)]) \cdots (\partial_{\bar{\theta}}^{i_{k+p}} \partial_{\bar{\rho}}^{j_{k+p}} [\bar{\rho}(\tau)]) \right) \right|.$$

In the $\sum_{2 \le k+p \le i^*+j^*} (\ldots)$, every non-zero term must have $i_l + j_l \ge 1$ for all $1 \le l \le k+p$, by Lemma 6.3.1. This implies $i_l + j_l \le N$, since $i^* + j^* = N + 1$, and hence $i_l + j_l = N + 1$ is impossible because both vectors **i** and **j** under consideration have $k + p$, thus at least 2, components. Therefore in every non-zero term in the above sum for all $1 \leq l \leq k$, by (6.54) and by induction hypotheses,

$$|\partial_{\bar{\theta}}^{i_l} \partial_{\bar{\rho}}^{j_l} [\bar{\theta}(\tau)]| \leq C(1 + \|A\|_{i_l + j_l} + \|B\|_{i_l + j_l}) \leq C(1 + \|A\|_N + \|B\|_N) \leq C.$$

Since analogously $|\partial_{\bar{\theta}}^{i_l} \partial_{\bar{\rho}}^{j_l} [\bar{\rho}(\tau)]| \leq C$ for $k + 1 \leq l \leq k + p$, we obtain from (6.31)

$$\left| \sum_{2 \leq k + p \leq i^* + j^*} (\ldots) \right| \leq C\bar{\delta},$$

and consequently, again by (6.31), for $F = f_1$ or $F = f_2$

$$\left| \partial_{\bar{\theta}}^{i^*} \partial_{\bar{\rho}}^{j^*} [F(\bar{\theta}(\tau), \bar{\rho}(\tau), \tau)] \right| \leq C\bar{\delta} \left(1 + |\partial_{\bar{\theta}}^{i^*} \partial_{\bar{\rho}}^{j^*} [\bar{\theta}(\tau)]| + |\partial_{\bar{\theta}}^{i^*} \partial_{\bar{\rho}}^{j^*} [\bar{\rho}(\tau)]| \right).$$

Then it follows from (6.52a), (6.52b), (6.53), observing also $\tau \in [0, 2\pi]$ and w.l.o.g. $\varepsilon \leq 1$, that in any case

$$\left| \partial_{\bar{\theta}}^{i^*} \partial_{\bar{\rho}}^{j^*} [F(\bar{\theta}(\tau), \bar{\rho}(\tau), \tau)] \right| \leq C\bar{\delta} \left(1 + \bar{\delta} [|\partial_{\bar{\theta}}^{i^*} \partial_{\bar{\rho}}^{j^*} A|_{\infty} + |\partial_{\bar{\theta}}^{i^*} \partial_{\bar{\rho}}^{j^*} B|_{\infty}] \right). \tag{6.55}$$

This in turn implies by (6.51b)

$$|\partial_{\bar{\theta}}^{i^*} \partial_{\bar{\rho}}^{j^*} B|_{\infty} \leq C \left(1 + \bar{\delta} [|\partial_{\bar{\theta}}^{i^*} \partial_{\bar{\rho}}^{j^*} A|_{\infty} + |\partial_{\bar{\theta}}^{i^*} \partial_{\bar{\rho}}^{j^*} B|_{\infty}] \right),$$

and thus $|\partial_{\bar{\theta}}^{i^*} \partial_{\bar{\rho}}^{j^*} B|_{\infty} \leq C(1 + \bar{\delta} |\partial_{\bar{\theta}}^{i^*} \partial_{\bar{\rho}}^{j^*} A|_{\infty})$ if $\bar{\delta}$ is sufficiently small. Inserting this and (6.55) with $F = f_1$ in (6.51a), we finally conclude that $|\partial_{\bar{\theta}}^{i^*} \partial_{\bar{\rho}}^{j^*} A|_{\infty} + |\partial_{\bar{\theta}}^{i^*} \partial_{\bar{\rho}}^{j^*} B|_{\infty} \leq C$ for sufficiently small $\bar{\delta}$, so that we have shown $\|A\|_{N+1} + \|B\|_{N+1} \leq C$, and thus in particular the claim of the lemma. $\square$

7. Planar non-smooth dynamical systems

We provide a survey of some results for the special case of planar non-smooth dynamical systems.

7.1 Lyapunov constants

Following COLL/GASULL/PROHENS [53] we investigate the number of small periodic solutions for real-analytic planar systems by means of the so-called Lyapunov constants. For smooth systems, a main source of motivation to study the number of periodic solutions of planar equations with polynomial right-hand sides is that this was stated as an important problem by Hilbert, as part of "Hilbert's 16th problem".

We consider the system

$$\dot{x} = f_1^+(x,y) \atop \dot{y} = f_2^+(x,y), \quad y > 0, \qquad \text{and} \qquad \dot{x} = f_1^-(x,y) \atop \dot{y} = f_2^-(x,y), \quad y < 0, \qquad (7.1)$$

where $f^{\pm} = (f_1^{\pm}, f_2^{\pm})$ are real-analytic functions defined in a neighborhood of $(x,y) = (0,0)$. We write their series expansion as

$$f_1^{\pm}(x,y) = a^{\pm} + b^{\pm}x + c^{\pm}y + d^{\pm}x^2 + e^{\pm}xy + f^{\pm}y^2 + \ldots \qquad (7.2a)$$
$$f_2^{\pm}(x,y) = k^{\pm} + l^{\pm}x + m^{\pm}y + n^{\pm}x^2 + o^{\pm}xy + p^{\pm}y^2 + \ldots \qquad (7.2b)$$

and since we need $(0,0)$ to be an equilibrium for both component equations, we will throughout assume that

$$a^{\pm} = 0 = k^{\pm}.$$

Definition 7.1.1. *A component equation $\pm$ is of focus type, if*

$$(b^{\pm} - m^{\pm})^2 + 4l^{\pm}c^{\pm} < 0 \quad \text{and} \quad l^{\pm} > 0. \qquad (7.3)$$

We call $(0,0)$ a singularity of focus-focus type, if both equations $+$ and $-$ are of focus type.

Note that since

$$(f^{\pm})'(0,0) = \begin{pmatrix} b^{\pm} & c^{\pm} \\ l^{\pm} & m^{\pm} \end{pmatrix}$$

are the linearizations at $(0,0)$, condition (7.3) means that the eigenvalues

$$\frac{1}{2}\left((b^{\pm} + m^{\pm}) + / - \sqrt{(b^{\pm} - m^{\pm})^2 + 4c^{\pm}l^{\pm}}\right)$$

are complex conjugate, and close to the origin the flows locally do rotate counterclockwise, the latter due to $l^{\pm} > 0$.

Next we introduce the Poincaré return maps

$$s \mapsto h^{+}(s) \mapsto h^{-}(h^{+}(s))$$

that are well-defined in a neighborhood of the origin; see Fig. 7.1. Note also that therefore it is not possible for a solution of (7.1) to stick to the discontinuity line $\{(x,y) : y = 0\}$ for some time, i.e., (7.1) needs not be considered as a differential inclusion.

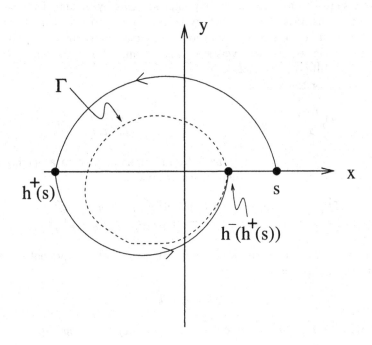

Fig. 7.1. The map $s \mapsto h^{+}(s) \mapsto h^{-}(h^{+}(s))$ and Γ from Ex. 7.1.1

In this situation one defines the Lyapunov constants to be the non-zero coefficients in the series expansion of the function $s \mapsto h^{-}(h^{+}(s)) - s$ for s close to zero; observe that $h^{-}(h^{+}(s)) = s$ corresponds to a periodic solution of (7.1).

Definition 7.1.2. *If $V_k \neq 0$ and*

$$h^-(h^+(s)) - s = V_k s^k + \mathcal{O}(s^{k+1})$$

for $s > 0$ close to zero, then V_k is called the kth Lyapunov constant.

Here we implicitly understand that V_k will only be considered in case that $V_1 = V_2 = \ldots = V_{k-1} = 0$. Thus in particular, if e.g. $V_1 = 0$, then the relations resulting from this equality can be used to simplify the expression for V_2, etc. As we are going to explain below in Ex. 7.1.1, information on the Lyapunov constants can be utilized to derive results about the number of small periodic solutions; see BAUTIN [22], BLOWS/LLOYD [26], GASULL/GUILLAMON/MAÑOSA [89], or GASULL/PROHENS [90] for more on the usage of Lyapunov constants for smooth systems.

Theorem 7.1.1. *Let $(0,0)$ be a singularity of focus-focus type, and*

$$\nu^\pm = \exp\left\{\frac{\pi(b^\pm + m^\pm)}{\sqrt{-[(b^\pm - m^\pm)^2 + 4l^\pm c^\pm]}}\right\}.$$

Then the first Lyapunov constant is

$$V_1 = \nu^+\nu^- - 1.$$

Proof: We will first deal with the equation $(\dot{x}, \dot{y}) = f^+(x, y)$ in the upper half-plane. It will be convenient to represent the solution starting at $(x(0), y(0)) = (s, 0)$ as

$$x(t) = R^+(\theta(t))\cos\theta(t), \quad y(t) = R^+(\theta(t))\sin\theta(t). \tag{7.4}$$

Differentiating this w.r. to t, it follows that

$$\frac{dR^+}{d\theta} = \frac{\operatorname{Re}(\bar{z}\dot{z})}{\operatorname{Im}(\bar{z}\dot{z})} R^+\Big|_{z=R^+e^{i\theta}}, \quad R^+(0; s) = s, \tag{7.5}$$

for the function $R^+ = R^+(\theta; s)$, with $\theta \in [0, \pi]$ and $s > 0$ small; here $z = x + iy$. Then $h^+(s) = R^+(\pi; s)$, and we are going to argue that

$$R^+(\pi; s) - s = (\nu^+ - 1)s + \mathcal{O}(s^2).$$

To see this, we expand

$$R^+(\theta; s) - s = \sum_{k=1}^{\infty} w_k^+(\theta)s^k, \quad \text{with} \quad w_k^+(0) = 0, \ k \in \mathbb{N}, \tag{7.6}$$

and derive a differential equation for w_1^+ using (7.5). We first write $(\dot{x}, \dot{y}) = f^+(x, y)$ in the complex form

$$\dot{z} = \sum_{k=1}^{\infty} F_k^+(z,\bar{z}), \quad \text{where} \quad F_k^+(z,\bar{z}) = \sum_{\substack{\alpha,\beta \in \mathbb{N}_0 \\ \alpha+\beta=k}} c_{\alpha\beta}^+ z^\alpha \bar{z}^\beta. \tag{7.7}$$

For simplicity we henceforth omit the superscript "+". Then

$$F_k(Re^{i\theta}, Re^{-i\theta}) = R^k F_k(e^{i\theta}, e^{-i\theta}),$$

and this shows that

$$\frac{\text{Re}(\bar{z}\dot{z})}{\text{Im}(\bar{z}\dot{z})} R \bigg|_{z=Re^{i\theta}} = \frac{\sum_{k=1}^{\infty} \text{Re}(S_k(\theta))R^k}{\sum_{k=1}^{\infty} \text{Im}(S_k(\theta))R^{k-1}}, \quad S_k(\theta) = e^{-i\theta} F_k(e^{i\theta}, e^{-i\theta}). \tag{7.8}$$

Next we observe that $\text{Im}(S_1(\theta)) > 0$ due to (7.3). Indeed, $F_1(z,\bar{z}) = c_{10}z + c_{01}\bar{z}$ implies upon decomposing c_{10} and c_{01} in real and imaginary part, and by comparing (7.7) to $(\dot{x}, \dot{y}) = f^+(x,y)$ and (7.2a), (7.2b), that

$$b^+ = \text{Re}(c_{10} + c_{01}), \quad c^+ = \text{Im}(c_{01} - c_{10}), \tag{7.9a}$$
$$l^+ = \text{Im}(c_{10} + c_{01}), \quad \text{and} \quad m^+ = \text{Re}(c_{10} - c_{01}), \tag{7.9b}$$

in particular

$$(b^+ - m^+)^2 + 4l^+c^+ = 4\left(|c_{01}|^2 - \text{Im}(c_{10})^2\right). \tag{7.10}$$

We have $S_1(\theta) = c_{10} + c_{01}e^{-2i\theta}$, whence $\text{Im}(S_1(\theta)) = \text{Im}(c_{10}) + \langle c_{01}, Je^{2\theta i}\rangle$, with $J = \begin{pmatrix} 0 & -1 \\ 1 & 0 \end{pmatrix}$ and the usual inner product $(\cdot, \cdot)$ in $\mathbb{R}^2 \times \mathbb{R}^2$. Assume $\text{Im}(c_{10}) < 0$. Then we get from (7.3), (7.9b), and (7.10) that $|\text{Im}(c_{10})| = -\text{Im}(c_{10}) < \text{Im}(c_{01}) \leq |c_{01}| < |\text{Im}(c_{10})|$, a contradiction. Therefore $\text{Im}(c_{10}) \geq 0$, and consequently $-\langle c_{01}, Je^{2\theta i}\rangle \leq |c_{01}| < |\text{Im}(c_{10})| = \text{Im}(c_{10})$ implies that in fact $\text{Im}(S_1(\theta)) > 0$. From this, (7.5), (7.6), and (7.8) we infer that with $R = R^+$

$$\sum_{k=1}^{\infty} w_k'(\theta)s^k = \frac{dR}{d\theta} = \frac{\text{Re}(S_1(\theta))}{\text{Im}(S_1(\theta))}R + \mathcal{O}(R^2)$$
$$= \frac{\text{Re}(S_1(\theta))}{\text{Im}(S_1(\theta))}(s + w_1(\theta)s) + \mathcal{O}(s^2),$$

hence

$$w_1'(\theta) = \frac{\text{Re}(S_1(\theta))}{\text{Im}(S_1(\theta))}(1 + w_1(\theta)), \quad \theta \in [0, 2\pi], \quad w_1(0) = 0,$$

by comparing the coefficients of s and recalling the initial condition from (7.6). After some calculation this may be integrated to yield $w_1(\pi) = \nu^+ - 1$, thus indeed

$$R^+(\pi; s) - s = (\nu^+ - 1)s + \mathcal{O}(s^2)$$

by evaluating (7.6) at $\theta = \pi$. Similarly it can be verified that

$$R^-(\pi; s) - s = (\nu^- - 1)s + \mathcal{O}(s^2)$$

for $s < 0$ small, where R^- is analogous to R^+ for the lower half-plane, in particular $h^-(s) = R^-(\pi; s)$. We therefore get

$$h^-(h^+(s)) - s = R^-(\pi; R^+(\pi; s)) - s = \nu^- R^+(\pi; s) + \mathcal{O}(s^2) - s$$
$$= (\nu^+ \nu^- - 1)s + \mathcal{O}(s^2),$$

as was to be shown. □

More precise information can be obtained if in addition

$$F_1^+(z, \bar{z}) = (i + \lambda^+)z \quad \text{and} \quad F_1^-(z, \bar{z}) = (i + \lambda^-)z, \tag{7.11}$$

with $\lambda^\pm \in \mathbb{R}$ and $F_1^\pm$ as in (7.7), which means that the linear parts are in Jordan form.

Theorem 7.1.2. If (7.11) holds and $(0,0)$ is a singularity of focus-focus type, then

$$V_1 = e^{\pi(\lambda^+ + \lambda^-)} - 1, \quad V_2 = w_2^+(\pi) + w_2^-(\pi)e^{3\lambda^+\pi},$$
$$V_3 = e^{\lambda^+\pi}w_3^+(\pi) - 2w_2^+(\pi)^2 + w_3^-(\pi)e^{5\lambda^+\pi}, \quad \text{and}$$
$$V_4 = e^{2\lambda^+\pi}w_4^+(\pi) - 5e^{\lambda^+\pi}w_2^+(\pi)w_3^+(\pi) + 5w_2^+(\pi)^3 + e^{7\lambda^+\pi}w_4^-(\pi).$$

Proof: Observe that using (7.9a) and (7.9b) we may alternatively write e.g.

$$\nu^+ = \exp\left\{ \frac{\pi \operatorname{Re}(c_{10})}{\sqrt{\operatorname{Im}(c_{10})^2 - |c_{01}|^2}} \right\}.$$

According to (7.11) we have

$$\operatorname{Re}(c_{10}) = \lambda^+, \quad \operatorname{Im}(c_{10}) = 1, \quad \text{and} \quad c_{01} = 0,$$

hence $\nu^+ = e^{\pi\lambda^+}$, and similarly $\nu^- = e^{\pi\lambda^-}$. By Thm. 7.1.1 we thus have $V_1 = e^{\pi(\lambda^+ + \lambda^-)} - 1$. Concerning the higher Lyapunov constants, in principle their explicit form can be found as we did for V_1, although this requires a lot of calculation and instead of the polar coordinates from (7.4) the using of so-called generalized polar coordinates

$$x(t) = R(\theta(t))^q \operatorname{Cs}(\theta(t)), \quad y(t) = R(\theta(t))^p \operatorname{Sn}(\theta(t)).$$

Here $q, p \in \mathbb{N}$ have to be chosen appropriately, and

$$\frac{d}{d\theta}\,\mathrm{Cs}\,(\theta) = -\mathrm{Sn}\,(\theta)^{2p-1}, \quad \frac{d}{d\theta}\,\mathrm{Sn}\,(\theta) = \mathrm{Cs}\,(\theta)^{2q-1},$$

$$\mathrm{Cs}(0) = \sqrt[2q]{\frac{1}{p}}, \quad \text{and} \quad \mathrm{Sn}(0) = 0;$$

see COLL/GASULL/PROHENS [53] for more details.   □

In the formulas for V_2, V_3, and V_4, the $w_k^{\pm}$ are the coefficients of $R^{\pm}$, cf. (7.6). We recall that V_2 is only relevant if $V_1 = 0$ and $V_2 \neq 0$, V_3 is used in case that $V_1 = V_2 = 0$ and $V_3 \neq 0$, and so on. The reference COLL/GASULL/PROHENS [53] also contains explicit formulas for the first four Lyapunov constants in case that one or both of the component equations do have a parabolic contact point at $(x, y) = (0, 0)$ rather than a focus.

The following example illustrates the calculation of Lyapunov constants and also their relevance for proving the existence of small periodic solutions in planar systems.

Example 7.1.1. We consider the problem

$$\dot{z} = \begin{cases} (i + \lambda)z + Az^2 + Bz\bar{z} + C\bar{z}^2 & : \quad \mathrm{Im}(z) > 0 \\ iz & : \quad \mathrm{Im}(z) < 0 \end{cases} \tag{7.12}$$

in complex form. Here

$$A = 2 + i\left(\varepsilon_1 + \frac{1}{3}\varepsilon_2 - 2\varepsilon_3 - \frac{\sqrt{6}}{4}\right), \quad B = 1 + i\left(\varepsilon_3 + \frac{\sqrt{6}}{8}\right), \quad \text{and}$$

$$C = i\left(\varepsilon_2 - 9\varepsilon_3 - 9\frac{\sqrt{6}}{8}\right).$$

Then for $\lambda > 0$, $\varepsilon_1 > 0$, $\varepsilon_2 < 0$, and $\varepsilon_3 < 0$ small such that $|\lambda| \ll |\varepsilon_1| \ll |\varepsilon_2| \ll |\varepsilon_3|$ there are at least three limit cycles for (7.12).

We first calculate the corresponding Lyapunov constants. Note that (7.12) is of the form required in (7.11), with $\lambda^+ = \lambda$ and $\lambda^- = 0$. Moreover, $(0, 0)$ is a singularity of focus-focus, as, writing $F_1^+(z, \bar{z}) = c_{10}^+ z + c_{01}^+ \bar{z}$ and $F_1^-(z, \bar{z}) = c_{10}^- z + c_{01}^- - \bar{z}$, we have

$$c_{10}^+ = i + \lambda, \quad c_{01}^+ = 0, \quad c_{10}^- = i, \quad \text{and} \quad c_{01}^- = 0,$$

and this in turn yields due to (7.10) and (7.9b)

$$(b^{\pm} - m^{\pm})^2 + 4l^{\pm}c^{\pm} = -4 \quad \text{and} \quad l^{\pm} = 1.$$

Hence Thm. 7.1.2 applies to give

$$V_1 = e^{\pi\lambda} - 1, \quad V_2 = -2\varepsilon_1, \quad V_3 = -\frac{\pi}{3}\varepsilon_2, \quad \text{and}$$

$$V_4 = 8(3 + 16\varepsilon_3^2 + 6\sqrt{6}\varepsilon_3)\varepsilon_3;$$

we omitted the explicit evaluation of V_2, V_3, and V_4.

To finally verify the existence of at least three limit cycles, the argument indicated in BLOWS/LLOYD [26, p. 220] may be employed. First we consider (7.12) with $\lambda = \varepsilon_1 = \varepsilon_2 = 0$. Then the first non-zero Lyapunov constant is $V_4 < 0$, the latter for $\varepsilon_3 < 0$ small. Hence

$$h^-(h^+(s)) - s = V_4 s^4 + \mathcal{O}(s^5) \tag{7.13}$$

for $s > 0$ small implies that the origin is stable. Thus we find a region encircled by a curve Γ such that the flow for (7.12) with $\lambda = \varepsilon_1 = \varepsilon_2 = 0$ strictly points inward across Γ; rigorously such Γ can be found as a level set of an appropriate Lyapunov function that is strictly decreasing along solutions. One may also imagine constructing Γ as follows: take some $s_0 > 0$ small and let $s_1 = h^-(h^+(s_0)) < s_0$. Then draw (in counterclockwise direction) a curve from s_1 to itself which, at each point (x, y), is a little more steepened than the solution trajectory through (x, y); see Fig. 7.1 on p. 186.

Having determined Γ, we note that perturbing (7.12) with $\lambda = \varepsilon_1 = \varepsilon_2 = 0$ a little will not affect the property of Γ that the flow strictly points inward across Γ. Hence we next choose $|\varepsilon_2| \ll 1$ such that $\varepsilon_2 < 0$, instead of $\varepsilon_2 = 0$, and we keep $\lambda = \varepsilon_1 = 0$. Then $V_3 > 0$ is the first non-zero Lyapunov constant, and

$$h^-(h^+(s)) - s = V_3 s^3 + \mathcal{O}(s^4)$$

for $s > 0$ in a neighborhood of $s = 0$ that will be smaller than the one where (7.13) has been valid. Consequently, the origin has become unstable, and similarly to the foregoing we find Γ_1 inside Γ such that the flow of (7.12), with $\lambda = \varepsilon_1 = 0$ and ε_4 fixed in the previous step, is strictly outward on Γ_1. Moreover, it is strictly inward on Γ. Continuing this way, for $\lambda > 0$, $\varepsilon_1 > 0$, $\varepsilon_2 < 0$, and $\varepsilon_3 < 0$ small such that $|\lambda| \ll |\varepsilon_1| \ll |\varepsilon_2| \ll |\varepsilon_3|$ there are curves Γ_3, Γ_2, Γ_1, and Γ, one contained inside the next, such that the flow of (7.12) is strictly inward on Γ and Γ_2, and strictly outward on Γ_1 and Γ_3. By the Poincaré–Bendixson theorem that is also valid for non-smooth planar systems, see FILIPPOV [82, Ch. 3.13, Thm. 6], we thus conclude that there is a limit cycle between Γ_3 and Γ_2, Γ_2 and Γ_1, and Γ_1 and Γ, respectively. Hence we find at least three limit cycles. $\Diamond$

We remark that this example also highlights a difference between smooth and non-smooth planar systems concerning Lyapunov constants, since $V_1 = V_2 = \ldots = V_{k-1} = 0$ and $V_k \neq 0$ implies k is odd for smooth systems. Choosing $\lambda = 0$ and $\varepsilon_1 \neq 0$, we however obtain that $V_1 = 0$ and $V_2 \neq 0$ for (7.12).

See COLL/GASULL/PROHENS [52] and COLL/PROHENS/GASULL [54] for further results in the same direction, mainly concerning the special case of a Liénard type system $f_1^\pm(x, y) = -y + f^\pm(x)$ and $f_2^\pm(x, y) = x$ in (7.1), where $f^\pm$ is a polynomial of degree at least two.

7.2 Hopf bifurcation

Closely related to the preceding section is the subject of Hopf bifurcation of periodic solutions for planar non-smooth systems. For simplicity we consider the problem in normal form

$$\begin{pmatrix} \dot{x} \\ \dot{y} \end{pmatrix} = \begin{pmatrix} \lambda & -\omega^+(\lambda) \\ \omega^+(\lambda) & \lambda \end{pmatrix} \begin{pmatrix} x \\ y \end{pmatrix} + g^+(x, y, \lambda), \quad y > 0, \quad \text{and} \quad (7.14a)$$

$$\begin{pmatrix} \dot{x} \\ \dot{y} \end{pmatrix} = \begin{pmatrix} \lambda & -\omega^-(\lambda) \\ \omega^-(\lambda) & \lambda \end{pmatrix} \begin{pmatrix} x \\ y \end{pmatrix} + g^-(x, y, \lambda), \quad y < 0, \quad (7.14b)$$

where $g^\pm = (g_1^\pm, g_2^\pm)$ are real-analytic functions defined in a neighborhood of $(x, y) = (0, 0)$ and $\lambda = 0$ such that $|g^\pm(x, y, \lambda)| \leq C(x^2 + y^2)$ in this neighborhood. The linearization matrices do have eigenvalues $\lambda \pm i\omega^+(\lambda)$ and $\lambda \pm i\omega^-(\lambda)$, respectively, and we suppose that $\omega^+(0) > 0$ as well as $\omega^-(0) > 0$; both functions $\omega^\pm(\cdot)$ are assumed to be real-valued and of class C^1 close to $\lambda = 0$.

Theorem 7.2.1. At $\lambda = 0$, system (7.14a), (7.14b) undergoes a Hopf bifurcation. More precisely, there exist $\delta > 0$ and a unique continuous function $\lambda^* :] - \delta, \delta[\to \mathbb{R}$ such that $\lambda^*(0) = 0$, and if $s \in] - \delta, \delta[$ and $s \neq 0$, then there is a periodic orbit of (7.14a), (7.14b) with $\lambda = \lambda^*(s)$ that passes through $(s, 0)$. The corresponding period is $T(s) > 0$ for a continuous function $T :] - \delta, \delta[\to]0, \infty[$ satisfying $T(0) = \frac{\pi}{\omega^+(0)} + \frac{\pi}{\omega^-(0)}$. In a neighborhood of $(x, y) = (0, 0)$ and $\lambda = 0$, all periodic solutions of (7.14a), (7.14b) are obtained this way.

Proof: The proof runs more or less analogously to the smooth case, see MARSDEN/MCCRACKEN [138]. Defining the return maps $h^\pm$ as in Sect. 7.1 and observing that both maps here do additionally depend on λ, we need to find a zero of $s \mapsto h^-(h^+(s, \lambda), \lambda) - s$. To exclude the trivial solution $s = 0$ one introduces

$$H(s, \lambda) = \begin{cases} \dfrac{h^-(h^+(s, \lambda), \lambda) - s}{s} & : \ s \neq 0 \\ V_1(\lambda) & : \ s = 0 \end{cases},$$

with $V_1(\lambda)$ the first Lyapunov constant. Observe that in the notation of (7.2a), (7.2b) we have

$$b^\pm = \lambda, \quad c^\pm = -\omega^\pm(\lambda), \quad l^\pm = \omega^\pm(\lambda), \quad \text{and} \quad m^\pm = \lambda,$$

whence $(0, 0)$ is a singularity of focus-focus type for λ close to zero, since $(b^\pm - m^\pm)^2 + 4l^\pm c^\pm = -4\omega^\pm(\lambda)^2$; recall Definition 7.1.1. Thus $\nu^\pm = \exp(\frac{\pi\lambda}{\omega^\pm(\lambda)})$, and Thm. 7.1.1 implies the explicit form

$$V_1(\lambda) = \exp\left(\frac{\pi\lambda}{\omega^+(\lambda)} + \frac{\pi\lambda}{\omega^-(\lambda)}\right) - 1 \tag{7.15}$$

for the first Lyapunov constant.

We wish to apply the implicit function theorem to solve $H(s,\lambda) = 0$ w.r. to $\lambda = \lambda^*(s)$ in a neighborhood of $s = 0$. By definition of the first Lyapunov constant, it may be shown that $(s,\lambda) \mapsto H(s,\lambda)$ is continuous close to $(s,\lambda) = (0,0)$. Moreover, with $H_1(s,\lambda) = h^-(h^+(s,\lambda),\lambda) - s$ we note that $H_1(0,\lambda) \equiv 0$, whence $\frac{\partial H_1}{\partial \lambda}(0,\lambda) \equiv 0$, and also $V_1(\lambda) = \frac{\partial H_1}{\partial s}(0,\lambda)$. This shows that

$$\frac{\frac{\partial H_1}{\partial \lambda}(s,\lambda)}{s} = \frac{\frac{\partial H_1}{\partial \lambda}(s,\lambda) - \frac{\partial H_1}{\partial \lambda}(0,\lambda)}{s} \longrightarrow \frac{\partial^2 H_1}{\partial s \partial \lambda}(0,\lambda) = \frac{\partial^2 H_1}{\partial \lambda \partial s}(0,\lambda) = V_1'(\lambda)$$

as $s \to 0$. Thus H is found to be continuously differentiable w.r. to λ. Next, $H(0,0) = 0$ due to (7.15), and $\frac{\partial H}{\partial \lambda}(0,0) = V_1'(0) = \frac{\pi}{\omega^+(0)} + \frac{\pi}{\omega^-(0)} > 0$. Hence a unique continuous function $\lambda^* :\,]-\delta,\delta[\to \mathbb{R}$ with $\lambda^*(0) = 0$ and such that $H(s,\lambda^*(s)) \equiv 0$ in $]-\delta,\delta[$ can be found; see DEIMLING [62, Thm. 15.1]. By definition of H this means that $h^-(h^+(s,\lambda^*(s)),\lambda^*(s)) \equiv s$, i.e., for each $s \in]-\delta,\delta[$ the system (7.14a), (7.14b) with $\lambda = \lambda^*(s)$ has a (small) periodic orbit. The fact that $T(0) = \frac{\pi}{\omega^+(0)} + \frac{\pi}{\omega^-(0)}$ follows by explicitly writing down the return maps. An alternative argument is that the corresponding periodic solution is pieced together by "half" a periodic solution of (7.14a) and "half" a periodic solution of (7.14b), both for $\lambda = 0$. From the (smooth) Hopf bifurcation theorem we know that those periodic solutions do have periods $\frac{2\pi}{\omega^+(0)}$ and $\frac{2\pi}{\omega^-(0)}$, respectively, whence we find $T(0) = \frac{1}{2}(\frac{2\pi}{\omega^+(0)}) + \frac{1}{2}(\frac{2\pi}{\omega^-(0)}) = \frac{\pi}{\omega^+(0)} + \frac{\pi}{\omega^-(0)}$; see also Ex. 7.2.1 below. The uniqueness assertion is a consequence of the uniqueness of the function $\lambda^*(\cdot)$; see MORITZ [146] for more details of the proof.   □

Note that $\lambda^*(\cdot)$ cannot be expected to be a C^1-function, since this would require that also H were C^1 w.r. to both s and λ. For the same reason, in general also T will only be continuous rather than differentiable. One may, however, formulate compatibility conditions on the coefficients of the real-analytic nonlinearities g^+ and g^- that will guarantee higher regularity of those functions.

Example 7.2.1. For illustration we investigate the simple system

$$\begin{pmatrix} \dot{x} \\ \dot{y} \end{pmatrix} = \begin{pmatrix} \lambda & -\omega^+ \\ \omega^+ & \lambda \end{pmatrix} \begin{pmatrix} x \\ y \end{pmatrix} - (x^2 + y^2)\begin{pmatrix} x \\ y \end{pmatrix}, \quad y > 0, \quad \text{and} \tag{7.16a}$$

$$\begin{pmatrix} \dot{x} \\ \dot{y} \end{pmatrix} = \begin{pmatrix} \lambda & -\omega^- \\ \omega^- & \lambda \end{pmatrix} \begin{pmatrix} x \\ y \end{pmatrix} - (x^2 + y^2)\begin{pmatrix} x \\ y \end{pmatrix}, \quad y < 0, \tag{7.16b}$$

with $\omega^\pm > 0$. Introducing polar coordinates $x(t) = r(t)\cos\theta(t)$, $y(t) = r(t)\sin\theta(t)$, this becomes

$$\begin{pmatrix} \dot{r} \\ \dot{\theta} \end{pmatrix} = \begin{pmatrix} r\lambda - r^3 \\ \omega^+ \end{pmatrix}, \quad \sin\theta > 0, \quad \text{and} \quad \begin{pmatrix} \dot{r} \\ \dot{\theta} \end{pmatrix} = \begin{pmatrix} r\lambda - r^3 \\ \omega^- \end{pmatrix}, \quad \sin\theta < 0,$$

$$(7.17)$$

From Thm. 7.2.1 we know that (7.16a), (7.16b) undergoes a Hopf bifurcation at $\lambda = 0$, and we will use (7.17) to obtain more precise information in this special case. Starting the "+"-system at $(r^+(0), \theta^+(0)) = (s, 0)$, we find $\theta^+(t) = \omega^+ t$ for $t \in [0, \frac{\pi}{\omega^+}]$, and the solution for $r = r^+$ is

$$r^+(t) = s\sqrt{\frac{\lambda}{s^2 + (\lambda - s^2)e^{-2\lambda t}}} \quad (\lambda \neq 0), \quad \text{and}$$

$$r^+(t) = \frac{s}{\sqrt{2s^2 t + 1}} \quad (\lambda = 0);$$

this can be shown by deriving an ODE for $\frac{1}{r^2}$, and also note that the denominator has the sign of λ. Then we start the "−"-system with data $(r^-(0), \theta^-(0)) = (r^+(\frac{\pi}{\omega^+}), \pi)$ to get $\theta^-(t) = \omega^- t + \pi$ for $t \in [0, \frac{\pi}{\omega^-}]$. Since the differential equation for r^- is the same as the one for r^+, we hence can directly consider the latter up to time $T = \frac{\pi}{\omega^+} + \frac{\pi}{\omega^-}$ to obtain

$$h^-(h^+(s,\lambda),\lambda) = s\sqrt{\frac{\lambda}{s^2 + (\lambda - s^2)e^{-2\lambda T}}} \quad (\lambda \neq 0), \quad \text{and}$$

$$h^-(h^+(s,0),0) = \frac{s}{\sqrt{2s^2 T + 1}}, \quad \text{where} \quad T = \frac{\pi}{\omega^+} + \frac{\pi}{\omega^-}.$$

Note that this is well-defined for all $s \in \mathbb{R}$. To find the non-trivial zeroes of $H_1(s,\lambda) = h^-(h^+(s,\lambda),\lambda) - s$, we observe that there is no such zero, hence no periodic solutions, in case that $\lambda = 0$. For $\lambda \neq 0$ the equation $H_1(s,\lambda) = 0$ simplifies to $\lambda = s^2$, whence $\lambda < 0$ is impossible, and $s = \sqrt{\lambda}$ for $\lambda > 0$. Therefore periodic solutions do exist exactly for $\lambda > 0$, all with equal period $T = \frac{\pi}{\omega^+} + \frac{\pi}{\omega^-}$, and the function $\lambda^*(\cdot)$ from Thm. 7.2.1 which solves $H_1(s,\lambda^*(s)) \equiv 0$ is found to be $\lambda^*(s) = s^2$. From the 1D phase portrait of $\dot{r} = \lambda r - r^3$, or else by calculating $\frac{d}{ds}h^-(h^+(s,\lambda),\lambda)|_{\lambda=s^2} = e^{-2\lambda T} < 1$, it moreover follows that the periodic orbits are asymptotically stable.   ◇

In MORITZ [146] some further results related to Hopf bifurcations are obtained, also on stability and for the case of real parts with opposite signs, i.e., when the linearizations in (7.14a) and (7.14b) are $\begin{pmatrix} \lambda & -\omega^+(\lambda) \\ \omega^+(\lambda) & \lambda \end{pmatrix}$ and $\begin{pmatrix} -\lambda & -\omega^-(\lambda) \\ \omega^-(\lambda) & -\lambda \end{pmatrix}$, respectively.

7.3 Piecewise linear planar systems

A particularly accessible special case of planar systems are piecewise linear systems of the form

$$\dot{q} = Aq + \mathrm{sgn}(w \cdot q)v, \qquad (7.18)$$

with $q = (x, y) \in \mathbb{R}^2$, A a real (2×2)-matrix, and given vectors $v, w \in \mathbb{R}^2$ with $w \neq 0$. Such systems play a role in electrical circuits with a twin triode, AN-DRONOV/VITT/KHAIKIN [9, p. 344], or in control systems with a two-point relay characteristic, LEFSCHETZ [127, p. 82]. Equation (7.18) has a discontinuity line $\{q \in \mathbb{R}^2 : w \cdot q = 0\}$, and the problem is to classify the dynamical behaviour of the system in dependence of the 8 parameters (4 for A, and 2 for w and v, respectively). Following GIANNAKOPOULOS/KAUL/PLIETE [92] or PLIETE [179], we make the following assumptions:

$$\mathrm{trace}(A) \neq 0, \quad \mathrm{trace}(A)^2 < 4 \det(A). \qquad (7.19)$$

It is then possible to determine the number of periodic solutions of (7.18) and their stability. Some of those solutions may stick to the discontinuity for some time (sliding motion), whence sgn has to be considered as the corresponding multi-valued Sgn, i.e., $\mathrm{Sgn}(0) = [-1, 1]$ as before.

As a first step, the number of parameters can be reduced to three, since by means of a suitable transformation (7.18) is seen to be equivalent to

$$\dot{q} = A_\sigma q + \mathrm{sgn}(y)b, \quad A_\sigma = \begin{pmatrix} 0 & -\sigma \\ 1 & -1 \end{pmatrix}, \quad \sigma = \det(A)/\mathrm{trace}(A)^2 > 1/4,$$

$$(7.20)$$

where

$$b_1 = -[(Aw^\perp) \cdot v^\perp]/\mathrm{trace}(A)^2, \quad b_2 = -(w \cdot v)/\mathrm{trace}(A),$$

with $w^\perp = (-w_2, w_1)$. In addition, this transformation rotates the discontinuity line to $\{(x, y) \in \mathbb{R}^2 : y = 0\}$. In (7.20) there are only three parameters left, σ, b_1, and b_2.

Then an analysis can be carried out for the resulting semiflow to find the critical points, i.e., such $q_0 = (x_0, y_0)$ with $(0, 0) \in A_\sigma q_0 + \mathrm{Sgn}(y_0)b$, in all the different possible cases; see GIANNAKOPOULOS/KAUL/PLIETE [92]. Concerning the existence of periodic solutions, $b_2 \geq 0$ is a necessary condition, as may be seen from the phase portraits. For $b_2 > 0$ the closed orbits that cross the discontinuity line transversely (no sticking) can be found as fixed points of the return map $h = h^- \circ h^+$, where e.g. h^+ maps a point $(x_0, 0)$ along a trajectory that is contained in $\{(x, y) : y > 0\}$ to the first return point $(x_1, 0)$ to the x-axis. Under certain circumstances, it may be verified that h, considered as a map on the x-axis, is strictly increasing and concave, and hence has at most two fixed points. Thus there are at most two closed orbits that do not stick to the discontinuity line. With some more effort, the precise number of periodic solutions can be detected, and also their stability. Moreover, the existence of periodic solutions with sliding motion and of homoclinic orbits can be studied in detail; cf. GIANNAKOPOULOS/KAUL/PLIETE [92] for the complete results. In particular, it turns out that there are at most three periodic solutions for (7.20).

8. Melnikov's method for non-smooth dynamical systems

A further tool to obtain analytical results for smooth dynamical systems is Melnikov's method, cf. ARROWSMITH/PLACE [16] or WIGGINS [225].

On the one hand, this method allows to detect subharmonic solutions of arbitrary period and homoclinic orbits in certain systems $\dot{y} = f(y) + \varepsilon g(y, t)$ that are small perturbations of systems possessing a homoclinic orbit. Although we will not go into further detail, we should mention that the corresponding techniques can be carried over to particular non-smooth systems which are of type $\dot{y} \in f(y) + \varepsilon G(y, t)$, with a multi-valued function $G(y, t)$ arising e.g. as a model of dry friction as in (1.2). Thus, the choice of $\varepsilon G(y, t)$ means that the friction effect has to be small; see FEČKAN [75, 76, 77, 79], and also AWREJCEWICZ/HOLICKE [19] for some analytical and numerical results.

A second situation where Melnikov's method can be used in non-smooth systems are particular planar systems, as in Chap. 7, where it is not possible for trajectories to stick to the discontinuity surface for some time. To be definite, we deal with an example taken from ZOU/KÜPPER [229]; the presentation follows KUNZE/KÜPPER [117]. For convenience we again denote $q = (x, y)$ instead of $y = (x, v)$ a typical point in $\mathbb{R}^2$. Define

$$f(x, y) = f^+(x, y), \quad y > 0, \quad \text{and} \quad f(x, y) = f^-(x, y), \quad y < 0,$$

where

$$f^+(x, y) = (1, -2x) \quad \text{and} \quad f^-(x, y) = (-y - 1, -x).$$

So the system we consider reads as

$$\begin{array}{ll} \dot{x} = 1 \\ \dot{y} = -2x \end{array}, \quad y > 0, \quad \text{and} \quad \begin{array}{ll} \dot{x} = -y - 1 \\ \dot{y} = -x \end{array}, \quad y < 0. \tag{8.1}$$

It has a homoclinic orbit

$$q_0(t) = \begin{cases} (-e^{t+1}, e^{t+1} - 1) & : \quad t \in]-\infty, -1], \\ (t, -t^2 + 1) & : \quad t \in [-1, 1], \\ (e^{-t+1}, e^{-t+1} - 1) & : \quad t \in [1, \infty[, \end{cases}$$

connecting the hyperbolic fixed point $p_0 = (0, -1)$ to itself. The solution has been adjusted in phase such that $q_0(0) = (0, 1)$, and it transversely intersects the discontinuity line $\{(x, y) \in \mathbb{R}^2 : y = 0\}$ twice, at time $t_d^- = -1$ in

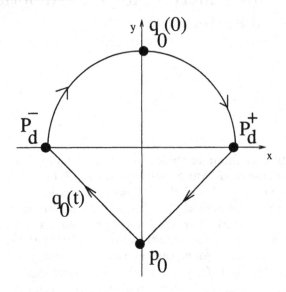

Fig. 8.1. The homoclinic orbit for (8.1)

$P_d^- = q_0(t_d^-) = (-1,0)$ and at $t_d^+ = 1$ in $P_d^+ = q_0(t_d^+) = (1,0)$; see Fig. 8.1 above.

We also note that here

$$\text{trace}(Df^+) = \text{trace}(Df^-) = 0 \tag{8.2}$$

in the respective domains of definition. It is for such situations and small perturbations

$$\dot{q} = f(q) + \varepsilon g(q,t) \tag{8.3}$$

that Melnikov's method can be used equally well for non-smooth system to prove e.g. the existence of a homoclinic orbit for ε small. Compared to the usual Melnikov function

$$M_0(\theta_0) = \int_{\mathbb{R}} f(q_0(t - \theta_0)) \wedge g(q_0(t - \theta_0), t)\, dt \tag{8.4}$$

for smooth systems, two additional terms will be picked up through crossing the discontinuity. To see this, we first have to re-develop some parts of the classical theory, cf. ARROWSMITH/PLACE [16, Sect. 3.8], always arguing on a quite heuristic rather than a rigorous level; in principle, the existence of all the stable and unstable manifolds etc. that we are going to use below should be verified first.

To begin with, we expect that the fixed point p_0 will perturb to a fixed point p_ε of (8.3) that is of distance $\mathcal{O}(\varepsilon)$ from p_0. For the augmented system

$$\dot{q} = f(q) + \varepsilon g(q,\theta), \quad \dot{\theta} = 1, \tag{8.5}$$

this means that there exists a hyperbolic periodic orbit γ_ε close to $\gamma_0 = \{(p_0, \theta) : \theta \in S^1)\}$. Fix $\theta_0 \in S^1 \cong [0, 2\pi]$ and denote L a line segment in the $\theta = \theta_0$ plane $\Sigma_{\theta_0} = \mathbb{R}^2 \times \{\theta_0\}$ that is perpendicular to the homoclinic orbit

$$\{q_0(t) : t \in \mathbb{R}\}$$

at $q_0(0)$, and therefore points in the direction of $f^\perp(q_0(0))$, with $f^\perp(q) = (-f_2(q), f_1(q))$ for $f(q) = (f_1(q), f_2(q))$. Denote p_{ε,θ_0} the intersection of γ_ε with Σ_{θ_0}, and let $q^{u,s}(t; \theta_0, \varepsilon)$ be the unique trajectories of (8.5) that lie in the unstable $(=u)$ resp. stable $(=s)$ manifold of p_{ε,θ_0} and cross L at shortest distance to $q_0(0)$; see Fig. 8.2.

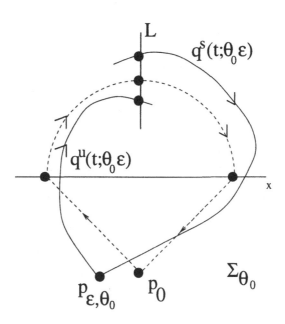

Fig. 8.2. Unstable and stable manifolds of p_{ε,θ_0}

Then we consider the function

$$\Delta_\varepsilon(t, \theta_0) = f(q_0(t - \theta_0)) \wedge [q^u(t; \theta_0, \varepsilon) - q^s(t; \theta_0, \varepsilon)] \qquad (8.6)$$

which for $t = \theta_0$ measures the distance between the unstable and stable manifolds of p_{ε,θ_0}. Hence to find a homoclinic orbit for (8.3) one needs to study the zeroes of $\theta_0 \mapsto \Delta_\varepsilon(\theta_0, \theta_0)$; see ARROWSMITH/PLACE [16, p. 173] or WIGGINS [225, Sect. 4.5]. We have

$$q^{u,s}(t; \theta_0, \varepsilon) = q_0(t - \theta_0) + \varepsilon q_1^{u,s}(t, \theta_0) + \mathcal{O}(\varepsilon^2), \qquad (8.7)$$

with $q_1^{u,s}$ solutions to the linearized equations

$$\dot{q}_1^{u,s}(t,\theta_0) = Df(q_0(t-\theta_0))q_1^{u,s}(t,\theta_0) + g(q_0(t-\theta_0),t). \qquad (8.8)$$

Let

$$\Delta_\varepsilon^{u,s}(t,\theta_0) = \varepsilon f(q_0(t-\theta_0)) \wedge q_1^{u,s}(t,\theta_0).$$

Since $f(q_0(t))$ is orthogonal to $q_0(t)$, it follows from (8.6) and (8.7) that

$$\Delta_\varepsilon(t,\theta_0) = \Delta_\varepsilon^u(t,\theta_0) - \Delta_\varepsilon^s(t,\theta_0) + \mathcal{O}(\varepsilon^2). \qquad (8.9)$$

According to the trace condition (8.2) it then can be shown as in the smooth case that

$$\dot{\Delta}_\varepsilon^u(t,\theta_0) = \varepsilon f(q_0(t-\theta_0)) \wedge g(q_0(t-\theta_0),t). \qquad (8.10)$$

To integrate this w.r. to t we have to know at which times $\theta_0 + T^{u,s}(\theta_0,\varepsilon)$ (on the level Σ_{θ_0}) the trajectories $q^{u,s}(t;\theta_0,\varepsilon)$ will cross the discontinuity surface. According to the above construction we expect

$$\theta_0 + T^{u,s}(\theta_0,\varepsilon) \cong \theta_0 + t_d^\pm + \mathcal{O}(\varepsilon),$$

and hence we will omit the error $\mathcal{O}(\varepsilon)$, since this should give a further term $\mathcal{O}(\varepsilon^2)$ in (8.9). The intersection points of $q^{u,s}(t;\theta_0,\varepsilon)$ with the discontinuity are ε-close to P_d^- and P_d^+, respectively. Integration of (8.10) over times $t \in \,]-\infty, \theta_0 + t_d^-]$ yields, taking into account that $\Delta_\varepsilon^u(-\infty,\theta_0) = 0$ due to $f(p_0) = 0$,

$$-\Delta_\varepsilon^{u,-}(\theta_0 + t_d^-,\theta_0) = \varepsilon \int_{-\infty}^{\theta_0+t_d^-} f^-(q_0(t-\theta_0)) \wedge g(q_0(t-\theta_0),t)\, dt;$$

here the "$-$" in $\Delta_\varepsilon^{u,-}$ is used to indicate that one needs to choose f^- in the definition of Δ_ε^u, also to determine q_1^u. Similarly, observing $t_d^- < 0$, we have

$$\Delta_\varepsilon^{u,+}(\theta_0,\theta_0) - \Delta_\varepsilon^{u,+}(\theta_0 + t_d^-,\theta_0)$$
$$= \varepsilon \int_{\theta_0+t_d^-}^{\theta_0} f^+(q_0(t-\theta_0)) \wedge g(q_0(t-\theta_0),t)\, dt,$$

and analogously, omitting the arguments in the integral,

$$\Delta_\varepsilon^{s,+}(\theta_0 + t_d^+,\theta_0) - \Delta_\varepsilon^{s,+}(\theta_0,\theta_0) = \varepsilon \int_{\theta_0}^{\theta_0+t_d^+} f^+ \wedge g\, dt,$$
$$-\Delta_\varepsilon^{s,-}(\theta_0 + t_d^+,\theta_0) = \varepsilon \int_{\theta_0+t_d^+}^{\infty} f^- \wedge g\, dt.$$

According to (8.9), and since $q_0(0)$ is in the upper half-plane, this leads to

$$\Delta_\varepsilon(\theta_0,\theta_0) = \Delta_\varepsilon^{u,+}(\theta_0,\theta_0) - \Delta_\varepsilon^{s,+}(\theta_0,\theta_0) + \mathcal{O}(\varepsilon^2)$$
$$= M_0(\theta_0) + \left(\Delta_\varepsilon^{u,+}(\theta_0 + t_d^-,\theta_0) - \Delta_\varepsilon^{u,-}(\theta_0 + t_d^-,\theta_0)\right.$$
$$\left. +\Delta_\varepsilon^{s,-}(\theta_0 + t_d^+,\theta_0) - \Delta_\varepsilon^{s,+}(\theta_0 + t_d^+,\theta_0)\right) + \mathcal{O}(\varepsilon^2),$$

with the standard Melnikov integral $M_0(\theta_0)$ from (8.4). The Δ-terms are contributions due to the discontinuity, and they are as well of order $\mathcal{O}(\varepsilon)$. Thus according to the definition of $\Delta_\varepsilon^{u,s}$ the relevant "non-smooth" Melnikov function is

$$M(\theta_0) = M_0(\theta_0) + \Big(f^+(q_0(t_d^-)) \wedge q_1^{u,+}(\theta_0 + t_d^-,\theta_0)$$
$$- f^-(q_0(t_d^-)) \wedge q_1^{u,-}(\theta_0 + t_d^-,\theta_0)$$
$$+ f^-(q_0(t_d^+)) \wedge q_1^{s,-}(\theta_0 + t_d^+,\theta_0)$$
$$- f^+(q_0(t_d^+)) \wedge q_1^{s,+}(\theta_0 + t_d^+,\theta_0)\Big).$$

The functions $q_1^{u/s,\pm}$ are obtained from (8.8), e.g. $\varphi(s) = q_1^{u,+}(\theta_0 + s,\theta_0)$ solves the equation $\frac{d}{ds}\varphi = Df^+(q_0(s))\varphi + g(q_0(s),\theta_0 + s)$ for s near $s = t_d^-$. Again the zeroes of $M(\theta_0)$ determine the (transversal) intersection of stable and unstable manifolds, whence a simple zero gives rise to a homoclinic orbit for (8.3).

Returning to the particular example (8.1) from the beginning of this section, with $g(q,t) = (0, \sin t)$ the corresponding Melnikov function was calculated in ZOU/KÜPPER [229] to be

$$M(\theta_0) = \alpha \sin\theta_0 + \beta \cos\theta_0 = \sqrt{\alpha^2 + \beta^2} \sin(\theta_0 + \phi),$$

with $\alpha = \frac{1}{2}(e + e^{-1}) - \sin 1 - 2\cos 1$, $\beta = \frac{1}{2}(e + e^{-1}) - \cos 1$, and $\phi = \arctan(\beta/\alpha)$. Numerically, two simple zeroes of M were found at $\theta_0 \approx 1.21$ and $\theta_0 \approx 4.35$, whence it was concluded that for the piecewise system

$$\begin{aligned} \dot{x} &= 1 \\ \dot{y} &= -2x + \varepsilon\sin t \end{aligned}, \quad y > 0, \quad \text{and} \quad \begin{aligned} \dot{x} &= -y - 1 \\ \dot{y} &= -x + \varepsilon\sin t \end{aligned}, \quad y < 0,$$

one may choose $\varepsilon \neq 0$ small enough to find periodic solutions of arbitrary high periods close to the homoclinic orbit q_0 of the system with $\varepsilon = 0$.

9. Further topics and notes

Without going into much detail, we review in this section some additional material and give a number of complementary remarks and references.

9.1 Numerical results

The calculation of solution trajectories for non-smooth systems is considered for instance in TAUBERT [215] and STEWART [209], a review paper is DONTCHEV/LEMPIO [73].

There are several works that, at least to some extent, deal with the numerical evaluation of Lyapunov exponents for non-smooth problems, mostly for particular example systems; cf. BERGER/ROKNI [23], BUDD/LAMBA [125], MICHAELI [140], NQI [160], or WIEDERHÖFT [223], and also many of the references to follow.

A great diversity of results has been obtained for friction oscillators of all kinds, e.g. one or several masses that are placed on a driven belt and that are coupled by means of springs; see Fig. 9.1.

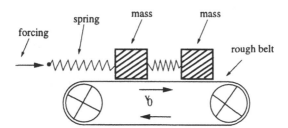

Fig. 9.1. A two-degrees-of-freedom oscillator

This system is called the Burridge–Knopoff model in seismology and thought of to describe earthquakes; the masses are representing the different tectonic plates. See BURRIDGE/KNOPOFF [41] or CARLSON/LANGER/SHAW [44]. A related model problem is the rocking block system, consisting of a block on a horizontal shake table. It has been used for quite a long time (since the 1880's) to study the motion of buildings during an earthquake; cf. in particular the papers cited in BROGLIATO [32, Sect. 6.4.1] and POPP [182, p. 234].

A selection of additional references containing mainly numerical results on the (often quite complicated) bifurcation behaviour of several types of such non-smooth oscillators is AWREJCEWICZ/DELFS [18], BALMER [21], BRANDL/HAJEK [30], BROMMUNDT [34], CHATTERJEE/MALLIK [49], FEENY [80], GALVANETTO/BISHOP [86, 87], GALVANETTO/BISHOP/BRISEGHELLA [88], KUNZE/KÜPPER [116], MICHAELI [140], OESTREICH/HINRICHS/POPP [164], PFEIFFER [171, 172, 173, 174], PFEIFFER/HAJEK [177], PFEIFFER/ SEYFFERTH [178], POPP/HINRICHS/OESTREICH [183], POPP/STELTER [184, 185], SCHNEIDER/POPP/IRRETIER [198], SHAW [200], SIKORA/BOGACZ [204], STELTER [207], STELTER/SEXTRO [208], STORZ [211], SZCZYGIELSKI [213], or WIEDERHÖFT [223].

In VOSSHAGE [217] a first step is undertaken towards the simulation of attractors and invariant measures for some non-smooth example systems by means of the program package GAIO, authored by DELLNITZ, HOHMANN, and co-workers, which allows for the effective approximation of such objects by a box sub-division algorithm; see e.g. DELLNITZ/HOHMANN [69].

9.2 Impact oscillator type problems

The impact oscillator is the system comprising a ball on a spring hitting a wall, as illustrated in Fig. 9.2.

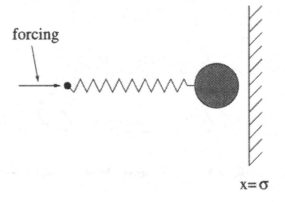

Fig. 9.2. The impact oscillator

The governing equations are

$$\ddot{x} + \delta\dot{x} + x = p(t) \quad (x < \sigma), \qquad \dot{x} \mapsto -r\dot{x} \quad (x = \sigma), \tag{9.1}$$

where $\delta > 0$ is the damping, p denotes some forcing of period $\omega > 0$, and $r \in]0, 1]$ is the coefficient of restitution at an impact with the wall which is fixed at $x = \sigma$. Thus there are four parameters σ, δ, ω, and r, and the goal is to study the bifurcations that can appear if those parameters are varied.

This problem is mostly approached by introducing a Poincaré map

$$P: \quad (\phi_i, v_i) \mapsto (\phi_{i+1}, v_{i+1}), \tag{9.2}$$

defined as follows: assume that an impact occurs at time $t_i > 0$ with positive velocity v_i, and denote $\phi_i = t_i \,(\mathrm{mod}\,\omega)$ the associated phase. Then the corresponding solution of (9.1) is calculated until the ball next hits the wall at time $t_{i+1} > t_i$, and then one lets $\phi_{i+1} = t_{i+1} \,(\mathrm{mod}\,\omega)$, whereas v_{i+1} is the velocity at the impact. Sometimes the return map alternatively is introduced as $P : (\phi_i, v_i^2) \mapsto (\phi_{i+1}, v_{i+1}^2)$, since this allows for an easier form of its Jacobian.

The simplest kind of interesting solutions to (9.1) are the periodic ones. If x is periodic with period $n\omega$, and if the motion repeats after m impacts (i.e., P^m has a fixed point), then the periodic orbit is addressed as being a (m, n) orbit. For instance, a $(1, n)$ orbit corresponds to a $n\omega$-periodic motion between every two consecutive impacts. It is possible to find explicit conditions for such periodic orbits and to analyze their linear stability; cf. WHISTON [220, 221] or BUDD/DUX/CLIFFE [40].

This is quite similar to periodically forced oscillators with a discontinuous and state-dependent stiffness, like

$$\ddot{x} + k(x)x = p(t), \quad k(x) = \begin{cases} k_1 & : \quad x < x_0 \\ k_2 & : \quad x > x_0 \end{cases},$$

as investigated e.g. in SHAW/HOLMES [201, 202], and it is also similar to bouncing ball problems, see PUSTYLNIKOV [186] and the corresponding references and remarks in BROGLIATO [32, p. 363 ff]. A great source of motivation to study bouncing ball type systems was the fact that they can be seen as simplified versions of the Fermi accelerator problem, FERMI [81], as described in e.g. LICHTENBERG/LIEBERMAN [131], being related to the origin of cosmic rays.

Returning to (9.2), the rigorous definition of P from (9.2) is more complicated (and often omitted), as the mass may hit the wall infinitely often during a finite time, and then stick to the wall before moving away again; cf. BUDD/DUX/CLIFFE [40] and LAMBA [123]. Another problem related to P is that not necessarily $v_{i+1} > 0$ at an impact, since the ball may just touch the wall with zero velocity to be then pulled away again. Hence those data that lead to an $v_i = 0$ for some $i \in \mathbb{N}$ do play a special role in the phase space cylinder $[0, 2\pi] \times]0, \infty[\,\cong\, \mathbb{S}^1 \times]0, \infty[$. Such zero velocity impacts are called grazes, and it is the presence of this discontinuity set that leads to a new kind of bifurcation, generally referred to as "grazing bifurcation", since a trajectory close to a reference trajectory that grazes at $x = \sigma$ with zero velocity will behave quite differently, depending on whether its velocity is slightly positive or "negative" (the latter means that there is no impact). For an extensive study of grazing bifurcations (also in other models) see PETERKA [169], WHISTON [220, 221, 222], NORDMARK [157, 158, 159], BUDD [35, 36, 37], BUDD/DUX [38, 39], CASAS/CHIN/GREBOGI/OTT [45], CHIN/OTT/NUSSE/GREBOGI [50, 51], DI BERNARDO/BUDD/CHAMPNEYS [71], FOALE/BISHOP [83, 84], FREDRIKSSON/NORDMARK [85], GONTIER/ TOULEMONDE [95], LAMBA [123, 124], LAMBA/BUDD [125], NUSSE/OTT/YORKE [161], NUSSE/YORKE [162, 163], TOULEMONDE/GONTIER [216], YUAN/BANERJEE/ODD/YORKE [227], and the references from those papers.

Often P is studied by introducing certain local coordinates (η, ξ) close to a point on $S = \{(\phi, v) : P(\phi, v) = (\varphi, 0)\}$, with η the coordinate in tangent direction to S, whereas ξ is the coordinate orthogonal to S, hence $\xi = 0$ locally on S; cf. NORDMARK [157], the above references, and also BROGLIATO [32, Sect. 7.2.3]. It turned out that P approximately has the form

$$\left\{ \begin{array}{rcll} P(\eta, \xi) & = & (-a\xi, b\xi + \eta \cdot & \vdots \leq 0, \\ P(\eta, \xi) & = & (-ad^2\xi, -\sqrt{\xi} & \vdots > 0, \end{array} \right. \tag{9.3}$$

for some parameters a, b, c, and $d \in [0, 1]$, a̶n̶d̶ ̶t̶h̶i̶s̶ ̶i̶n̶f̶o̶r̶mation can be used to further study the bifurcation behaviour of (9.1); note that P has a square-root type singularity at $\xi = 0$. Additional simplification is introduced by using only one-dimensional maps, like

$$\bar{P}_\varepsilon(\xi) = \left\{ \begin{array}{lll} a\xi + \varepsilon & : & \xi \leq 0 \\ b\xi^\gamma + \varepsilon & : & \xi > 0 \end{array} \right. ,$$

with $a \in]0, 1[$, $b < -1$, a certain power $\gamma \in]0, 1[$ (e.g. $\gamma = \frac{1}{2}$ corresponding to (9.3)), and the bifurcation parameter ε. For $\varepsilon < 0$, $\bar{P}_\varepsilon$ has an attractive fixed point, whereas for $\varepsilon > 0$ small the fixed point is of flip (saddle) type; see the references already mentioned before. Some material on the impact oscillator is reviewed in PETERKA [170].

Another class of models that is closely related to impact oscillators are billiard (or dual billiard) type systems which also can be described in terms of two-dimensional return maps: the angle and velocity at one impact of the billiard ball at the boundary is mapped to the angle and velocity at the next impact. There is a large literature concerning properties of billiards, and in particular with regard to ergodicity; see SINAI [205], KOZLOV/TRESHCHEV [110], or TABACHNIKOV [214]. The general theory of measure-preserving hyperbolic maps which are smooth except for some small set is developed in detail in KATOK/STRELCYN [106]. The connection of impact oscillators to dual billiards is investigated in BOYLAND [29].

9.3 General systems with impacts and/or friction

In this section we are concerned with general mechanical systems consisting of $l \geq 1$ rigid bodies which are subject to impacts and/or friction. Our aim is to describe the basic model and to comment on different issues related to this model.

The equations of motion in its most general form for such multi-body systems may be written as

$$M(q,t)\ddot{q} = f(q,\dot{q},t) \quad \text{for} \quad t \geq 0, \tag{9.4}$$

where $q = q(t) \in \mathbb{R}^m$ denotes the coordinates of one of the $l \geq 1$ bodies which is thought of to be distinguished. The nonlinearity $f : \mathbb{R}^{2m+1} \to \mathbb{R}^m$ contains all kind of forces acting on the system, like driving forces, inertia forces, contact forces, friction forces, etc., and $M : \mathbb{R}^{m+1} \to \mathcal{L}(\mathbb{R}^m, \mathbb{R}^m)$ is the symmetric and positive definite mass matrix. Besides (9.4), a solution also has to satisfy some constraint conditions, given through functions $\phi_i : \mathbb{R}^m \to \mathbb{R}$ describing the distances from the marked body to all other bodies, so that we require

$$\phi_i(q(t)) \geq 0 \quad \text{for} \quad t \geq 0 \quad \text{and} \quad 1 \leq i \leq p. \tag{9.5}$$

Since we allow measurement of multiple distances from the marked body to the others, we have $p \geq l - 1$. (As an example, one can think of a slab falling down to the floor, where we can measure simultaneously e.g. the distance of the top and of the bottom of the slab to the floor.) Following LÖTSTEDT [135], PFEIFFER/GLOCKER [175], and others, we let

$$J_N(q) = \{i \in \{1,\ldots,p\} : \phi_i(q) = 0\} \quad \text{for} \quad q \in \mathbb{R}^m, \tag{9.6}$$

so that $i_0 \in J_N(q(t_0))$ means that at time $t = t_0$ there was an impact between the distinguished body and the body labeled with i_0. In this case of an impact, there may be also additional friction forces, due to sticking or sliding. Thus if we denote

$$v_i(q,\dot{q}) = g_{F_i}(q)^T \dot{q} + w_i(q), \quad q, \dot{q} \in \mathbb{R}^m, \tag{9.7}$$

the corresponding relative velocities in tangential directions, with $g_{F_i} : \mathbb{R}^m \to \mathbb{R}^m$ and $w_i : \mathbb{R}^m \to \mathbb{R}^m$, then the index sets

$$\begin{aligned}
J_{F_0}(q,\dot{q}) &= \{i \in J_N(q) : v_i(q,\dot{q}) = 0\} \\
&= \{i \in \{1,\ldots,p\} : \phi_i(q) = 0 \text{ and } v_i(q,\dot{q}) = 0\}
\end{aligned} \tag{9.8}$$

and

$$\begin{aligned}
J_{FW}(q,\dot{q}) &= \{i \in J_N(q) : v_i(q,\dot{q}) \neq 0\} \\
&= \{i \in \{1,\ldots,p\} : \phi_i(q) = 0 \text{ and } v_i(q,\dot{q}) \neq 0\}
\end{aligned} \tag{9.9}$$

are useful in an analogous way as J_N is: if q is a solution of (9.4), then $i_0 \in J_F(q(t_0),\dot{q}(t_0))$ [resp. $i_0 \in J_{FW}(q(t_0),\dot{q}(t_0))$] is equivalent to stating that at time $t = t_0$ there is a sticking [resp. slipping] contact between the body labeled with i_0 and the distinguished body.

By specifying at every time $t = t_0$ the forces arising due to contacts, sticking, or slipping, (9.4) reads as

$$\ddot{q}(t) = f(q(t), \dot{q}(t), t) \quad + \sum_{i \in J_N(q(t))} \lambda_{N_i}(t)\, g_{N_i}(q(t))$$

$$+ \sum_{i \in J_{F_0}(q(t), \dot{q}(t))} \lambda_{F_i}(t)\, g_{F_i}(q(t)) \quad + \sum_{i \in J_{FW}(q(t), \dot{q}(t))} \lambda_{F_i}(t)\, g_{F_i}(q(t)),$$

$$(9.10)$$

cf. e.g. PFEIFFER [172, Sect. 2.2] or PFEIFFER/SEYFFERTH [178, Sect. 2]. Here $g_{N_i} = \phi_i'$, and $\lambda_{N_i} \geq 0$ resp. λ_{F_i} are the Lagrange multipliers corresponding to the constraints. (They are proportional to the normal force resp. the friction force, both with the same nonnegative factor.) In contrast to (9.4), in (9.10) we chose $M(\dot{q}, t)$ to be the identity matrix (this may be achieved by a suitable change of coordinate system), and f in (9.10) is assumed to contain only the external forces and the inertia forces.

To further rewrite (9.10), note that for $i \in J_{FW}(q(t), \dot{q}(t))$ we have $v_i(q(t), \dot{q}(t)) \neq 0$. Now Coulomb's law in its general form says (cf. KILMISTER/REEVE [108]), if N, F resp. v denote the normal force, the friction force resp. the relative velocity at an impact point, then

$$(N \geq 0, \quad v \neq 0 \Longrightarrow |F| = \mu N) \text{ and } (v F \leq 0, \quad v = 0 \Longrightarrow |F| \leq \mu N).$$
$$(9.11)$$

Thus, if we drop the argument t for the moment,

$$v_i(q, \dot{q})\, \lambda_{F_i} \leq 0 \quad \text{and} \quad \lambda_{F_i} = s_i\, \mu_i\, \lambda_{N_i} \quad \text{with} \quad s_i = \begin{cases} 1 & \text{for } v_i(q, \dot{q}) < 0 \\ -1 & \text{for } v_i(q, \dot{q}) > 0 \end{cases}$$
$$(9.12)$$

and suitable friction coefficients $\mu_i > 0$. Inserting this into (9.10) we finally obtain the basic equation

$$\ddot{q}(t) = f(q(t), \dot{q}(t), t) \quad + \sum_{i \in J_N(q(t))} \lambda_{N_i}(t)\, g_{N_i}(q(t))$$

$$+ \sum_{i \in J_{F_0}(q(t), \dot{q}(t))} \lambda_{F_i}(t)\, g_{F_i}(q(t))$$

$$+ \sum_{i \in J_{FW}(q(t), \dot{q}(t))} s_i(t)\, \mu_i\, \lambda_{N_i}(t)\, g_{F_i}(q(t)), \qquad (9.13)$$

where we of course let $\sum_{\emptyset} = 0$.

From (9.13), the inherent problem in numerical simulations of the dynamics is obvious, cf. PFEIFFER [174] or PFEIFFER/GLOCKER [175]: in multi-body systems with multiple contacts the change of one contact may lead to a change in all other contact points, i.e., at every time $t = t_0$ the "active" index sets $J_N(q(t))$, $J_{F_0}(q(t), \dot{q}(t))$, and $J_{FW}(q(t), \dot{q}(t))$ have to be determined anew. Therefore, to model the dynamics, the following questions arise:

(a) Suppose that $(q_0, \dot{q}_0) = (q(t_0), \dot{q}(t_0))$ is known at some time $t = t_0$, and hence also $J_N(q_0)$, $J_{F_0}(q_0, \dot{q}_0)$, and $J_{FW}(q_0, \dot{q}_0)$ are known. How is it possible to modify these active index sets in such a way that they (at best) do not change in infinitesimal time, i.e., that they remain constant on some $[t_0, t_0 + h]$?

(b) How to incorporate the effects of possible changes of the active set? The following possibilities have to be taken into account:
  (i) two bodies impact, i.e., J_N has to be enlarged;
  (ii) two bodies loose contact, i.e., J_N has to be decreased;
  (iii) there is a transition from slipping to sticking, i.e., an index from J_{FW} is changing to J_{F_0};
  (iv) there is a transition from sticking to slipping, i.e., an index from J_{F_0} is changing to J_{FW};

(c) How to solve (9.13) in time-intervals $[t_0, t_0 + h]$ where the active sets are constant?

While (c) is standard, problems (a) and (b) may be transformed to solving so-called (extended) linear complementarity problems (LCPs), either on the force level or on the momentum level; cf. e.g. LÖTSTEDT [135, 136, 137], MOREAU [145], GLOCKER/PFEIFFER [93, 94], PFEIFFER [174], PFEIFFER/GLOCKER [175], PFEIFFER/SEYFFERTH [178], MEITINGER/PFEIFFER [139], or ALSCHER [8], HEEMELS [98], KUNZE/NEUMANN [121], and NEUMANN [156]. For general information on LCPs, see COTTLE/DANTZIG [58] or MURTY [155].

9.4 Concluding remarks

Of course there is still a lot to do for non-smooth systems. Since the models that are mathematically accessible are often by far too simplified, it will be a challenge to rigorously study systems that are closer to real-world problems than those investigated in these notes. One could start this programme e.g. by introducing in (1.2) some more general friction laws that are relevant for practical applications, as described e.g. in IBRAHIM [100]; see also the references already given on p. 39.

Finally we should emphasize that currently there is growing interest in non-smooth dynamical systems, multi-body systems, sweeping processes, and related issues, as is reflected by the large number of papers, reviews, and books that appeared during the last few years, much too much to be comprehensively reviewed here. The interested reader may further consult BROGLIATO [32, 33], KUNZE/KÜPPER [117], MONTEIRO MARQUES [144], PFEIFFER/GLOCKER [176], STEWART [210], WIERCIGROCH/DE KRAKER [224], and the references therein.

Notation

Basic Notation

$\mathbb{N}$	$= \{1, 2, 3, \ldots\}$
$\mathbb{N}_0$	$= \{0, 1, 2, 3, \ldots\}$
$\mathcal{L}(\mathbb{R}^d)$	linear operators on $\mathbb{R}^d$, i.e., the $(d \times d)$-matrices, with norm $\|\cdot\|_d$ induced by the Euclidean norm on $\mathbb{R}^d$
$\langle x, y \rangle$, $x \cdot y$	standard Euclidean inner product of $x, y \in \mathbb{R}^d$
$B_r(x_0)$	$= \{x \in X : |x - x_0| < r\}$; open ball of radius $r > 0$ centered at $x_0 \in X$, where $(X, |\cdot|)$ is a normed space
$\overline{B}_r(x_0) = \overline{B_r(x_0)}$	$= \{x \in X : |x - x_0| \le r\}$; closure of $B_r(x_0)$
$A+B$	$= \{x + y : x \in A, y \in B\}$ for $A, B \subset X$
λA	$= \{\lambda x : x \in A\}$ for $\lambda \in \mathbb{R}$ and $A \subset X$
$\overline{A}$	closure of $A \subset X$
A^0	interior of $A \subset X$
$\mathrm{co}A$	$= \left\{ \sum_{i=1}^m \lambda_i x_i : m \in \mathbb{N}, \lambda_i \in [0,1], \sum_{i=1}^m \lambda_i = 1, x_i \in A \right\}$; convex hull of $A \subset X$
$\overline{\mathrm{co}}A$	$= \overline{\mathrm{co}A}$
$\mathrm{dist}(x, A)$	$= \inf\{|x - a| : a \in A\}$; distance from $x \in X$ to $A \subset X$
$d_{\mathrm{H}}(A, B)$	$= \max\left\{ \sup_{x \in B} \mathrm{dist}(x, A),\ \sup_{x \in A} \mathrm{dist}(x, B) \right\}$; Hausdorff-distance of $A, B \subset X$

$A \dot{\cup} B$	disjoint union of $A, B \subset X$, i.e., $A \cap B = \emptyset$
2^A	power set of $A \subset X$, i.e., collection of all subsets of A
id	identity on X, i.e., $\mathrm{id}(x) = x$ for $x \in X$
$\mathbf{1}_A$	characteristic function of A, i.e., $\mathbf{1}_A(x) = 1$ for $x \in A$ and $\mathbf{1}_A(x) = 0$ for $x \in X \setminus A$
$a \wedge b$	$= \min\{a, b\}$ for $a, b \in \mathbb{R}$
$a \vee b$	$= \max\{a, b\}$ for $a, b \in \mathbb{R}$
$Y_1 \vee Y_2$	sum of the pointed spaces Y_1 and Y_2; see p. 146
$\mathrm{Sgn}\, v$	$= \{v/|v|\}$ for $v \in \mathbb{R}$ with $v \neq 0$, and $\mathrm{Sgn}(0) = [-1, 1]$
$F(A)$	$= \bigcup_{y \in A} F(y)$ for a multi-valued mapping F
$F^{-1}(A)$	$\{y : F(y) \cap A \neq \emptyset\}$ for a multi-valued mapping F
$\|F(y)\|$	$= \sup\{|z| : z \in F(y)\}$ for a multi-valued mapping F
$\sup(F)\ [\inf(F)]$	$= \sup\{z : z \in F(y)\}\ [= \inf\{z : z \in F(y)\}]$ for a multi-valued mapping F
$T_D(y)$	$= \left\{ z \in \mathbb{R}^n : \liminf_{\lambda \to 0+} \lambda^{-1}\mathrm{dist}(y + \lambda z, D) = 0 \right\}$; tangent cone to $D \subset \mathbb{R}^n$ at $y \in D$
$L^1_{\mathrm{loc}}(I; \mathbb{R}^n)$	space of Lebesgue-measurable functions $u : I \to \mathbb{R}^n$ such that $\int_J |u(t)|\, dt < \infty$ for every compact interval $J \subset I$
$C^k_b(I)$	space of functions $u \in C^k(I)$ such that $|u|_{C^k_b(I)} = \sum_{j=0}^k \sup_{t \in I} |u^{(j)}(t)| < \infty$
$\mathcal{E}_p(t; \xi_0, t_0),\ \mathcal{E}(\tau; \xi_0, \tau_0)$	see p. 20, p. 111
$\Delta_\mathcal{E}(t)$	usually denotes $(x(t) - \bar{x}(t))^2 + (\dot{x}(t) - \dot{\bar{x}}(t))^2$ for two solutions $x, \bar{x}$
$\langle p, q \rangle_{\mathrm{ap}}$	$= \lim_{T \to \infty} \frac{1}{T} \int_0^T p(t)\, \overline{q(t)}\, dt$ for almost periodic functions $p, q : \mathbb{R} \to \mathbb{C}$
$\lambda^+(y_0, z_0)$	upper Lyapunov exponents, see p. 63 and p. 118
$\lambda^{(i)},\ \lambda_y^{(i)}$	Lyapunov exponents, see p. 64 and p. 136
λ^d	Lebesgue measure on $\mathbb{R}^d$

$T(t,y)$	cocycle, see p. 63 and p. 136
$\Phi_f(\cdot,y_0)$ [or $\pi_f(\cdot,y_0)$]	global flow generated by f, i.e., $y(t) = \Phi_f(t,y_0)$ solves $\dot{y} = f(y)$, $y(0) = y_0$
$\mathcal{P}_{\mathrm{inv}}(\varphi)$ [$\mathcal{P}_{\mathrm{erg}}(\varphi)$]	invariant [ergodic] probability measures for a measurable semiflow φ
$h(I) = [B/b^+]$	equivalence class of B/b^+ modulo homotopy invariance in the category of pointed spaces; see p.145
Σ^k	equivalence class of the pointed k-sphere
$\mathrm{inv}(N;F)$	see p. 147
V_k	kth Lyapunov constant; see p.187
a.e.	almost everywhere
cf.	confer
e.g.	for example
i.e.	that is
MET	Oseledets' multiplicative ergodic theorem
w.l.o.g.	without loss of generality
w.r. to	with respect to
usc	upper semicontinuous
$\square$	marks end of a proof
$\diamond$	marks end of a remark or an example

More specific notation

M_k, M_∞, E_k, E_∞, f_k, f_∞, $\xi_i(y)$, $t_i(y)$	see p. 70
D_k, D_∞	see p. 69, p. 71 in $(A7)$
$G_{<\infty}$, G_∞, G	see p. 73, p. 74, p. 75
$\Gamma_l^i(t,y)$	see p. 78
$A_{i+1}(y_0)$	see p. 78, p. 80

D_k, $D_{I \cup \{\infty\}}$	see p. 86
$\mathbb{R}^d_\omega$	see p. 88
$\varphi^t_{\text{per}}(\xi_0, e^{2\pi i \tau_0/\omega})$	see p. 89
$T_{\text{per}}(t, y_0)$	see p. 89
S_γ	$= [\gamma - 1, -\gamma + 1]$
$b^\pm(\tau)$	$= \gamma \sin(\eta\tau) \pm 1$
τ_{mod}	$= \tau \operatorname{mod}(2\pi/\eta)$
M_+, M_-, M_{left}, M_{right}, M_∞	see p. 91
f_+, f_-, f_{left}, f_{right}, f_∞	see p. 92
E_+, E_-, E_{left}, E_{right}, E_∞	see p. 92
$s_1(y_0)$	see p. 94
h_+, h_-, h_{left}, h_{right}, h_∞	see p. 102
D_+, D_-, D_{left}, D_{right}	see p. 103
C_+, C_-, C_{left} C_{right}, C_∞, $C_{I \cup \{\infty\}}$	see p. 103
$A(t, \bar{t}, \alpha)$	see p. 120
$A(\alpha = 0)$	$= A(t, t, 0)$
$T(h)$	$= 4 \arccos (2h)^{-1/2}$
$I(h)$	$= hT(h) - 2\sqrt{2h - 1}$
$M_0(\theta_0)$	see p. 198
$\Delta_\varepsilon(t, \theta_0)$, $\Delta_\varepsilon^{u,s}(t, \theta_0)$	see p. 199, p. 200

References

1. Abraham R., Robbin J. (1967) Transversal Mappings and Flows. W.A. Benjamin, New York
2. Adams R.A. (1975) Sobolev Spaces. Academic Press, New York London
3. Aizerman M.A., Gantmakher F.R. (1958) On the stability of periodic motions. J Appl Math Mech 22: 1065-1078
4. Aizerman M.A., Gantmakher F.R. (1960) On the stability of the equilibrium position for discontinuous systems. J Appl Math Mech 24: 406-421
5. Aizerman M.A., Gantmakher F.R. (1963) Stability in a linear approximation of the periodic solution of a system of differential equations with discontinuous right sides. Amer Math Soc Transl 26: 339-351
6. Alfsen E.M. (1971) Compact Convex Sets and Boundary Integrals. Springer, Berlin Heidelberg New York
7. Alonso J.M., Ortega R. (1996) Unbounded solutions of semilinear equations at resonance. Nonlinearity 9: 1099-1111
8. Alscher Ch. (2000) Equations of Constrained Motion. Perturbation Analysis and Application to the Hydrostatic Skeleton. Fortschritt-Berichte VDI, Reihe 20: Rechnerunterstützte Verfahren Nr. 307, VDI Verlag, Düsseldorf
9. Andronov A.A., Vitt A.A., Khaikin S.E. (1966) Theory of Oscillators. Dover Publications, New York
10. Arnold L. (1998) Random Dynamical Systems. Springer, Berlin Heidelberg New York
11. Arnold L., Wihstutz V. (1986) Lyapunov exponents: a survey. In: Arnold L., Wihstutz V. (Eds.) Lyapunov Exponents, Proceedings, Bremen 1984. Lecture Notes in Mathematics Vol 1186. Springer, Berlin Heidelberg New York, 1-26
12. Arnold V.I. (1962) On the behaviour of an adiabatic invariant under slow periodic variation of the Hamiltonian. Translated in: Soviet Math Dokl 3: 136-140
13. Arnold V.I. (1963) Proof of A.N. Kolmogorov's theorem on the preservation of quasiperiodic motions under small perturbations of the Hamiltonian. Translated in: Russ Math Surv 18: 9-36
14. Arnold V.I. (1983) Geometric Methods in the Theory of Ordinary Differential Equations. Springer, Berlin Heidelberg New York
15. Arnold V.I. (1989) Mathematical Methods of Classical Mechanics, 2nd edn. Springer, Berlin Heidelberg New York
16. Arrowsmith D.K., Place C.M. (1990) An Introduction to Dynamical Systems. Cambridge University Press, Cambridge New York
17. Aubin J.P., Cellina A. (1984) Differential Inclusions. Springer, Berlin Heidelberg New York
18. Awrejcewicz J., Delfs J. (1990) Dynamics of a self-excited stick-slip oscillator with two degrees of freedom. Part II: slip-stick, slip-slip, stick-slip transitions, periodic and chaotic orbits. European J Mech A/Solids 9: 397-418

19. Awrejcewicz J., Holicke M.M. (1999) Melnikov's method and stick-slip chaotic oscillations in very weakly forced mechanical systems. Int J Bifurcation and Chaos 9: 505-518

20. Bainov D.D., Zabreiko P.P., Kostadinov S.T. (1988) Stability of the general exponent of nonlinear impulsive differential equations in a Banach space. Int J Theoretical Phys 27: 373-380

21. Balmer B. (1992) Ein Reibungsmodell zur Berechnung reibungsgedampfter Schaufelschwingungen. Z Angew Math Mech 72: T117-T119

22. Bautin N.N. (1954) On the number of limit cycles which appear with variation of coefficients from an equilibrium position of focus or center type. American Math Soc Transl 100: 397-413

23. Berger B.S., Rokni M. (1990) Lyapunov exponents for discontinuous differential equations. Quarterly Appl Math 48: 549-553

24. Besicovitch A.S. (1954) Almost Periodic Functions. Dover Publications, New York

25. Birkhoff G.D. (1927) Dynamical Systems. American Mathematical Society Publications, Providence/RI

26. Blows T.R., Lloyd N.G. (1984) The number of limit cycles of certain polynomial differential equations. Proc Roy Soc Edinburgh Sect A 98: 215-239

27. Bockman S.F. (1991) Lyapunov exponents for systems described by differential equations with discontinuous right-hand sides. In: Proc American Control Conference, Vol 2, 1673-1678

28. Bothe D. (1999) Periodic solutions of non-smooth friction oscillators. Z Angew Math Phys 50: 779-808

29. Boyland P. (1996) Dual billiards, twist maps and impact oscillators. Nonlinearity 9: 1411-1438

30. Brandl M., Hajek M. (1988) Mechanische Systeme mit Trockenreibung. Z Angew Math Mech 68: T59-T61

31. Brezis H., Nirenberg L. (1978) Characterization of the ranges of some nonlinear operators and applications to boundary value problems. Ann Scuola Norm Sup Pisa 5: 225-326

32. Brogliato B. (1999) Nonsmooth Mechanics. 2nd edn. Springer, Berlin Heidelberg New York

33. Brogliato B. (Ed.) (2000) Impacts in Mechanical Systems. Analysis and Modelling. Lecture Notes in Physics Vol 551. Springer, Berlin Heidelberg New York

34. Brommundt E. (1995) Ein Reibschwinger mit Selbsterregung ohne fallende Reibkennlinie. Z Angew Math Mech 75: 811-820

35. Budd C. (1995) The global dynamics of impact oscillators. In: Branner B., Hjorth P. (Eds.) Real and Complex Dynamical Systems, Hillerød 1993. NATO Adv Sci Inst Ser C Math Phys Sci 464. Kluwer, Dordrecht Boston London, 27-46

36. Budd C. (1995) Grazing in impact oscillators. In: Branner B., Hjorth P. (Eds.) Real and Complex Dynamical Systems, Hillerød 1993. NATO Adv Sci Inst Ser C Math Phys Sci 464. Kluwer, Dordrecht Boston London, 47-63

37. Budd C. (1996) Non-smooth dynamical systems and the grazing bifurcation. In: Aston Ph. (Ed.) Nonlinear Mathematics and its Applications, Guildford 1995. Cambridge University Press, Cambridge New York, 219-235

38. Budd C., Dux F. (1994) Chattering and related behaviour in impact oscillators. Phil Trans Roy Soc London A 347: 365-389

39. Budd C., Dux F. (1994) Intermittency in impact oscillators close to resonance. Nonlinearity 7: 1191-1224

40. Budd C., Dux F., Cliffe K.A. (1995) The effect of frequency and clearance variations on single-degree-of-freedom impact oscillators. J Sound Vibration 184: 475-502

41. Burridge R., Knopoff L. (1967) Model and theoretical seismicity. Bull Seism Soc Am 57: 341-371

42. Buttazzo G., Giaquinta M., Hildebrandt S. (1998) One-Dimensional Variational Problems. Oxford University Press, Oxford

43. Capecchi D. (1991) Periodic response and stability of hysteretic oscillators. Dynamics and Stability of Systems 6: 89-106

44. Carlson J.M., Langer J.S., Shaw B.E. (1994) Dynamics of earthquakes faults. Rev Mod Phys 66: 657-670

45. Casas F., Chin W., Grebogi C., Ott W. (1996) Universal grazing bifurcations in impact oscillators. Phys Rev E 53: 134-139

46. Caughey T.K., Masri S.F. (1966) On the stability of the impact damper. J Applied Mech 33: 586-592

47. Caughey T.K., Vijayaraghavan A. (1970) Free and forced oscillations of a dynamical system with "linear hysteretic damping" (non-linear theory). Int J Non-Linear Mech 5: 533-555

48. Caughey T.K., Vijayaraghavan A. (1976) Stability analysis of the periodic solution of a piecewise-linear non-linear dynamic system. Int J Non-Linear Mech 11: 127-134

49. Chatterjee S., Mallik A. (1996) Bifurcations and chaos in autonomous self-excited oscillators with impact damping. J Sound Vibration 191: 539-562

50. Chin W., Ott E., Nusse H.E., Grebogi C. (1994) Grazing bifurcations in impact oscillators. Phys Rev E 50: 4427-4444

51. Chin W., Ott E., Nusse H.E., Grebogi C. (1995) Universal behaviour of impact oscillators near grazing incidence. Phys Lett A 201: 197-204

52. Coll B., Gasull A., Prohens R. (1995) First Lyapunov constants for non-smooth Liénard differential equations. In: Proceedings of the 2nd Catalan Days on Applied Mathematics, Odeillo 1995. Presses Univ Perpignan, Perpignan, 77-83

53. Coll B., Gasull A., Prohens R. (1999) Degenerate Hopf bifurcations in discontinuous planar systems. Preprint

54. Coll B., Prohens R., Gasull A. (1999) The center problem for discontinuous Liénard differential equation. Discrete dynamical systems. Int J Bifurcation and Chaos 9: 1751-1761

55. Conley C. (1978) Isolated Invariant Sets and the Morse Index. CBMS 38, American Mathematical Society Publications, Providence/RI

56. Conley C., Zehnder E. (1984) Morse type index theory for flows and periodic solutions for Hamiltonian equations. Comm Pure Appl Math 37: 207-254

57. Corduneanu C. (1968) Almost Periodic Functions. John Wiley & Sons, New York London

58. Cottle R.W., Dantzig G.B. (1968) Complementary pivot theory of mathematical programming. Linear Algebra Appl 1: 103-125

59. Dancer E.N. (1984) Degenerate critical points, homotopy indices and Morse inequalities. J Reine Angew Math 350: 1-22

60. Dankowicz H. (1999) On the modeling of dynamic friction phenomena. Z Angew Math Mech 79: 399-409

61. Dankowicz H., Nordmark A. (2000) On the origin and bifurcations of stick-slip oscillations. Physica D 136: 280-302

62. Deimling K. (1985) Nonlinear Functional Analysis. Springer, Berlin Heidelberg New York

63. Deimling K. (1986) Fixed points of weakly inward multis. Nonlinear Anal 10: 1261-1262

64. Deimling K. (1992) Multivalued differential equations and dry friction problems. In: Fink A.M., Miller R.K., Kliemann W. (Eds.) Proc Conf Differential and Delay Equations, Ames/Iowa 1991. World Scientific, Singapore New York, 99-106

65. Deimling K. (1992) Multivalued Differential Equations. de Gruyter, Berlin New York

66. Deimling K. (1994) Resonance and Coulomb friction. Differential Integral Equations 7: 759-765

67. Deimling K., Hetzer G., Shen W.X. (1996) Almost periodicity enforced by Coulomb friction. Adv Differential Equations 1: 265-281

68. Deimling K., Szilágyi P. (1994) Periodic solutions of dry friction problems. Z Angew Math Phys 45: 53-60

69. Dellnitz M., Hohmann A. (1997) A subdivision algorithm for the computation of unstable manifolds and global attractors. Numer Mathematik 75: 293-317

70. Den Hartog J.P. (1930) Forced vibrations with combined viscous and Coulomb damping. Philos Mag VII, Ser 9: 801-817

71. Di Bernardo M., Budd C., Champneys A. (1998) Grazing, skipping and sliding: analysis of the non-smooth dynamics of the DC/DC buck converter. Nonlinearity 11: 859-890

72. Dieckerhoff R., Zehnder E. (1987) Boundedness of solutions via the twist-theorem. Ann Sc Norm Sup Pisa 14: 79-95

73. Dontchev A., Lempio F. (1992) Difference methods for differential inclusions: a survey. SIAM Review 34: 263-294

74. Eckmann J.P., Ruelle D. (1985) Ergodic theory of chaos and strange attractors. Reviews of Modern Phys 57: 617-656

75. Fečkan M. (1996) Bifurcations from homoclinic to periodic solutions in ordinary differential equations with multivalued perturbations. J Differential Equations 130: 415-450

76. Fečkan M. (1997) Bifurcation of periodic solutions in differential inclusions. Appl Math 42: 369-393

77. Fečkan M. (1997) Bifurcations from homoclinic to periodic solutions in singularly perturbed differential inclusions. Proc Roy Soc Edinburgh Sect A 127: 727-753

78. Fečkan M. (1998) Differential inclusions at resonance. Bull Belg Math Soc Simon Stevin 5: 483-495

79. Fečkan M. (1999) Chaotic solutions in differential inclusions: chaos in dry friction problems. Trans Amer Math Soc 351: 2861-2873

80. Feeny B. (1992) A nonsmooth Coulomb friction oscillator. Physica D 59: 25-38

81. Fermi E. (1949) On the origin of cosmic radiation. Phys Rev 75: 1169-1174

82. Filippov A.F. (1988) Differential Equations with Discontinuous Right-Hand Sides. Kluwer, Dordrecht Boston London

83. Foale S., Bishop S.R. (1992) Dynamical complexities of forced impacting systems. Phil Trans Roy Soc London A 338: 547-556

84. Foale S., Bishop S.R. (1994) Bifurcations in impact oscillations. Nonlinear Dynamics 6: 285-299

85. Fredriksson M., Nordmark A. (1997) Bifurcations caused by grazing incidence in many degrees of freedom impact oscillators. Proc Roy Soc London Ser A 453: 1261-1276

86. Galvanetto U., Bishop S.R. (1994) Stick-slip vibrations of a two degree-of-freedom geophysical fault model. Int J Mech Sci 36: 683-698

87. Galvanetto U., Bishop S.R. (1995) Characterization of the dynamics of a four-dimensional stick-slip system by a scalar variable. Chaos, Solitons and Fractals 5: 2171-2179
88. Galvanetto U., Bishop S.R., Briseghella L. (1995) Mechanical stick-slip vibrations. Int J Bifurcation and Chaos 5: 637-651
89. Gasull A., Guillamon A., Mañosa V. (1997) An explicit expression of the first Liapunov and period constants with applications. J Math Anal Appl 211: 190-212
90. Gasull A., Prohens R. (1997) Effective computation of the first Lyapunov constant for a planar differential equation. Applicationes Mathematicae 24: 243-250
91. Georgii H.O. (1988) Gibbs Measures and Phase Transitions. de Gruyter, Berlin New York
92. Giannakopoulos F., Kaul A., Pliete K. (1999) Qualitative analysis of a planar system of piecewise linear differential equations with a line of discontinuity. Preprint
93. Glocker Ch., Pfeiffer F. (1992) An LCP approach for multibody systems with planar friction. In: Proc International Symposium on Contact Mechanics Lausanne, 13-30
94. Glocker Ch., Pfeiffer F. (1993) Complementary problems in multibody systems with planar friction. Arch Appl Mech 63: 452-463
95. Gontier C., Toulemonde C. (1997) Approach to the periodic and chaotic behaviour of the impact oscillator by a continuation method. Eur J Mech A Solids 16: 141-163
96. Guran A., Feeny F., Hinrichs N., Popp K. (1998) A historical review on dry friction and stick-slip phenomena. Appl Mech Rev 51: 321-341
97. Hale J.K., Koçak H. (1991) Dynamics and Bifurcations. Springer, Berlin Heidelberg New York
98. Heemels M. (1999) Linear Complementarity Systems. A Study in Hybrid Dynamics. Ph D Thesis, Technische Universiteit Eindhoven
99. Herman M.R. (1982) Sur les Courbes Invariantes par les Difféomorphismes de l'Anneau. Astérisque 103-104, Soc Math de France, Paris
100. Ibrahim R.A. (1994) Friction-induced vibration, chatter, squeal and chaos. Part I: Mechanics of contact and friction, Part II: Dynamics and modeling. Appl Mech Rev 47: 209-226 and 227-253
101. Inaudi J.A., Kelly J.M. (1995) Mass damper using friction-dissipating devices. J Eng Mech 121: 142-149
102. Irisov A.E., Tonkova V.S., Tonkov E.L. (1978) Periodic solutions of differential inclusions (in Russian). Nonlinear Oscillations and Control Theory (Izhevsk) 2: 3-14
103. Kaczynski T., Mrozek M. (1995) Conley index for discrete multi-valued dynamical systems. Topology Appl 65: 83-96
104. Kato T. (1980) Perturbation Theory for Linear Operators. Springer, Berlin Heidelberg New York
105. Katok A., Hasselblatt B. (1995) Introduction to the Modern Theory of Dynamical Systems. Cambridge University Press, Cambridge
106. Katok A., Strelcyn V. (1986) Invariant Manifolds, Entropy and Billiards; Smooth Maps with Singularities. Lecture Notes in Mathematics Vol 1222. Springer, Berlin Heidelberg New York
107. Kauderer H. (1958) Nichtlineare Mechanik. Springer, Berlin Heidelberg New York
108. Kilmister C.W., Reeve J.E. (1966) Rational Mechanics. Longmans, London

109. Klotter K. (1980) Technische Schwingungslehre IB: Nichtlineare Schwingungen. 3rd edn. Springer, Berlin Heidelberg New York
110. Kozlov V.V., Treshchev D.V. (1991) Billiards. A Genetic Introduction to the Dynamics of Systems with Impacts. American Mathematical Society, Providence/RI
111. Krengel U. (1985) Ergodic Theorems. de Gruyter, Berlin New York
112. Kunze M. (1998) Unbounded solutions in non-smooth dynamical systems at resonance. Z Angew Math Mech 78, Supplement 3: S985-S986
113. Kunze M. (1999) Periodic solutions of conservative non-smooth dynamical systems. Z Angew Math Mech 79, Supplement 1: S97-S100
114. Kunze M. (1999) Non-Smooth Dynamical Systems. Habilitation Thesis, Universität Köln
115. Kunze M. (2000) On Lyapunov exponents for non-smooth dynamical systems with an application to a pendulum with dry friction. J Dynamics Differential Equations 12: 31-116
116. Kunze M., Küpper T. (1997) Qualitative analysis of a non-smooth friction-oscillator model. Z Angew Math Phys 48: 1-15
117. Kunze M., Küpper T. (1999) Non-smooth dynamical systems: an overview. Submitted to: Fiedler B. (Ed.) DANSE book, final book of the DFG piority research program "Dynamical Systems", Springer, Berlin Heidelberg New York
118. Kunze M., Küpper T., Li J. (2000) On the application of Conley index theory to non-smooth dynamical systems. Differential Integral Equations 13: 479-502
119. Kunze M., Küpper T., You J. (1997) On the application of KAM theory to discontinuous dynamical systems. J Differential Equations 139: 1-21
120. Kunze M., Michaeli B. (1995) On the rigorous applicability of Oseledets' ergodic theorem to obtain Lyapunov exponents for non-smooth dynamical systems. To appear in: Arino O. (Ed.) Proc 2nd Marrakesh International Conference on Differential Equations
121. Kunze M., Neumann J. (1997) Linear complementary problems and the simulation of the motion of rigid body systems subject to Coulomb friction. Z Angew Math Mech 77: 833-838
122. Laederich S., Levi M. (1991) Invariant curves and time-dependent potentials. Ergodic Th Dynamical Systems 11: 365-378
123. Lamba H. (1993) Impacting Oscillators and Non-Smooth Dynamical Systems. Ph D Thesis, University of Bristol
124. Lamba H. (1995) Chaotic, regular and unbounded behaviour in the elastic impact oscillator. Physica D 82: 117-135
125. Lamba H., Budd C. (1994) Scaling of Lyapunov exponents at nonsmooth bifurcations. Phys Rev E 50: 84-90
126. Ledrappier F. (1984) Quelques proprietés des exposants characteristiques. In: Hennequin P.L. (Ed.) École d'Été de Probabilités de Saint-Flour XII-1982. Lecture Notes in Mathematics Vol 1097. Springer, Berlin Heidelberg New York, 305-396
127. Lefschetz S. (1965) Stability of Nonlinear Control Systems. Academic Press, New York London
128. Levi M. (1991) Quasiperiodic motions in superquadratic time-periodic potentials. Comm Math Phys 143: 43-83
129. Levi M. (1992) On the Littlewood's counterexample of unbounded motions in superquadratic potentials. In: Jones Ch., Kirchgraber U., Walther H.O. (Eds.) Dynamics Reported 1, 113-124
130. Levi M., You J. (1997) Oscillatory escape in a Duffing equation with polynomial potential. J Differential Equations 140: 415-426

131. Lichtenberg A.J., Lieberman M.A. (1991) Regular and Chaotic Dynamics. 2nd edn. Springer, Berlin Heidelberg New York
132. Littlewood J. (1966) Unbounded solutions of an equation $\ddot{y} + g(y) = p(t)$ with $p(t)$ periodic and bounded and $g(y)/y \to \infty$ as $y \to \pm\infty$. J London Math Soc 41: 497-507
133. Littlewood J. (1968) Some Problems in Real and Complex Analysis. Heath. Lexington, Massachusetts
134. Liu B. (1999) Boundedness in nonlinear oscillations at resonance. J Differential Equations 153: 142-174
135. Lötstedt P. (1981) Coulomb friction in rigid body systems. Z Angew Math Mech 61: 605-615
136. Lötstedt P. (1982) Mechanical systems of rigid bodies subject to unilateral constraints. SIAM J Appl Math 42: 281-296
137. Lötstedt P. (1982) Numerical simulation of time-dependent contact and friction problems in rigid body mechanics. Dep of Numerical Anal and Comp Sc, Royal Inst of Technology Stockholm. Report TRITA-NA-7914
138. Marsden J.E., McCracken M. (1976) The Hopf Bifurcation and its Applications. Springer, Berlin Heidelberg New York
139. Meitinger T., Pfeiffer F. (1995) Dynamic simulation of assembly processes. In: Proceedings of the 1995 IEEE/RSJ International Conference on Intelligent Robots and Systems, Pittsburgh/PA, Vol 2. IDEE Computer Society Publications Office, Los Alamitos, 298-304
140. Michaeli B. (1998) Lyapunov-Exponenten bei nichtglatten dynamischen Systemen. Ph D Thesis, Universität Köln
141. Mischaikow K. (1995) Conley Index Theory. In: Johnson R. (Ed.) Dynamical Systems, Montecatini Terme 1994. Lecture Notes in Mathematics Vol 1609. Springer, Berlin Heidelberg New York, 119-207
142. Mischaikow K., Mrozek M. (1995) Chaos in the Lorenz equation: a computer-assisted proof. Bull Amer Math Soc (N.S.) 32: 66-72
143. Mischaikow K., Mrozek M. (1998) Chaos in the Lorenz equation: a computer-assisted proof. II. Details. Math Comp 67: 1023-1046
144. Monteiro Marques M.D.P. (1993) Differential Inclusions in Nonsmooth Mechanical Problems-Shocks and Dry Friction. Birkhäuser, Basel Boston
145. Moreau J.J. (1988) Nonsmooth Mechanics and Applications. CISM Lectures Vol. 203. Springer, Berlin Heidelberg New York
146. Moritz S. (2000) Hopf-Verzweigung bei unstetigen planaren Systemen. MA Thesis, Universität Köln
147. Morris G.R. (1976) A case of boundedness in Littlewood's problem on oscillatory differential equations. Bull Austral Math Soc 14: 71-93
148. Moser J. (1962) On invariant curves of area preserving mappings of an annulus. Nachr Akad Wiss Göttingen Math Phys Kl II: 1-20
149. Moser J. (1973) Stable and Random Motions in Dynamical Systems. Princeton University Press, Princeton
150. Moser J. (1989) Quasi-periodic solutions of nonlinear elliptic partial differential equations. Boll Soc Bras Math 20: 29-45
151. Mrozek M. (1990) A cohomological index of Conley type for multi-valued admissible flows. J Differential Equations 84: 15-51
152. Mrozek M. (1995) From the theorem of Ważewski to computer assisted proofs in dynamics. In: Panoramas of Mathematics, Banach Center Publications 34, Polish Acad Science, Warsaw, 105-120
153. Müller A. (1994) Lyapunov Exponenten in nicht-glatten dynamischen Systemen. MA Thesis, Universität Köln

222   References

154. Müller P.C. (1995) Calculation of Lyapunov exponents for dynamic systems with discontinuities. Chaos, Solitons and Fractals 5: 1671-1681
155. Murty K.G. (1988) Linear Complementarity, Linear and Nonlinear Programming. Heldermann Verlag, Berlin
156. Neumann J. (1995) Die Dynamik starrer Körper in ebenen Systemen mit Coulombscher Reibung. MA Thesis, Universität Köln
157. Nordmark A. (1991) Non-periodic motion caused by grazing incidence in an impact oscillator. J Sound Vibration 145: 279-297
158. Nordmark A. (1992) Effects due to low velocity impacts in mechanical oscillators. Int J Bifurcation and Chaos 2: 597-605
159. Nordmark A. (1993) Non-smooth bifurcations in mappings with square-root singularities. Preprint
160. Nqi F. (1997) Etude Numérique de Divers Problèmes Dynamiques avec Impact et de Leurs Propriétés Qualitatives. Ph D Thesis, Université Claude Bernard Lyon
161. Nusse H., Ott E., Yorke J. (1994) Border-collision bifurcations: an explanation for observed bifurcation phenomena. Phys Rev E 49: 1073-1076
162. Nusse H., Yorke J. (1992) Border-collision bifurcations including "period two to period three" for piecewise smooth systems. Physica D 57: 39-57
163. Nusse H., Yorke J. (1995) Border-collision bifurcations for piecewise smooth one-dimensional maps. Int J Bifurcation and Chaos 5: 189-207
164. Oestreich M., Hinrichs N., Popp K. (1996) Bifurcation and stability analysis for a non-smooth friction oscillator. Archive of Applied Mechanics 66: 301-314
165. Ortega R. (1996) Asymmetric oscillators and twist mappings. J London Math Soc 53: 325-342
166. Ortega R. (1999) Boundedness in a piecewise linear oscillator and a variant of the small twist theorem. Proc London Math Soc 79: 381-413
167. Oseledets V.I. (1968) A multiplicative ergodic theorem. Ljapunov characteristic numbers for dynamical systems. Trans Moscow Math Soc 19: 197-231
168. Parnes R. (1984) Response of an oscillator to a ground motion with Coulomb friction slippage. J Sound Vibration 94: 469-482
169. Peterka F. (1992) Transition to chaotic motion in mechanical systems with impacts. J Sound Vibration 154: 95-115
170. Peterka F. (2000) Impact oscillator. In: WIERCIGROCH/DE KRAKER [224]
171. Pfeiffer F. (1988) Seltsame Attraktoren in Zahnradgetrieben. Ingenieur Archiv 58: 113-125
172. Pfeiffer F. (1991) Dynamical systems with time-varying or unsteady structure. Z Angew Math Mech 71: T6-T22
173. Pfeiffer F. (1992) On stick-slip vibrations in machine dynamics. Machine Vibrations 1: 20-28
174. Pfeiffer F. (1994) Methoden zur Nichtlinearen Antriebstechnik. VDI Berichte 1153: 599-624
175. Pfeiffer F., Glocker Ch. (1996) Multibody Dynamics with Unilateral Contacts. John Wiley & Sons, New York London
176. Pfeiffer F., Glocker Ch. (Eds.) (1999) Unilateral Multibody Contacts. Proceedings IUTAM Symposium, München 1998, Kluwer, Dordrecht Boston London
177. Pfeiffer F., Hajek M. (1992) Slip-stick motions of turbine blade dampers. Phil Trans Roy Soc London A 338: 503-517
178. Pfeiffer F., Seyfferth W. (1992) Dynamics of assembly processes with a manipulator. In: Proc IEEE/RSJ Intern Conf Intelligent Robots and Systems, 1303-1310

179. Pliete K. (1998) Über die Anzahl der geschlossenen Orbits bei unstetigen stückweise linearen dynamischen Systemen in der Ebene. MA Thesis, Universität Köln
180. Poincaré H. (1899) Les Méthodes Nouvelles de la Mécanique Céleste. 3 Volumes, Gauthier-Villars, Paris
181. Pollicott M. (1993) Lectures on Ergodic Theory and Pesin Theory on Compact Manifolds. London Math Soc Lecture Notes Series Vol 180. Cambridge University Press, Cambridge
182. Popp K. (1999) Examples of non-smooth dynamical systems-an overview. In: PFEIFFER/GLOCKER [175], 233-242
183. Popp K., Hinrichs N., Oestreich M. (1995) Dynamical behaviour of a friction oscillator with simultaneous self and external excitation. Sādhanā 20: 627-654
184. Popp K., Stelter P. (1990) Nonlinear oscillations of structures induced by dry friction. In: Schiehlen W. (Ed.) Nonlinear Dynamics in Engineering Systems-IUTAM Symposium Stuttgart 1989. Springer, Berlin Heidelberg New York, 233-240
185. Popp K., Stelter P. (1990) Stick-slip vibrations and chaos. Phil Trans Roy Soc London A 332: 89-105
186. Pustylnikov L. (1978) Stable and oscillating modes in nonautonomous dynamical systems. Trans Moscow Math Soc 2: 1-101
187. Rabinowitz P. (1971) Some global results for nonlinear eigenvalue problems. J Functional Anal 7: 487-513
188. Reissig R. (1953) Über die Differentialgleichung $\frac{d^2x}{d\tau^2} + 2D \cdot \frac{dx}{d\tau} + \mu \cdot \text{sgn} \frac{dx}{d\tau} + x = \Phi(\eta\tau)$, wo $\Phi(\eta\tau + 2\pi) \equiv \Phi(\eta\tau)$ ist. Das Verhalten der Lösungen für $\tau \to \infty$. Abhandlungen der Deutschen Akademie der Wissenschaften zu Berlin, Klasse für Mathematik und Naturwissenschaften 1
189. Reissig R. (1954) Erzwungene Schwingungen mit zäher und trockner Reibung, Math Nachrichten 11: 345-384
190. Reissig R. (1954) Erzwungene Schwingungen mit zäher Reibung und starker Gleitreibung II. Math Nachrichten 12: 119-128
191. Reissig R. (1954) Erzwungene Schwingungen mit zäher und trockner Reibung: Ergänzung. Math Nachrichten 12: 249-252
192. Reissig R. (1954) Erzwungene Schwingungen mit zäher und trockner Reibung: Abschätzung der Amplituden. Math Nachrichten 12: 283-300
193. Reithmeier E. (1991) Periodic Solutions of Nonlinear Dynamical Systems. Lecture Notes in Mathematics Vol 1483. Springer, Berlin Heidelberg New York
194. Reitmann V. (1996) Reguläre und chaotische Dynamik. Teubner, Stuttgart Leipzig
195. Rudin W. (1973) Functional Analysis. McGraw-Hill, New Delhi
196. Ruelle D. (1979) Ergodic theory of differentiable dynamical systems. Publ Math IHES 50: 275-306
197. Rybakowski K.P. (1987) The Homotopy Index and Partial Differential Equations. Springer, Berlin Heidelberg New York
198. Schneider E., Popp K., Irretier H. (1988) Noise generation in railway wheels due to rail-wheel contact forces. J Sound Vibration 120: 227-244
199. Schwartz J.T. (1969) Nonlinear Functional Analysis. Gordon and Breach, New York
200. Shaw S.W. (1986) On the dynamic response of a system with dry friction. J Sound Vibration 108: 305-325
201. Shaw S.W., Holmes P.J. (1983) A periodically forced piecewise linear oscillator. J Sound Vibration 90: 129-155
202. Shaw S.W., Holmes P.J. (1983) Periodically forced linear oscillator with impacts: chaos and long-period motions. Phys Rev Lett 51: 623-626

203. Siegel C.L., Moser J. (1971) Lectures on Celestial Mechanics. Springer, Berlin Heidelberg New York
204. Sikora J., Bogacz R. (1993) On dynamics of several degrees of freedom. Z Angew Math Mech 73: T118-T122
205. Sinai Y. (1976) Introduction to Ergodic Theory. Princeton University Press, Princeton
206. Smoller J. (1994) Shock Waves and Reaction-Diffusion Equations, 2nd edn. Springer, Berlin Heidelberg New York
207. Stelter P. (1990) Nichtlineare Schwingung reibungserregter Strukturen. Ph D Thesis, Universität Hannover
208. Stelter P., Sextro W. (1991) Bifurcations in dynamic systems with dry friction. In: Bifurcations and Chaos: Analysis, Algorithms, Applications, Würzburg 1990. Birkhäuser, Basel Boston, 343-347
209. Stewart D.E. (1990) A high accuracy method for solving ODEs with discontinuous right-hand side. Numer Mathematik 58: 299-328
210. Stewart D.E. (2000) Rigid body dynamics with friction and impact. SIAM Review 42: 3-39
211. Storz M. (1992) Ein ungefesseltes, harmonisch erregtes System mit Coulombscher Reibung und Stoß als Modell für die Vibrationsramme. Z Angew Math Mech 72: T123-T127
212. Szablewski W. (1954) Einfluß der Coulombschen Reibung auf Schwingungsvorgänge. Math Nachrichten 12: 183-208
213. Szczygielski W.M. (1986) Dynamisches Verhalten eines schnell drehenden Rotors bei Anstreifvorgängen. Ph D Thesis, ETH Zürich
214. Tabachnikov S. (1995) Billiards. Panoramas et Synthèses 1, Société Mathématique de France, Paris
215. Taubert K. (1976) Differenzenverfahren für Schwingungen mit trockener und zäher Reibung für Regelungssysteme. Numer Mathematik 26: 379-395
216. Toulemonde C., Gontier C. (1998) Sticking motions of impact oscillators. Eur J Mech A Solids 17: 339-366
217. Voßhage Ch. (2000) Visualisierung von Attraktoren und invarianten Maßen in nichtglatten dynamischen Systemen. MA Thesis, Universität Köln
218. Ward J.R., Jr. (1996) Bifurcating continua in infinite dimensional dynamical systems and applications to differential equations. J Differential Equations 125: 117-132
219. Ward J.R., Jr. (1998) Global bifurcation of periodic solutions to ordinary differential equations. J Differential Equations 142: 1-16
220. Whiston G.S. (1987) The vibro-impact response of a harmonically excited and preloaded one-dimensional linear oscillator. J Sound Vibration 115: 303-324
221. Whiston G.S. (1987) Global dynamics of a vibro-impacting linear oscillator. J Sound Vibration 118: 395-429
222. Whiston G.S. (1992) Singularities in vibro-impact dynamics. J Sound Vibration 152: 427-460
223. Wiederhöft A. (1994) Der periodisch erregte Einmassenreibschwinger. MA Thesis, Universität Köln
224. Wiercigroch M., de Kraker A. (2000) Applied Nonlinear Dynamics and Chaos of Mechanical Systems with Discontinuities. World Scientific, Singapore New York
225. Wiggins S. (1990) Introduction to Applied Nonlinear Dynamical Systems and Chaos. Springer, Berlin Heidelberg New York
226. Yosida K. (1980) Functional Analysis. Springer, Berlin Heidelberg New York

227. Yuan G., Banerjee S., Odd E., Yorke J. (1988) Border-collision bifurcations in the Buck converter. IEEE Trans Circuits Systems I Fund Theory Appl 45: 707-716

228. Zeidler E. (1986) Nonlinear Functional Analysis and its Applications I: Fixed-Point Theorems. Springer, Berlin Heidelberg New York

229. Zou Y.-K., Küpper T. (2000) Melnikov method and detection of chaos for non-smooth systems. Submitted to SIAM J Math Anal

Index

Printing: Weihert-Druck GmbH, Darmstadt
Binding: Buchbinderei Schäffer, Grünstadt

Vol. 1706: S. Yu. Pilyugin, Shadowing in Dynamical Systems. XVII, 271 pages. 1999.

Vol. 1707: R. Pytlak, Numerical Methods for Optimal Control Problems with State Constraints. XV, 215 pages. 1999.

Vol. 1708: K. Zuo, Representations of Fundamental Groups of Algebraic Varieties. VII, 139 pages. 1999.

Vol. 1709: J. Azéma, M. Émery, M. Ledoux, M. Yor (Eds), Séminaire de Probabilités XXXIII. VIII, 418 pages. 1999.

Vol. 1710: M. Koecher, The Minnesota Notes on Jordan Algebras and Their Applications. IX, 173 pages. 1999.

Vol. 1711: W. Ricker, Operator Algebras Generated by Commuting Projections: A Vector Measure Approach. XVII, 159 pages. 1999.

Vol. 1712: N. Schwartz, J. J. Madden, Semi-algebraic Function Rings and Reflectors of Partially Ordered Rings. XI, 279 pages. 1999.

Vol. 1713: F. Bethuel, G. Huisken, S. Müller, K. Steffen, Calculus of Variations and Geometric Evolution Problems. Cetraro, 1996. Editors: S. Hildebrandt, M. Struwe. VII, 293 pages. 1999.

Vol. 1714: O. Diekmann, R. Durrett, K. P. Hadeler, P. K. Maini, H. L. Smith, Mathematics Inspired by Biology. Martina Franca, 1997. Editors: V. Capasso, O. Diekmann. VII, 268 pages. 1999.

Vol. 1715: N. V. Krylov, M. Röckner, J. Zabczyk, Stochastic PDE's and Kolmogorov Equations in Infinite Dimensions. Cetraro, 1998. Editor: G. Da Prato. VIII, 239 pages. 1999.

Vol. 1716: J. Coates, R. Greenberg, K. A. Ribet, K. Rubin, Arithmetic Theory of Elliptic Curves. Cetraro, 1997. Editor: C. Viola. VIII, 260 pages. 1999.

Vol. 1717: J. Bertoin, F. Martinelli, Y. Peres, Lectures on Probability Theory and Statistics. Saint-Flour, 1997. Editor: P. Bernard. IX, 291 pages. 1999.

Vol. 1718: A. Eberle, Uniqueness and Non-Uniqueness of Semigroups Generated by Singular Diffusion Operators. VIII, 262 pages. 1999.

Vol. 1719: K. R. Meyer, Periodic Solutions of the N-Body Problem. IX, 144 pages. 1999.

Vol. 1720: D. Elworthy, Y. Le Jan, X-M. Li, On the Geometry of Diffusion Operators and Stochastic Flows. IV, 118 pages. 1999.

Vol. 1721: A. Iarrobino, V. Kanev, Power Sums, Gorenstein Algebras, and Determinantal Loci. XXVII, 345 pages. 1999.

Vol. 1722: R. McCutcheon, Elemental Methods in Ergodic Ramsey Theory. VI, 160 pages. 1999.

Vol. 1723: J. P. Croisille, C. Lebeau, Diffraction by an Immersed Elastic Wedge. VI, 134 pages. 1999.

Vol. 1724: V. N. Kolokoltsov, Semiclassical Analysis for Diffusions and Stochastic Processes. VIII, 347 pages. 2000.

Vol. 1725: D. A. Wolf-Gladrow, Lattice-Gas Cellular Automata and Lattice Boltzmann Models. IX, 308 pages. 2000.

Vol. 1726: V. Marić, Regular Variation and Differential Equations. X, 127 pages. 2000.

Vol. 1727: P. Kravanja, M. Van Barel, Computing the Zeros of Analytic Functions. VII, 111 pages. 2000.

Vol. 1728: K. Gatermann, Computer Algebra Methods for Equivariant Dynamical Systems. XV, 153 pages. 2000.

Vol. 1729: J. Azéma, M. Émery, M. Ledoux, M. Yor, Séminaire de Probabilités XXXIV. VI, 431 pages. 2000.

Vol. 1730: S. Graf, H. Luschgy, Foundations of Quantization for Probability Distributions. X, 230 pages. 2000.

Vol. 1731: T. Hsu, Quilts: Central Extensions, Braid Actions, and Finite Groups,. XII, 185 pages. 2000.

Vol. 1732: K. Keller, Invariant Factors, Julia Equivalences and the (Abstract) Mandelbrot Set. X, 206 pages. 2000.

Vol. 1733: K. Ritter, Average-Case Analysis of NumericalProblems. IX, 254 pages. 2000.

Vol. 1734: M. Espedal, A. Fasano, A. Mikelić, Filtration in Porous Media and Industrial Applications. Cetraro 1998. Editor: A. Fasano. 2000.

Vol. 1735: D. Yafaev, Scattering Theory: Some Old and New Problems. XVI, 169 pages. 2000.

Vol. 1736: B. O. Turesson, Nonlinear Potential Theory and Weighted Sobolev Spaces. XIV, 173 pages. 2000.

Vol. 1737: S. Wakabayashi, Classical Microlocal Analysis in the Space of Hyperfunctions. VIII, 367 pages. 2000.

Vol. 1738: M. Emery, A. Nemirovski, D. Voiculescu, Lectures on Probability Theory and Statistics. XI, 356 pages. 2000.

Vol. 1739: R. Burkard, P. Deuflhard, A. Jameson, J.-L. Lions, G. Strang, Computational Mathematics Driven by Industrial Problems. Martina Franca. 1999. Editors: V. Capasso, H. Engl, J. Periaux. VI, 414 pages. 2000.

Vol. 1740: B. Kawohl, O. Pironneau, L. Tartar, J.-P. Zolesio, Optimal Shape Design. Troia 1999. Editors: A. Cellina, A. Ornelas. IX, 393 pages. 2000.

Vol. 1741: E. Lombardi, Oscillatory Integrals and Phenomena Beyond all Algebraic Orders. XV, 413 pages. 2000.

Vol. 1742: A. Unterberger, Quantization and Non-holomorphic Modular Forms. VIII, 253 pages. 2000.

Vol. 1743: L. Habermann, Riemannian Metrics of Constant Mass and Moduli Spaces of Conformal Structures. XII, 116 pages. 2000.

Vol. 1744: M Kunze, Non-Smooth Dynamical Systems. X, 228 pages. 2000.

4. Lecture Notes are printed by photo-offset from the master-copy delivered in camera-ready form by the authors. Springer-Verlag provides technical instructions for the preparation of manuscripts. Macro packages in T_EX, L^AT_EX2e, $L^AT_EX2.09$ are available from Springer's web-pages at

http://www.springer.de/math/authors/b-tex.html.

Careful preparation of the manuscripts will help keep production time short and ensure satisfactory appearance of the finished book.

The actual production of a Lecture Notes volume takes approximately 12 weeks.

5. Authors receive a total of 50 free copies of their volume, but no royalties. They are entitled to a discount of 33.3 % on the price of Springer books purchase for their personal use, if ordering directly from Springer-Verlag.

Commitment to publish is made by letter of intent rather than by signing a formal contract. Springer-Verlag secures the copyright for each volume. Authors are free to reuse material contained in their LNM volumes in later publications: A brief written (or e-mail) request for formal permission is sufficient.

Addresses:

Professor F. Takens, Mathematisch Instituut,
Rijksuniversiteit Groningen, Postbus 800,
9700 AV Groningen, The Netherlands
E-mail: F.Takens@math.rug.nl

Professor B. Teissier
Université Paris 7
UFR de Mathématiques
Equipe Géométrie et Dynamique
Case 7012
2 place Jussieu
75251 Paris Cedex 05
E-mail: Teissier@math.jussieu.fr

Springer-Verlag, Mathematics Editorial, Tiergartenstr. 17,
D-69121 Heidelberg, Germany,
Tel.: *49 (6221) 487-701
Fax: *49 (6221) 487-355
E-mail: lnm@Springer.de